Lecture Notes in Mathematics

Edited by A. Dold and B. Eckmann

446

Partial Differential Equations and Related Topics

Ford Foundation Sponsored Program
at Tulane University, January to May, 1974

Edited by Jerome A. Goldstein

Springer-Verlag
Berlin · Heidelberg · New York 1975

Prof. Jerome A. Goldstein
Dept. of Mathematics
Tulane University
New Orleans, LA 70118/USA

Library of Congress Cataloging in Publication Data

Main entry under title:

Partial differential equations and related topics.

(Lecture notes in mathematics ; 446)
1. Differential equations, Partial--Addresses, essays, lectures. I. Goldstein, Jerome A., 1941- II. Series: Lecture notes in mathematics (Berlin) ; 446.
QA3.L28 no. 446 [QA377] 510'.85 [515'.353]
75-6604

AMS Subject Classifications (1970): 34G05, 35-02, 35A22, 35B35, 35J60, 35K55, 35L05, 35L45, 35L60, 35L65, 35L99, 35Q10, 44-02, 45G05, 47M05, 49E99, 55C20, 60G50, 76G05, 92A10

ISBN 3-540-07148-2 Springer-Verlag Berlin · Heidelberg · New York
ISBN 0-387-07148-2 Springer-Verlag New York · Heidelberg · Berlin

Offsetdruck: Julius Beltz, Hemsbach/Bergstr.

TABLE OF CONTENTS

LIST OF PARTICIPANTS

Carlos Berenstein, University of Maryland, College Park, Maryland

Haïm Brézis, Université de Paris, Paris, France

Felix E. Browder, University of Chicago, Chicago, Illinois

David Goldstein Costa, Universidade Federal do Rio de Janeiro, Rio de Janeiro, Brasil

Hector O. Fattorini, Universidad de Buenos Aires, Buenos Aires, Argentina and University of California, Los Angeles, California

Djairo G. de Figueiredo, Universidade de Brasilia, Brasilia, Brasil

Ciprian Foiaş, University of Bucharest, Bucharest, Romania and Courant Institute of Mathematical Sciences, New York, New York

James M. Greenberg, State University of New York, Buffalo, New York and Courant Institute of Mathematical Sciences, New York, New York

F. Alberto Grünbaum, California Institute of Technology, Pasadena, California

Chaitan P. Gupta, Northern Illinois University, DeKalb, Illinois

Reuben Hersh, University of New Mexico, Albuquerque, New Mexico

Robert G. Kuller, Northern Illinois University, DeKalb, Illinois

Jacques-Louis Lions, Collège de France, I.R.I.A., and Université de Paris, Paris, France

Luiz Adauto Medeiros, Universidade Federal do Rio de Janeiro, Rio de Janeiro, Brasil

Beatriz P. Neves, Universidade Federal do Rio de Janeiro, Rio de Janeiro, Brasil

Louis Nirenberg, Courant Institute of Mathematical Sciences, New York, New York

Jeffrey Rauch, University of Michigan, Ann Arbor, Michigan

Pedro H. Rivera, Universidade Federal do Rio de Janeiro, Rio de Janeiro, Brasil and Universidad Nacional Mayor de San Marcos, Lima, Peru

Joel Spruck, Courant Institute of Mathematical Sciences, New York, New York

Hans Weinberger, University of Minnesota, Minneapolis, Minnesota

Calvin H. Wilcox, University of Utah, Salt Lake City, Utah

PREFACE

During the spring 1974 semester (January to May) Tulane University organized a special program in partial differential equations and related topics. The program was made possible solely through a grant from the Ford Foundation. One of the goals of the program was to promote contact with Latin American mathematicians, and accordingly all of the long-term visitors were South Americans. In addition there were many short-term visitors from the United States, Europe, and South America.

This volume consists of contributions of many of our visitors. These papers contain much new research, but the emphasis is on expository work, and many of the contributions have a good deal of pedagogical value. An accompanying volume, "Lectures on Scattering Theory for the d'Alembert Wave Equation in Exterior Domains" is an expanded version of Calvin Wilcox's lectures.

It is a pleasure to thank Tom Beale, Ed Conway, and Steve Rosencrans for their many contributions to the planning and other aspects of the program. The typing of this volume was done efficiently and patiently by Deborah Casey and Meredith Mickel of the Tulane University Mathematics Department secretarial staff. Deborah Casey also typed the accompanying volume of Wilcox's lecture notes. Finally we wish to express our deep gratitude to the Ford Foundation for its generous support and cooperation.

J. A. G.

NONLINEAR DIFFUSION IN POPULATION GENETICS, COMBUSTION, AND NERVE PULSE PROPAGATION

by

D. G. ARONSON* and H. F. WEINBERGER*

School of Mathematics
University of Minnesota
Minneapolis, Minnesota 55455

1. INTRODUCTION

In this paper we shall investigate the behavior of solutions of the semilinear diffusion equation

$$\frac{\partial u}{\partial t} = \frac{\partial^2 u}{\partial x^2} + f(u) \qquad (1.1)$$

for large values of the time t. Throughout this work we shall assume that $f(0) = f(1) = 0$ and consider only solutions $u(x,t)$ with values in $[0,1]$. The problems which we consider are the pure initial value problem in the half-space $\mathbb{R} \times \mathbb{R}^+$ and the initial-boundary value problem in the quarter-space $\mathbb{R}^+ \times \mathbb{R}^+$.

The equation (1.1) occurs in various applications, and we shall consider forms of the function $f(u)$ which are suggested by some of these applications.

---

* This work was supported through grants AFOSR71-2098 and NSF GP37660X.

The classical application is to the following problem in population genetics, which was formulated by R. A. Fisher [4].

Consider a population of diploid individuals. Suppose that the gene at a specific locus in a specific chromosome pair occurs in two forms, called alleles, which we denote by a and A. Then the population is divided into three classes or genotypes. Two of these classes consist of individuals called homozygotes which carry only one kind of allele. The members of these classes are denoted by aa or AA, depending on the alleles they carry. The third class consists of individuals, called heterozygotes, which carry one of each allele. We denote these individuals by aA.

Let the population be distributed in a one-dimensional habitat. The linear densities of the genotypes aa, aA, and AA at the point x of the habitat at time t are denoted by $\rho_1(x,t)$, $\rho_2(x,t)$, and $\rho_3(x,t)$, respectively. We assume that the population mates at random, thereby producing offspring with a birthrate denoted by r, and that the population diffuses through the habitat with diffusion constant 1. We further assume that the death rate depends only on the genotype with respect to the alleles a and A, and denote the death rates of the genotypes aa, aA, and AA by τ_1, τ_2, and τ_3, respectively. In general, these death rates differ slightly, so that some genotypes are more viable than others. Reproduction by cell division can be incorporated into this model by adding negative quantities to the death rates. Therefore we make no assumption about the signs of the τ_i.

Under the assumptions stated above the population densities satisfy the system of partial differential equations

$$
\left.
\begin{aligned}
\frac{\partial \rho_1}{\partial t} &= \frac{\partial^2 \rho_1}{\partial x^2} - \tau_1 \rho_1 + \frac{r}{\rho}\left(\rho_1 + \frac{1}{2}\rho_2\right)^2 \\
\frac{\partial \rho_2}{\partial t} &= \frac{\partial^2 \rho_2}{\partial x^2} - \tau_2 \rho_2 + \frac{2r}{\rho}\left(\rho_1 + \frac{1}{2}\rho_2\right)\left(\rho_3 + \frac{1}{2}\rho_2\right) \\
\frac{\partial \rho_3}{\partial t} &= \frac{\partial^2 \rho_3}{\partial x^2} - \tau_3 \rho_3 + \frac{r}{\rho}\left(\rho_3 + \frac{1}{2}\rho_2\right)^2
\end{aligned}
\right\} \qquad (1.2)
$$

where

$$\rho(x,t) \equiv \rho_1(x,t) + \rho_2(x,t) + \rho_3(x,t) .$$

In the Appendix we show that if the derivatives of the initial data are small, if r is very large, and if the quantity

$$\varepsilon = |\tau_1 - \tau_2| + |\tau_3 - \tau_2|$$

is very small, then for times which are small relative to ε^{-1} the relative density

$$u(x,t) = \frac{\rho_3 + \frac{1}{2}\rho_2}{\rho_1 + \rho_2 + \rho_3} \qquad (1.3)$$

can be expected to be close to the solution with the same initial values of the equation (1.1) with

$$f(u) = u(1-u)\{(\tau_1 - \tau_2)(1-u) - (\tau_3 - \tau_2)u\}. \qquad (1.4)$$

Other heuristic derivations of this equation are given in [4] and [11]. In general, the equation (1.1) should be regarded as a highly idealized and simplified model of some qualitative features of the genetic processes rather than as a strict quantitative model. It is therefore of interest to study the relation between the qualitative form of the function $f(u)$ and the qualitative behavior of solutions of the equation (1.1).

Regardless of the values of the τ_i, the function $f(u)$ given by (1.4) has the properties

$$f \in C^1[0,1], \quad f(0) = f(1) = 0. \tag{1.5}$$

We shall always deal with functions $f(u)$ which satisfy these conditions. Additional assumptions on $f(u)$ which depend on the relative values of the τ_i are also suggested by the function (1.4). Since we can always interchange the labels of a and A and hence the values of τ_1 and τ_3, there is no loss of generality in assuming that $\tau_1 \geq \tau_3$, so that AA is at least as viable as aa. There are then three cases.

CASE 1. If $\tau_3 \leq \tau_2 < \tau_1$, the viability of the heterozygote is between the viabilities of the homozygotes, and we call this the <u>heterozygote intermediate</u> case. The relevant properties of the function (1.4) are

$$f'(0) > 0, \quad f(u) > 0 \text{ in } (0,1). \tag{1.6}$$

This is the case which was considered in the classical studies of Fisher [4] and Kolmogoroff, Petrovsky, and Piscounoff [11].

CASE 2. If $\tau_2 < \tau_3 \leq \tau_1$, we have <u>heterozygote superiority</u>. The relevant features of $f(u)$ are

$$\left.\begin{array}{l} f'(0) > 0 \;\; f'(1) > 0, \text{ and } f(u) > 0 \\ \text{in } (0,\alpha), \;\; f(u) < 0 \text{ in } (\alpha,1) \\ \text{for some } \alpha \in (0,1). \end{array}\right\} \tag{1.7}$$

CASE 3. If $\tau_3 \le \tau_1 < \tau_2$, we have heterozygote inferiority. The relevant features of $f(u)$ are

$$\left.\begin{array}{l} f'(0) < 0, \quad f(u) < 0 \text{ in } (0,\alpha), \quad f(u) > 0 \\ \text{in } (\alpha,1) \text{ for some } \alpha \in (0,1), \\ \int_0^1 f(u)du > 0. \end{array}\right\} \qquad (1.8)$$

There are various other applications which lead to similar models. For example certain flame propagation problems in chemical reactor theory lead to equations of the form (1.1) with a function $f(u)$ which satisfies (1.5) and the generalization

$$\left.\begin{array}{l} f(u) \le 0 \text{ in } (0,\alpha), \quad f(u) > 0 \text{ in } (\alpha,1) \\ \text{for some } \alpha \in (0,1), \int_0^1 f(u)du > 0 \end{array}\right\} \qquad (1.8')$$

of (1.8). (See, for example, the article of Gelfand [5]). Here u represents a normalized temperature and α represents a critical temperature at which an exothermic reaction starts.

A model for the propagation of a voltage pulse through the nerve axon of a squid has been proposed by Hodgkin and Huxley [6]. The voltage u satisfies an equation of the form

$$\frac{\partial u}{\partial t} = \frac{\partial^2 u}{\partial x^2} + F[u]$$

where F is a certain rather complicated nonlinear functional. An electrical analogue which exhibits the qualitative features of the Hodgkin-Huxley model was proposed by Nagumo, Arimoto, and Yoshizawa [12]. This model leads to the equation

$$\frac{\partial u}{\partial t} = \frac{\partial^2 u}{\partial x^2} + f(u) - \varepsilon \int_0^t u(x,\tau)\,d\tau \qquad (1.9)$$

where ε is a nonnegative parameter and

$$f(u) = u(1 - u)(u - \alpha)$$

for some $\alpha \in (0, \frac{1}{2})$. Note that this function satisfies the conditions (1.5) and (1.8). It has been suggested by Cohen [2, p. 35] that (1.9) with $\varepsilon = 0$ is a model for a nerve which has been treated with certain toxins. Moreover, a rescaled version of (1.9) with $\varepsilon = 0$ has been used by Nagumo, Yoshizawa, and Arimoto [13] as a model for a bistable active transmission line.

In their classical paper [11] Kolmogoroff, Petrovsky, and Piscounoff considered equation (1.1) in the heterozygote intermediate case. They proved the existence of a number $c^* > 0$ such that (1.1) possesses travelling wave solutions $u(x,t) = q(x - ct)$ for all velocities c with $|c| \geq c^*$. (These travelling wave solutions in the heterozygote intermediate case were also discussed by Fisher in [4].) Moreover, they proved that the solution of the special initial value problem with

$$u(x,0) = \begin{cases} 1 & \text{for } x < 0 \\ 0 & \text{for } x > 0 \end{cases}$$

converges (in a certain sense) to a travelling wave solution with speed c^*.

Kanel' [7, 8, 9, 10] has extended and generalized these results in the heterozygote inferior case (1.8) and the case of flame propagation (1.8'). Moreover, Kanel' has observed the occurrence of a threshold behavior with respect to the initial values $u(x,0)$ in these cases.

We study solutions of the equation (1.1) with $f(u)$ subject to (1.5) and (1.6), (1.7), (1.8) or (1.8'). In the applications to flame propagation and voltage pulse propagation it is natural to consider the initial-boundary value problem on the quarter plane $\mathbb{R}^+ \times \mathbb{R}^+$ as well as the pure initial value problem. We shall deal with both of these problems under rather mild restrictions on the data.

In the various cases under consideration we derive the limit behavior of the solution $u(x,t)$ as $t \to \infty$. We study the stability properties of the equilibrium states $u \equiv 0$, $u \equiv \alpha$, and $u \equiv 1$ in the initial value problem in Section 3. We show in Section 4 that in every case there exists a $c^* > 0$ with the property that in the pure initial value problem every disturbance which is initially confined to a bounded set and which is propagated at all is propagated at the asymptotic speed c^*. These results are extended to solutions of the initial-boundary value problem in Section 5.

Many of the results which we obtain here for functions $f(u)$ which satisfy (1.6), (1.7), (1.8), or (1.8') are valid in more general circumstances. These generalizations will be published elsewhere [1].

2. A MAXIMUM PRINCIPLE AND ITS APPLICATIONS

All the forcing functions $f(u)$ described in Section 1 satisfy the conditions

$$f(0) = f(1) = 0, \quad f \in C^1[0,1]. \tag{2.1}$$

In the remainder of this paper these conditions will be understood to hold even if they are not mentioned explicitly.

We begin our study of the equation (1.1) with a version of the maximum principle.

PROPOSITION 2.1. *Let* $u(x,t) \in [0,1]$ *and* $v(x,t) \in [0,1]$ *satisfy the inequalities*

$$u_t - u_{xx} - f(u) \geq v_t - v_{xx} - f(v) \quad \textit{in} \quad (a,b) \times (0,T],$$

$$0 \leq v(x,0) \leq u(x,0) \leq 1 \quad \textit{in} \quad (a,b)$$

where $-\infty \leq a < b \leq \infty$ *and* $0 < T \leq \infty$. *Moreover, if* $a > -\infty$, *assume that*

$$0 \leq v(a,t) \leq u(a,t) \leq 1 \quad \textit{on} \quad [0,T]$$

and if $b < \infty$ *assume that*

$$0 \leq v(b,t) \leq u(b,t) \leq 1 \quad \textit{on} \quad [0,T].$$

Then $u \geq v$, *and if* $u(x,0) > v(x,0)$ *in an open subinterval of* (a,b) *then* $u > v$, *in* $(a,b) \times (0,T]$.

PROOF. By the theorem of the mean we find that

$$(u-v)_t - (u-v)_{xx} \geq f(u) - f(v) = f'(v+\theta(u-v))(u-v)$$

for some $\theta \in (0,1)$. Let $\alpha = \max_{[0,1]} f'(u)$, and define

$$w(x,t) = (u-v)e^{-\alpha t}.$$

Then

$$w_t - w_{xx} \geq \{f'(v + \theta(u-v)) - \alpha\}w.$$

Since the coefficient of w is nonpositive, our result

follows from the strong maximum principle for linear parabolic inequalities. (See, for example, [14, p. 172].)

We now derive the principal tool for our investigation.

PROPOSITION 2.2. *Let* $q(x) \in [0,1]$ *be a solution of the ordinary differential equation*

$$q'' + f(q) = 0 \quad in \quad (a,b)$$

where $-\infty \leq a < b \leq \infty$. *If* $a > -\infty$ *assume that* $q(a) = 0$ *and if* $b < \infty$ *assume that* $q(b) = 0$. *Let* $v(x,t)$ *denote the solution of the initial value problem*

$$v_t = v_{xx} + f(v) \; in \; \mathbb{R} \times \mathbb{R}^+, \tag{2.2}$$

$$v(x,0) = \begin{cases} q(x) & in \quad (a,b) \\ 0 & in \quad \mathbb{R}\setminus(a,b). \end{cases}$$

Then $v(x,t)$ *is a nondecreasing function of* t *for each* x. *Moreover,*

$$\lim_{t\to\infty} v(x,t) = \tau(x)$$

uniformly in each bounded interval, where $\tau(x)$ *is the smallest nonnegative solution of the differential equation*

$$\tau'' + f(\tau) = 0 \tag{2.3}$$

on the whole real line $\mathbb{R}$ *which satisfies the inequality*

$$\tau(x) \geq q(x) \quad in \quad (a,b). \tag{2.4}$$

PROOF. By Proposition 2.1, $v(x,t) \geq 0$ in $\mathbb{R} \times \mathbb{R}^+$. Thus we can apply Proposition 2.1 with u replaced by the present $v(x,t)$ and v replaced by $q(x)$ to find that $v(x,t) \geq q(x)$ in $(a,b) \times \mathbb{R}^+$. We then see that for any $h > 0$ we have

$$v(x,h) \geq v(x,0) \quad \text{in } \mathbb{R}.$$

We now apply Proposition 2.1 to $v(x,t+h)$ and $v(x,t)$ to conclude that for any $h > 0$,

$$v(x,t+h) \geq v(x,t) \quad \text{in } \mathbb{R} \times \mathbb{R}^+.$$

Since $u \equiv 1$ is a solution of (1.1), Proposition 2.1 shows that $v(x,t) \leq 1$. Thus for each x, $v(x,t)$ is nondecreasing in t and bounded above. Therefore the limit $\tau(x)$ exists.

By applying the inverse of the heat operator to the equation (2.2) it is easy to show that v_x, v_{xx}, and v_t are uniformly bounded in $\mathbb{R} \times [1,\infty)$. It then follows from the Schauder-type theory for parabolic equations (see [3, p. 92]) that on each bounded x-interval the families of functions v_x, v_{xx}, and v_t, parametrized by t, are equicontinuous in x. Therefore on each bounded x-interval, v converges to τ and v_x, v_{xx}, and v_t converge to the corresponding derivatives of τ uniformly. It follows that τ satisfies the steady-state equation (2.3) in $\mathbb{R}$. Since $v(x,t) \in [0,1]$, $\tau(x) \in [0,1]$, and since $v(x,t) \geq q(x)$ in $(a,b) \times \mathbb{R}^+$, τ satisfies the inequality (2.4).

Finally, if $\sigma(x)$ is any nonnegative solution of (2.3) in all of $\mathbb{R}$ which satisfies the inequality (2.4), then $v(x,0) \leq \sigma(x)$. Hence by Proposition 2.1 $v(x,t) \leq \sigma(x)$ and therefore also $\tau(x) \leq \sigma(x)$. This shows that $\tau(x)$ is the minimal nonnegative solution of (2.3) with the property (2.4).

Note that Proposition 2.2 establishes the existence of a unique nonnegative solution of (2.3) which is minimal with respect to the condition (2.4).

3. STABILITY AND THRESHOLD RESULTS

Our first result establishes the stability of the equilibrium state $u \equiv 1$ in the heterozygote intermediate case and the state $u \equiv \alpha$ in the heterozygote superior case.

THEOREM 3.1. *Let* $u(x,t) \in [0,1]$ *be a solution of* (1.1) *in* $\mathbb{R} \times \mathbb{R}^+$.

(i) If $f(u)$ *satisfies* (2.1) *and* (1.6), *then either* $u(x,t) \equiv 0$ *or*

$$\lim_{t\to\infty} u(x,t) = 1.$$

(ii) If $f(u)$ *satisfies* (2.1) *and* (1.7), *then*

$$u(x,t) \equiv 0, \quad u(x,t) \equiv 1, \quad \textit{or}$$

$$\lim_{t\to\infty} u(x,t) = \alpha.$$

PROOF. The differential equation

$$q'' + f(q) = 0$$

has the first integral

$$\frac{1}{2} q'^2 + F(q) = k \qquad (3.1)$$

where k is an arbitrary constant and

$$F(q) \equiv \int_0^q f(u)\,du.$$

Suppose first that $f(u)$ satisfies the conditions (2.1) and (1.7) of the heterozygote superior case. Then for any $\varepsilon \in (0,\alpha)$, $F(q)$ is increasing in $(0,\varepsilon)$ and in particular $F(\varepsilon) > 0$. Since $f(\varepsilon) > 0$, it is easily seen that $\{F(\varepsilon) - F(u)\}^{-1/2}$ is integrable on the interval $[0,\varepsilon]$. It follows that for each $\varepsilon \in (0,\alpha)$ the problem

$$\frac{1}{2} q'^2 + F(q) = F(\varepsilon)$$

$$q(0) = 0$$

$$q'(0) = \{2F(\varepsilon)\}^{1/2}$$

has a solution $q_\varepsilon(x)$ which is positive in the interval $(0,b_\varepsilon)$, where

$$b_\varepsilon = 2\int_0^\varepsilon [2\{F(\varepsilon) - F(u)\}]^{-1/2}\,du.$$

Moreover, $q_\varepsilon(x) \le q_\varepsilon(\frac{1}{2} b_\varepsilon) = \varepsilon$, and

$$q_\varepsilon(0) = q_\varepsilon(b_\varepsilon) = 0.$$

As ε decreases to zero, q_ε approaches the solution $\varepsilon \sin\{f'(0)\}^{1/2} x$ of the corresponding linearized problem, and b_ε approaches $\pi/\{f'(0)\}^{1/2}$.

In view of Proposition 2.1, if $u(x,t) \not\equiv 0$, $u(x,h) > 0$ for any $h > 0$. Since $b_\varepsilon < 2\pi/\{f'(0)\}^{1/2}$ when ε is sufficiently small, and since $q_\varepsilon(x) \le \varepsilon$, we can choose $\varepsilon > 0$ so small that $u(x,h) \ge q_\varepsilon(x)$ in $(0,b_\varepsilon)$. It then follows from Proposition 2.2 that

$$\liminf_{t\to\infty} u(x,t) = \liminf_{t\to\infty} u(x, t+h) \ge \tau(x)$$

where $\tau(x)$ is the smallest nonnegative solution of

$q'' + f(q) = 0$ which satisfies $q(x) \geq q_\varepsilon(x)$ in $(0,b_\varepsilon)$.

To show that $\tau(x) \geq \alpha$, we assume the contrary and show that a contradiction results. Suppose that there is an x_0 such that $\beta \equiv \tau(x_0) \in (0,\alpha)$. Then $\tau(x)$ satisfies the first order equation (3.1) with some $k \geq F(\beta)$. Hence $\{k - F(u)\}^{-1/2}$ is integrable on the interval $[0,\beta]$. Therefore $\tau(x)$ is implicitly determined by the equation

$$x = x_0 \mp \int_\tau^\beta [2\{k - F(u)\}]^{-1/2}\, du,$$

where the sign is determined by the sign of $\tau'(x_0)$. It follows that $\tau(x)$ becomes zero with $\tau' \neq 0$ at a finite value of x, so that τ cannot be a nonnegative solution of $q'' + f(q) = 0$ for all x. This contradiction shows that $\tau(x) \geq \alpha$ and hence that

$$\liminf_{t\to\infty} u(x,t) \geq \alpha.$$

If we apply this proof with $\alpha = 1$ when f satisfies (1.6) and recall that $u \leq 1$, we obtain the statement (i) of the Theorem.

We now let $v = 1 - u$ and note that v satisfies the equation (1.1) with $f(u)$ replaced by $-f(1 - v)$. If $f(u)$ satisfies (1.7), then $-f(1 - v) > 0$ for $v \in (0, 1-\alpha)$. Hence the same proof shows that if $u(x,t) \not\equiv 1$, then

$$\liminf_{t\to\infty} (1 - u(x,t)) \geq 1 - \alpha.$$

Hence we have proved statement (ii), and the Theorem is proved.

We remark that Theorem 3.1 not only proves the stability of the state $u \equiv 1$ in the heterozygote intermediate case but also the very strong instability of the state $u \equiv 0$. Similarly, statement (ii) shows that in the heterozygote superior case both the states $u \equiv 0$ and $u \equiv 1$ are very unstable.

We now turn to the case (1.8) of heterozygote inferiority. In this case we shall show that the equilibrium states $u \equiv 0$ and $u \equiv 1$ are stable while $u \equiv \alpha$ is unstable. As a consequence, we can expect threshold phenomena to be associated with this case.

We begin with the following elementary lemma.

LEMMA. *Let* $u(x,t) \in [0,1]$ *be a solution of* (1.1) *in* $\mathbb{R} \times \mathbb{R}^+$ *and let* $f(u) < 0$ *in the interval* $(0,\gamma]$. *If* $u(x,0) \in [0,\gamma]$, *then*

$$\lim_{t \to 0} u(x,t) = 0$$

uniformly on $\mathbb{R}$.

PROOF. Let v be the solution of the initial value problem

$$v_t = v_{xx} + f(v)$$
$$v(x,0) = \gamma.$$

Then v is independent of x and satisfies the relation

$$t = \int_v^\gamma [-f(u)]^{-1} \, du.$$

Hence v goes to zero as $t \to \infty$.

Since $v(x,0) \geq u(x,0)$, the Lemma follows from Proposition 2.1.

Our next theorem concerns the stability of the equilibrium state $u \equiv 0$ in the heterozygote inferior case. It is a generalization of a result proved by Kanel'[10]. In stating the theorem we shall use the following notation: For any $\rho \in [0,\alpha)$ we define

$$s(\rho) \equiv \sup\left\{\frac{f(u)}{u-\rho} : u \in (\alpha,1)\right\}.$$

Moreover, we shall use the notation

$$[\mu]^+ \equiv \max\{\mu, 0\}.$$

THEOREM 3.2. *Let* $u(x,t) \in [0,1]$ *be a solution of equation* (1.1) *in* $\mathbb{R} \times \mathbb{R}^+$ *where* $f(u)$ *satisfies* (2.1) *and* (1.8). *If for some* $\rho \in [0,\alpha)$

$$\int_{-\infty}^{\infty} [u(x,0) - \rho]^+ \, dx < \left\{\frac{2\pi}{s(\rho)e}\right\}^{1/2} (\alpha-\rho), \qquad (3.2)$$

then

$$\lim_{t\to\infty} u(x,t) = 0$$

uniformly on $\mathbb{R}$.

PROOF. Fix ρ and write s for $s(\rho)$. Let $w(x,t)$ denote the solution of the problem

$$w_t = w_{xx} + sw$$

$$w(x,0) = [u(x,0) - \rho]^+.$$

By Proposition 2.1, $w \geq 0$ so that $w \equiv [w]^+$. Since $f(u) \leq 0$ on $[0,\alpha]$, it follows from the definition of $s(\rho)$ that

$$f(u) \leq s[u-\rho]^+.$$

Let

$$v(x,t) \equiv u(x,t) - \rho .$$

Then

$$\begin{aligned} v_t - v_{xx} - s[v]^+ &\le u_t - u_{xx} - f(u) \\ &= 0 \\ &= w_t - w_{xx} - s[w]^+ . \end{aligned}$$

In view of Proposition 2.1, $v(x,t) \le w(x,t)$ so that

$$u(x,t) \le w(x,t) + \rho .$$

The function we^{-st} satisfies the equation of heat conduction. Therefore

$$\begin{aligned} w(x,t) &= \frac{1}{2\sqrt{\pi t}} e^{st} \int_{-\infty}^{\infty} e^{-\frac{(x-\xi)^2}{4t}} [u(\xi,0) - \rho]^+ d\xi \\ &\le \frac{1}{2\sqrt{\pi t}} e^{st} \int_{-\infty}^{\infty} [u(\xi,0) - \rho]^+ d\xi . \end{aligned}$$

In particular, it follows from (3.2) that $u(x,(2s)^{-1})$ is bounded above by a constant $\gamma < \alpha$. Then the preceding Lemma proves Theorem 3.2.

Theorem 3.2 shows that the state $u \equiv 0$ is locally stable while the Lemma proves that the state $u \equiv \alpha$ is unstable in the case of heterozygote inferiority. We shall now show that the state $u \equiv 0$ is not globally stable, even with respect to disturbances of bounded support.

We observe that the hypotheses (1.8) or even (1.8') imply the existence of a unique $\kappa \in [\alpha,1)$ for which

$$F(\kappa) \equiv \int_0^\kappa f(u)du = 0. \tag{3.3}$$

Moreover, $F(q) > 0$ and $F'(q) = f(q) > 0$ for $q \in (\kappa,1)$. For any $\beta \in (\kappa,1)$ we define the length

$$b_\beta = 2\int_0^\beta \{2F(\beta) - 2F(u)\}^{-1/2} du,$$

and the solution $q_\beta(x)$ of $q'' + f(q) = 0$ which has the first integral

$$\frac{1}{2}q'^2 + F(q) = F(\beta)$$

and which satisfies $q(0) = 0$, $q'(0) = \{2F(\beta)\}^{1/2}$. Then $q_\beta > 0$ in $(0,b_\beta)$, $q_\beta(0) = q_\beta(b_\beta) = 0$, and

$$q_\beta(x) \le q_\beta(\tfrac{1}{2}b_\beta) = \beta \quad \text{on} \quad [0,b_\beta].$$

With the aid of this function we state the following theorem.

THEOREM 3.3. *Let* $u(x,t) \in [0,1]$ *be a solution in the half-plane* $\mathbb{R} \times \mathbb{R}^+$ *of the equation* (1.1) *where* $f(u)$ *satisfies the conditions* (2.1) *and* (1.8) *or* (1.8').

If for some $\beta \in (\kappa,1)$ *and some* x_0

$$u(x,0) \ge q_\beta(x-x_0) \quad \text{on} \quad (x_0,x_0+b_\beta),$$

then

$$\lim_{t\to\infty} u(x,t) = 1.$$

PROOF. We apply Proposition 2.2 with $q(x) = q_\beta(x)$.

The proof of the fact that the minimal nonnegative solution τ of the equation $q'' + f(q) = 0$ on $\mathbb{R}$ which satisfies $q \ge q_\beta(x-x_0)$ in (x_0,x_0+b_β) is identically one is the same as that of the fact that $\tau \equiv \alpha$ in the proof of Theorem 3.1. Thus the theorem is proved.

We note that Theorem 3.3 not only shows that the state $u \equiv 0$ is unstable with respect to disturbances with bounded support but also that the state $u \equiv 1$ is globally stable with respect to such disturbances.

Theorems 3.2 and 3.3 together exhibit a threshold phenomenon. A disturbance of bounded support of the state $u \equiv 0$ which is sufficiently large on a sufficiently large interval grows to one, while a disturbance which is not sufficiently large on a sufficiently large interval dies out.

If $f(u)$ satisfies only (1.8'), the Lemma does not apply. However, if (3.2) holds, then from the proof of Theorem 3.2 we find that $u < \alpha$ for $t \geq 1/2s(\rho)$ and that $[u(\cdot,1/2s(\rho)) - \rho]^{+} \in L^{1}(\mathbb{R})$. A comparison with the equation of heat conduction then yields $\limsup u(x,t) \leq \rho$ as $t \to \infty$. Since Theorem 3.3 is valid when (1.8') holds, there are also threshold effects in this case.

4. PROPAGATION

In this section we investigate how the solution $u(x,t)$ of (1.1) behaves as a function of time. For this purpose we introduce the moving coordinate

$$\xi = x - ct, \quad c > 0.$$

If we define

$$v(\xi,t) \equiv u(\xi+ct,t),$$

the equation (1.1) becomes

$$v_t = v_{\xi\xi} + cv_\xi + f(v). \tag{4.1}$$

We note that the maximum principle, Proposition 2.1, and the convergence result, Proposition 2.2, are immediately extendable to this equation. Since the proofs are identical to those given in Section 2, we shall simply use these results without further comment.

The steady state equation which corresponds to (4.1) is, of course,

$$q'' + cq' + f(q) = 0. \tag{4.2}$$

This equation is equivalent to the system

$$q' = p$$
$$p' = -cp - f(q).$$

The functions $p(\xi)$, $q(\xi)$ corresponding to a solution of (4.2) give a trajectory in the q-p plane or, as it is usually called, the phase plane. Such a trajectory has the slope

$$\frac{dp}{dq} = -c - \frac{f(q)}{p} \tag{4.3}$$

for $p \neq 0$.

When $c = 0$, each trajectory satisfies an equation of the form

$$\frac{1}{2} p^2 + F(q) = \text{constant}.$$

Under our hypotheses on $f(u)$ there is an $\eta \in (0,1)$ such that $F(\eta) > 0$. For any ν such that $0 < \nu < [2F(\eta)]^{1/2}$ the trajectory through $(0,-\nu)$ lies in the strip $q \in [0,1)$ and contains a point of the positive p-axis. By continuity there is a $\tilde{c} = \tilde{c}(\nu) > 0$ such that the same is true for all $c \in [0,\tilde{c})$. Hence for $c \in [0,\tilde{c})$ there is no trajectory joining the origin and the line

$q = 1$.

We now consider $c > 0$. If $c^2 > 4f'(0)$, there is a nontrivial trajectory from the origin [15,§56]. The unique trajectory in the strip $q \in [0,1]$ that goes to the point $(0,-\nu)$ with $\nu > 0$ cannot cross any trajectory that goes to the origin. Hence if we take the limit as $\nu \downarrow 0$ of the trajectory that goes to $(0,-\nu)$ and if $c^2 > 4f'(0)$, we obtain a nontrivial extremal trajectory going to the origin. We denote this extremal trajectory by T_c.

We define

$$\sigma = \sup_{u \in [0,1]} \frac{f(u)}{u} ,$$

so that

$$f(u) \le \sigma u \quad \text{for} \quad u \in [0,1].$$

It follows that if T is any trajectory of (4.3), then

$$\frac{dp}{dq} \le -c - \sigma \frac{q}{p}$$

at every point of T where $q \in [0,1]$ and $p < 0$. On the other hand if $c^2 > 4\sigma$, the line through the origin

$$p = -\frac{1}{2}\left(c + \sqrt{c^2 - 4\sigma}\right) q \qquad (4.4)$$

satisfies the differential equation

$$\frac{dp}{dq} = -c - \sigma \frac{q}{p} .$$

Consequently, the trajectory through $(0,-\nu)$ with $\nu > 0$ cannot cross this line for $q \in [0,1]$. It must therefore lie below it. Taking the limit as $\nu \downarrow 0$, we see that for $c^2 > 4\sigma$, T_c is bounded above by the line (4.4). In particular, T_c connects the origin with a point of the form $(1,-\nu)$ with $\nu > 0$.

In view of the above observations, the number

$$c^* = \inf\{c: c^2 > 4f'(0), \text{ there exists } \nu > 0 \text{ such that } (1,-\nu) \in T_c\}$$

is well-defined and positive. In the remainder of this section we shall exhibit various properties of c^*. In particular, we shall show that c^* is the asymptotic speed of propagation associated with the equation (1.1).

THEOREM 4.1. *Let* $u(x,t) \in [0,1]$ *be a solution of equation* (1.1), *where* $f(u)$ *satisfies* (1.6), (1.7), (1.8), *or* (1.8'), *in* $\mathbb{R} \times \mathbb{R}^+$. *If for some* x_0

$$u(x,0) \equiv 0 \quad \textit{in} \quad (x_0, \infty), \tag{4.5}$$

then for each ξ *and each* $c > c^*$,

$$\lim_{t\to\infty} u(\xi+ct,t) = 0. \tag{4.6}$$

PROOF. Let $q_c(x)$ denote the solution of the steady state equation (4.2) in $\mathbb{R}^+$ which corresponds to the trajectory T_c and for which $q_c(0) = 1$. q_c is decreasing and approaches zero as $x \to \infty$.

We observe that the function $w \equiv 1 - u$ satisfies an equation like (1.1) but with $f(u)$ replaced by $-f(1 - w)$. We apply the extension of Proposition 2.1 with $q = 1 - q_c(x - x_0)$ in (x_0, ∞) to the equation

$$w_t = w_{\xi\xi} + cw_\xi - f(1 - w).$$

We note that $1 - u(x,0) \geq 1 - q_c(x - x_0)$. Hence by the extensions of Propositions 2.1 and 2.2

$$\liminf_{t \to \infty} (1 - u(\xi+ct,t)) \geq 1 - \tau(\xi) \tag{4.7}$$

where $\tau(\xi)$ is the solution of equation (4.2) which is maximal with respect to the properties

$$\tau(\xi) \leq 1 \text{ in } \mathbb{R} \tag{4.8}$$

and

$$\tau(\xi) \leq q_c(\xi - x_o) \text{ in } (x_o, \infty). \tag{4.9}$$

We must now show that $\tau(\xi) \equiv 0$.

For any $c > 0$ such that $c^2 > 4f'(0)$ the trajectory T_c has slope S^- at the origin, where

$$S^{\pm} = \frac{1}{2}\{-c \pm \sqrt{c^2 - 4f'(0)}\} .$$

Moreover, T_c is the unique trajectory with this slope at the origin. Any other trajectory which approaches the origin with $q > 0$ must do so with the slope S^+. These statements can be proved by the methods used by Petrovski [15, §56].

Since $c > c^*$, the trajectory T_c lies in the half-plane $p < 0$ for $q \in (0,1]$ and contains a point $(1,-\nu)$ with $\nu > 0$. If $\tau(\xi) \not\equiv 0$, then the corresponding trajectory T has slope S^+ or S^- at the origin. Since the slope of T_c at $(0,0)$ is S^-, it follows from (4.9) that T cannot have slope S^+ at $(0,0)$. Therefore $T \equiv T_c$, and there exists $\zeta \in \mathbb{R}$ such that $\tau(\zeta) = 1$ and $\tau'(\zeta) = -\nu < 0$. Hence $\tau(\xi) > 1$ for some $\xi < \zeta$. This contradicts (4.8), and we conclude that $\tau \equiv 0$. Since $u(x,t) \geq 0$ the assertion of the theorem follows from (4.7).

In view of (4.3) and the uniqueness property of T_c it can be shown that for $\bar{c}^2 > 4f'(0)$, T_c approaches $T_{\bar{c}}$ as $c \to \bar{c}$. That is, T_c is continuous in c.

REMARKS. 1. Because the trajectory T_c has slope $-\frac{1}{2}\left(c + \sqrt{c^2 - 4f'(0)}\right)$, the function q_c has the property

$$\lim_{x\to\infty} q_c(x)^{1/x} = e^{-\frac{1}{2}\left(c+\sqrt{c^2-4f'(0)}\right)}.$$

One can then see from the proof that the condition (4.5) can be replaced by

$$\lim_{x\to\infty} u(x,0)^{1/x} = 0.$$

2. Since the equation (1.1) is invariant when x is replaced by $-x$, the conclusion (4.6) holds for $c < -c^*$ if (4.5) is replaced by

$$u(x,0) \equiv 0 \text{ in some interval } (-\infty, x_0)$$

or by

$$\lim_{x\to-\infty} u(x,0)^{-1/x} = 0.$$

3. If $u(x,0) \equiv 0$ outside a bounded interval, we have (4.6) for $|c| > c^*$. Moreover, it can be shown that the convergence is uniform in this case.

Next we show that there always exists a travelling

wave solution with velocity c^*.

THEOREM 4.2. *If* $f(u)$ *satisfies* (1.6), (1.7), (1.8), *or* (1.8'), *there exists a travelling wave solution* $u = q^*(x - c^*t)$ *of* (1.1). *Moreover,* $q^{*\prime}(\xi) < 0$,

$$\lim_{\xi \to \infty} q^*(\xi) = 0$$

and

$$\lim_{\xi \to -\infty} q^*(\xi) = \begin{cases} 1 & \text{if } f(u) \text{ satisfies (1.6), (1.8), or (1.8')} \\ \alpha & \text{if } f(u) \text{ satisfies (1.7).} \end{cases}$$

PROOF. If ${c^*}^2 > 4f'(0)$, then the trajectory T_{c^*} exists and lies in the half-strip $q \in [0,1]$, $p \le 0$ at least in a relative neighborhood of the origin. T_{c^*} is minimal in the sense that there is no other trajectory which lies below T_{c^*} and approaches the origin.

If ${c^*}^2 > 4f'(0)$ and T_{c^*} does not intersect the positive q-axis, then by continuity the same will be true of T_c for a slightly smaller value of c, which contradicts the definition of c^*. Therefore T_{c^*} intersects the q-axis at a point $(\eta, 0)$ with $\eta \in (0,1]$. If $f(\eta) \neq 0$, then since T_{c^*} must go in the negative q-direction for $p < 0$, (4.3) implies that $f(\eta) > 0$. But then there is a number $\eta_1 > \eta$ such that $f(\eta_1) > 0$. The part of the trajectory through $(\eta_1, 0)$ on which $p < 0$ lies below T_{c^*}, and must go to the negative p-axis. By continuity, the same will be true for sufficiently small $c > c^*$, and the resulting trajectory through $(\eta_1, 0)$ bounds T_c away from $q = 1$. This again contradicts the definition of c^*. We conclude that T_{c^*} must hit the q-axis at a point $(\eta,0)$ where $f(\eta) = 0$.

According to (4.3), if $f(q) < 0$, the slope of T_{c^*} is negative. Therefore, T_{c^*} cannot hit the q-axis at a zero η of $f(u)$ which is the right endpoint of an interval where $f(u)$ is negative. Thus if $c^{*2} > 4f'(0)$, then T_{c^*} must hit the q-axis at $(1,0)$ in the cases (1.6), (1.8), and (1.8') and at $(\alpha,0)$ in the case (1.7).

If, on the other hand, $c^{*2} = 4f'(0)$, then since $c^* > 0$, $f'(0)$ must be positive. Hence $f(u)$ satisfies (1.6) or (1.7). In particular, $f(u) > 0$ in an interval $(0,\alpha)$, where we set $\alpha = 1$ in the case (1.6). The trajectories through the points of the interval $(0,\alpha)$ of the q-axis go downward and to the left in the half-strip $q \in [0,1]$, $p < 0$. Hence they cannot recross the positive q-axis. On the other hand, suppose that a trajectory $S_{c^*}(\eta)$ through some point $(\eta,0)$ with $\eta \in (0,\alpha)$ went to a point $(0, -\nu)$ on the negative p-axis. By continuity the trajectory $S_c(\eta)$ through $(\eta,0)$ would still go to the negative p-axis for any sufficiently small $c > c^*$. Since $S_c(\eta)$ would bound T_c away from the q-axis, we would again find a contradiction with the definition of c^*. We conclude that every trajectory $S_{c^*}(\eta)$ through a point $(\eta,0)$ with $\eta \in (0,\alpha)$ must go to the origin. By continuity the same is true of the limit of these trajectories as η approaches α. This limiting trajectory connects the point $(\alpha,0)$ with the origin.

We have shown that there is always a trajectory in the phase plane that connects $(1,0)$ to the origin in the cases (1.6), (1.8), and (1.8') and that connects $(\alpha,0)$ to the origin in the case (1.7). Any solution $q^*(\xi)$ corresponding to this trajectory clearly has the properties stated in the theorem.

REMARKS. 1. If $c^{*2} = 4f'(0)$ the trajectory corresponding to q^* may not be the minimal trajectory through the origin, which we have called T_{c^*}. For example, if $f(u)$ has the property $f(u) \le f'(0)u$, then $\sigma = f'(0)$. Hence T_{c^*} lies below the line (4.4) with $c = c^*$. Thus T_{c^*} goes to the negative half-line $q = 1$. Since $f'(0) > 0$ in this case, the proof of Theorem 4.1 works with $c = c^*$.

2. If $f'(0) > 0$, the above proof can be extended to show that there is a travelling wave solution with the properties stated in Theorem 4.2 for every $c \ge c^*$. The problem treated by Kolmogoroff, Petrovsky, and Piscounoff [11] has the properties of Remarks 1 and 2.

3. The function $q^*(-x - c^*t)$ gives a travelling wave with velocity $-c^*$.

Finally, we consider the behavior of $u(\xi+ct,t)$ for $|c| < c^*$. Here we shall have to consider the three cases separately.

THEOREM 4.3. *Let* $u(x,t) \in [0,1]$ *be a solution of* (1.1) *in* $\mathbb{R} \times \mathbb{R}^+$ *where* $f(u)$ *satisfies* (1.6). *If* $u(x,t) \not\equiv 0$, *then for each* c *with* $|c| < c^*$ *and each* ξ

$$\lim_{t\to\infty} u(\xi+ct,\, t) = 1.$$

PROOF. If $c \in (0,\{4f'(0)\}^{1/2})$, then the origin in the phase plane is a spiral point. This means that there are trajectories in the strip $q \in [0,1]$ which connect the positive p-axis to the negative p-axis.

If $c^{*2} > 4f'(0)$, the proof of Theorem 4.2 shows that T_{c^*} goes from $(1,0)$ to $(0,0)$ in the lower half plane. Consider any $c \in (\{4f'(0)\}^{1/2}, c^*)$. Because of equation (4.3), the trajectory T_c lies above T_{c^*}. Hence T_c crosses the q-axis at a point $(\eta,0)$ with $\eta \in (0,1)$. Then if $\beta \in (\eta,1)$ the lower part of the trajectory T through $(\beta,0)$ stays below T_c. Therefore, T goes to the negative p-axis. Since $f(u) > 0$ in $(0,1)$, we see from (4.3) that the slope of T is negative in the upper half-plane. Moreover, the slope is bounded below when p is bounded away from zero. Therefore, T goes from a point on the positive p-axis to $(\beta,0)$ and from there to a point on the negative p-axis.

We have shown that for each $c \in (0, c^*)$ there is a trajectory T which connects the positive p-axis to the negative p-axis. T crosses the q-axis at a point $(\beta,0)$ with $\beta \in (0,1)$, and lies in the strip $q \in [0,\beta]$. Let q_β be the corresponding solution of $q'' + cq' + f(q) = 0$ for which $q_\beta(0) = 0$, $q_\beta'(0) > 0$. This solution is positive on a finite interval $(0,b)$ and vanishes at its ends. Moreover, $q_\beta(x) \le \beta < 1$.

According to Theorem 3.1, $u(x,t)$ converges to 1 as $t \to \infty$. Moreover, this theorem was proved by using Proposition 2.2, which states that the convergence is uniform on every bounded x-interval. In particular, there is a time T so that

$$u(x,T) \geq \beta \geq q_\beta(x) \text{ on } [0,b].$$

Theorem 4.3 for $c \in [0,c^*)$ now follows from applying the extensions of Propositions 2.1 and 2.2 to the solution v of (4.1) and recalling that $v(\xi,t) = u(\xi+ct,t)$. Since replacing x by $-x$ replaces c by $-c$, the Theorem is also true for $c \in (-c^*,0]$.

In exactly the same manner we can prove:

THEOREM 4.4. *Let* $u(x,t) \in [0,1]$ *be a solution of* (1.1) *in* $\mathbb{R} \times \mathbb{R}^+$ *where* $f(u)$ *satisfies* (1.7). *If* $u(x,0) \not\equiv 0$ *then for each* c *such that* $|c| < c^*$ *and each* ξ

$$\liminf_{t \to \infty} u(\xi+ct,\, t) \geq \alpha.$$

We remark that in this case there will in general be another propagation speed at which the decrease of u to α travels.

In the heterozygote inferior case we have seen that $u(x,t) \to 1$ if and only if the initial conditions exceed some threshold value. Thus we cannot expect the analogue of Theorems 4.3 and 4.4 to hold without some conditions such as those of Theorem 3.3 on $u(x,0)$. With this in mind we can carry through the argument used above to obtain the following result.

THEOREM 4.5. *Let* $u(x,t) \in [0,1]$ *be a solution of* (1.1) *in* $\mathbb{R} \times \mathbb{R}^+$ *where* $f(u)$ *satisfies* (1.8) *or*

(1.8'). *Suppose that*

$$\lim_{t\to\infty} u(x,t) \equiv 1.$$

Then for every c *with* $|c| < c^*$ *and each* ξ

$$\lim_{t\to\infty} u(\xi+ct,\ t) = 1.$$

We see from Theorems 4.1, 4.3, 4.4, and 4.5, that a disturbance which is initially confined to a half-line $x < x_0$ and which increases to either 1 or α is propagated with the asymptotic speed c^*. More precisely, if β is any constant such that $\beta \in (0,1)$ in cases (1.6), (1.8), or (1.8') or $\beta \in (0,\alpha)$ in case (1.7), and if we define

$$\overline{x}(t) = \max\ \{x: u(x,t) = \beta\},$$

$$\underline{x}(t) = \min\ \{x > 0: u(x,t) = \beta\},$$

then

$$\lim_{t\to\infty} \overline{x}/t = \lim_{t\to\infty} \underline{x}/t = c^*.$$

5. THE INITIAL-BOUNDARY VALUE PROBLEM

We now consider the initial-boundary value problem

$$\left.\begin{aligned} &u_t = u_{xx} + f(u) && \text{in } \mathbb{R}^+ \times \mathbb{R}^+, \\ &u(x,0) = 0 && \text{in } \mathbb{R}^+, \\ &u(0,t) = \psi(t) && \text{in } \mathbb{R}^+, \end{aligned}\right\} \qquad (5.1)$$

where $\psi(t)$ is a given function with values on the interval $[0,1]$. Since both the results for this problem and their derivations are very similar to those for the initial value problem, we shall only sketch the proofs.

The analog of Proposition 2.2 is the following proposition, which is proved in the same manner.

PROPOSITION 5.1. *Let* $q(x) \in [0,1]$ *be a solution of the equation* $q'' + f(q) = 0$ *in* (a,b) *with* $a > 0$, *and let* $q(a) = q(b) = 0$.

Let $v(x,t)$ *denote the solution of the initial-boundary value problem*

$$v_t = v_{xx} + f(v) \quad \text{in} \quad \mathbb{R}^+ \times \mathbb{R}^+,$$

$$v(x,0) = \begin{cases} q(x) & \text{in} \quad (a,b) \\ 0 & \text{in} \quad \mathbb{R}^+ \setminus (a,b), \end{cases}$$

$$v(0,t) = \phi(t) \quad \text{in} \quad \mathbb{R}^+.$$

Suppose that $\phi(t)$ *is nondecreasing,* $\phi(0) = 0$, *and* $\phi(t) \in [0,1]$.

Then $v(x,t)$ *is nondecreasing in* t *and*

$$\lim_{t\to\infty} v(x,t) = \tau(x)$$

where $\tau(x)$ *is the smallest nonnegative solution of the equation*

$$\tau'' + f(\tau) = 0 \quad \text{in} \quad \mathbb{R}^+$$

which satisfies the inequalities

$$\tau(0) \geq \lim_{t\to\infty} \phi(t)$$

and

$$\tau(x) \geq q(x) \quad \text{in} \quad (a,b).$$

Moreover, the convergence of v *to* τ *is uniform on each closed bounded interval in the interior of* $\mathbb{R}^+$.

If $f(u)$ satisfies the conditions (1.6) of the heterozygote intermediate case, we see from the first integral (3.1) that the initial value problem

$$\left.\begin{aligned} q'' + f(q) &= 0 \quad \text{in } \mathbb{R}^+ \\ q(0) &= \beta \end{aligned}\right\} \qquad (5.2)$$

has a unique solution in $[0,1]$ for each $\beta \in (0,1]$ and two such solutions for $\beta = 0$. All these solutions other than $q \equiv 0$ approach 1 as $x \to \infty$. By employing Proposition 2.1 and Proposition 5.1 in a proof like that of Theorem 3.1 we find the following result:

THEOREM 5.1. *Let* $u(x,t) \in [0,1]$ *be the solution of the problem* (5.1) *where* $f(u)$ *satisfies* (1.6). *If* $u(x,t) \not\equiv 0$, *then*

$$\liminf_{t \to \infty} u(x,t) \geq \tau(x)$$

where $\tau(x)$ *is the unique positive solution of the problem* (5.2) *with*

$$\beta = \liminf_{t \to \infty} \psi(t) .$$

In particular,

$$\lim_{x\to\infty} \liminf_{t \to \infty} u(x,t) = 1.$$

Thus if $\psi(t) \not\equiv 0$, $u(x,t)$ approaches values near one far from the boundary regardless of the behavior of $\psi(t)$.

In the same manner we find that if $f(u)$ satisfies the conditions (1.7) of the heterozygote superior case, then

$$\lim_{x\to\infty} \liminf_{t\to\infty} u(x,t) = \lim_{x\to\infty} \limsup_{t\to\infty} u(x,t) = \alpha$$

unless $\psi(t) \equiv 0$.

In the heterozygote inferior case (1.8) or the combustion case (1.8'), it is easily seen from (3.1) that for $\beta \in [0,\kappa)$, with κ defined by (3.3), there is a solution $q_\beta(x)$ of the problem (5.2) such that

$$\lim_{x\to\infty} q_\beta(x) = 0 .$$

(There is another solution which approaches 1, but we shall not use it.)

We then find the following result from Proposition 2.1.

THEOREM 5.2. *Let* $u(x,t) \in [0,1]$ *be the solution of the problem* (5.1) *and let* $f(u)$ *satisfy* (1.8). *If*

$$\beta = \sup_{t\in\mathbb{R}^+} \psi(t) < \kappa,$$

then $u(x,t) \le q_\beta(x)$. *In particular,*

$$\lim_{x\to\infty} \limsup_{t\to\infty} u(x,t) = 0 .$$

One can, in fact, obtain the following analogue of Theorem 3.2.

THEOREM 5.3. *Let* $u(x,t) \in [0,1]$ *be the solution of the problem* (5.1) *with* $f(u)$ *subject to the conditions* (1.8).

Suppose that for some $T \in \mathbb{R}^+$ *and some* $\rho \in (0,\alpha)$

$$\psi(t) \leq \rho \quad \textit{in} \quad (T,\infty)$$

and

$$\int_0^T e^{s(\rho)(T-t)} [\psi(t) - \rho]^+ \, dt < \sqrt{2\pi/e}\,(\alpha - \rho)/s(\rho),$$

where $s(\rho)$ *and* $[\mu]^+$ *are defined as in Theorem* 3.2. *Then*

$$\lim_{x\to\infty} \limsup_{t\to\infty} u(x,t) = 0.$$

REMARK. If $f(u)$ only satisfies (1.8'), we can still show that under the conditions of Theorem 5.3 $\limsup_{t\to\infty} u(x,t) \leq \rho$.

The following result, together with the two preceding theorems, shows that there is a threshold effect in the initial-boundary value problem.

THEOREM 5.4. *Let* $u(x,t) \in [0,1]$ *be the solution of the initial-boundary value problem* (5.1) *and let* $f(u)$ *satisfy* (1.8) *or* (1.8'). *Let* κ *be defined by* (3.3).

For any $\beta \in (\kappa,1)$ *there is a positive time* T_β *with the property that the condition*

$$\psi(t) \geq \beta \quad \textit{on} \quad (t_0, t_0 + T_\beta) \tag{5.3}$$

for some nonnegative t_0 *implies*

$$\lim_{x\to\infty} \liminf_{t\to\infty} u(x,t) = 1. \tag{5.4}$$

PROOF. Let $\chi(t)$ be a smooth nondecreasing function

which satisfies the conditions

$$\chi(t) = \begin{cases} 0 & \text{in } (-\infty,0) \\ \beta & \text{in } (1,\infty). \end{cases}$$

Let $w(x,t)$ denote the solution of the problem

$$w_t = w_{xx} + f(w) \quad \text{in } \mathbb{R}^+ \times \mathbb{R}^+,$$

$$w(x,0) = 0 \qquad \text{in } \mathbb{R}^+,$$

$$w(0,t) = \chi(t) \qquad \text{in } \mathbb{R}^+.$$

By Proposition 5.1

$$\lim_{t\to\infty} w(x,t) = \tau(x)$$

where $\tau(x)$ is the smallest nonnegative solution of the problem (5.2). Since $\beta > \kappa$, the problem (5.2) has only one nonnegative solution, and this solution is increasing, so that

$$\tau(x) > \beta \quad \text{in } \mathbb{R}^+.$$

Moreover, the convergence of $w(x,t)$ to $\tau(x)$ is uniform on each bounded interval.

We recall the solution $q_\beta(x)$ of $q'' + f(q) = 0$ which was used in the proof of Theorem 3.3. It is defined and positive in an interval $(0,b_\beta)$, vanishes at the ends of this interval, and satisfies the inequality

$$q_\beta(x) \le q_\beta(\tfrac{1}{2} b_\beta) = \beta \quad \text{on} \quad (0,b_\beta).$$

Thus $q_\beta(x - 1) < \tau(x)$ on $(1,b_\beta + 1)$.

Since $w(x,t)$ converges to $\tau(x)$ uniformly on $[1,b_\beta + 1]$, there is a time T_β for which

$$w(x,T_\beta) \ge q_\beta(x - 1) \quad \text{on} \quad [1,b_\beta + 1].$$

We now apply Proposition 2.1 to see that because of (5.3),

$$u(x, t+t_0) \geq w(x,t) \quad \text{in} \quad \mathbb{R}^+ \times [0,T_\beta].$$

Hence

$$u(x, T_\beta+t_0) \geq q_\beta(x - 1) \quad \text{in} \quad (1, b_\beta+1).$$

Because of Propositions 2.1 and 5.1, $\liminf_{t \to \infty} u(x,t)$ is bounded below by a nonnegative solution τ^* in $\mathbb{R}^+$ of $q'' + f(q) = 0$ which, in turn, is bounded below by $q_\beta(x - 1)$ in $(1, b_\beta+1)$. In particular $\tau^*(\frac{1}{2} b_\beta + 1) \geq \beta > \kappa$. The first integral (3.1) then shows that $\tau(x) \to 1$ as $x \to \infty$, which proves (5.4).

If, as in Section 4, we introduce the coordinate $\xi = x - ct$, the set $\mathbb{R}^+ \times \mathbb{R}^+$ is mapped onto the set $\{(\xi,t) : \xi > -ct,\ t > 0\}$. If $c > 0$, we can prove an extension of Proposition 2.2 for solutions $v(\xi,t)$ of $v_{\xi\xi} + cv_\xi + f(v) = 0$ which vanish on the boundary $\xi = -ct$. The limit $\tau(\xi)$ is a nonnegative solution of $\tau'' + c\tau' + f(\tau) = 0$ in all of $\mathbb{R}$. The proofs of Theorems 4.1, 4.3, 4.4, and 4.5 now yield the following result, with c^* defined as before.

THEOREM 5.5. *Let* $u(x,t) \in [0,1]$ *be the solution of the problem* (5.1) *and let* $f(u)$ *satisfy one of the conditions* (1.6), (1.7), (1.8), *or* (1.8').

(a) Then for any $c > c^*$ *and any real* ξ

$$\lim_{t\to\infty} u(\xi+ct,\ t) = 0.$$

(b) If $\lim_{x\to\infty} \liminf_{t \to \infty} u(x,t) = 1$, *then for any*

$c \in (0,c^*)$ *and any real* ξ

$$\lim_{t\to\infty} u(\xi+ct,\, t) = 1.$$

(c) If $f(u)$ *satisfies* (1.7) *and* $u(x,t) \not\equiv 0$, *then for any* $c \in (0,c^*)$ *and any real* ξ

$$\liminf_{t\to\infty} u(\xi+ct,\, t) \geq \alpha.$$

Thus c^* is also the asymptotic propagation velocity associated with the initial-boundary value problem (5.1).

6. APPENDIX: REDUCTION OF THE SYSTEM (1.2) TO A SINGLE EQUATION

In this section we shall indicate how the initial value problem for the equation (1.1) with $f(u)$ given by (1.4) is related to the initial value problem for the system (1.2).

We first consider the initial value problem which consists of finding a solution of the system (1.2) in $\mathbb{R}\times\mathbb{R}^+$ subject to the initial conditions

$$\rho_j(x,0) = \gamma_j(x) \quad \text{in } \mathbb{R}, \quad (j = 1,2,3). \tag{6.1}$$

The functions γ_j are assumed to be smooth and nonnegative. Moreover, we assume that there exist positive constants a and b such that

$$0 < a \leq \gamma(x) \equiv \sum_{j=1}^{3} \gamma_j(x) \leq b. \tag{6.2}$$

The solution of the problem (1.2), (6.1) is obtained by inverting the linear part of the operator and applying the method of successive approximations. Let $\rho_j^{(\ell)}$ denote the j-th component of the ℓ-th iterate, where $\rho_j^{(0)}$ is the j-th component of the solution of the uncoupled system

$$\frac{\partial \rho_j}{\partial t} = \frac{\partial^2 \rho_j}{\partial x^2} - \tau_j \rho_j \quad (j = 1,2,3)$$

with the initial data (6.1). By the maximum principle [14], $\rho_j^{(0)} \geq 0$. Let

$$\nu = \max (\tau_1, \tau_2, \tau_3)$$

and

$$\rho^{(\ell)} = \rho_1^{(\ell)} + \rho_2^{(\ell)} + \rho_3^{(\ell)} .$$

Using (6.2) in a standard comparison argument, we obtain the estimate

$$\rho^{(0)}(x,t) \geq ae^{-\nu t} \quad \text{in } \mathbb{R} \times \mathbb{R}^+ .$$

Since the nonlinear terms in (1.2) are nonnegative when the ρ_j are nonnegative and their sum ρ is positive, it follows that $\rho_j^{(\ell)} \geq \rho_j^{(0)}$ and hence also that $\rho^{(\ell)} \geq \rho^{(0)} \geq ae^{-\nu t}$ for all ℓ. It is then a routine matter to show that the $\rho_j^{(\ell)}$ converge to the unique bounded solution of the problem (1.2), (6.1) in $\mathbb{R} \times (0,T]$ for any $T \in \mathbb{R}^+$. In particular, the components ρ_j of the solution are nonnegative and

$$\rho(x,t) \geq ae^{-\nu t} \quad \text{in } \mathbb{R} \times \mathbb{R}^+ . \tag{6.3}$$

Moreover, in carrying through the details of the successive approximations, one also obtains the bounds

$$\rho(x,t) \leq be^{(r-\lambda)t}$$

and

$$\left|\frac{\partial\rho}{\partial x}(x,t)\right| \leq e^{-\lambda t}\sum_{j=1}^{3}\sup_{\mathbb{R}}|\gamma_j'| + 2br\sqrt{\frac{t}{\pi}}\,e^{(r-\lambda)t} \tag{6.4}$$

where

$$\lambda = \min(\tau_1, \tau_2, \tau_3).$$

If we introduce the new dependent variables

$$\left.\begin{aligned} v &= \frac{1}{\rho}(\rho_3 + \frac{1}{2}\rho_2) \\ \sigma &= \frac{1}{\rho^2}(\rho_2^2 - 4\rho_1\rho_3) \\ \mu &= \frac{\partial}{\partial x}(\log\rho), \end{aligned}\right\} \tag{6.5}$$

the system (1.2) becomes

$$\frac{\partial v}{\partial t} - \frac{\partial^2 v}{\partial x^2} - f(v) = 2\mu\frac{\partial v}{\partial x} + \frac{1}{4}\{(\tau_2-\tau_1)v - (\tau_2-\tau_3)(1-v)\}\sigma \tag{6.6a}$$

$$\left.\begin{aligned} &\frac{\partial\sigma}{\partial t} - \frac{\partial^2\sigma}{\partial x^2} - 2\mu\frac{\partial\sigma}{\partial x} + \{r - (\tau_1-\tau_3)(1-2v) + \frac{\tau^*}{4}\sigma\}\sigma \\ &\qquad = 4\tau^* v^2(1-v)^2 - 8\left(\frac{\partial v}{\partial x}\right)^2 \end{aligned}\right\} \tag{6.6b}$$

$$\frac{\partial\mu}{\partial t} - \frac{\partial^2\mu}{\partial x^2} = \frac{\partial}{\partial x}\{\mu^2 + \frac{\tau^*}{4}\sigma + (\tau_2-\tau_3)v^2 + (\tau_2-\tau_1)(1-v)^2\} \tag{6.6c}$$

where f is defined by (1.4) and

$$\tau^* = \tau_1 - 2\tau_2 + \tau_3.$$

Note that v represents the relative density of the allele A in the population. The quantity σ measures the deviation of the system from the Hardy-Weinberg equilibrium, while μ measures its deviation from uniform population density.

To establish a relationship between the equation (1.1) and the system (1.2) we must find conditions which guarantee that the right-hand side of (6.6a) is negligible relative to f. In the usual derivations of equation (1.1) in population genetics it is tacitly assumed that $\sigma \equiv \mu \equiv 0$. However, (6.6c) shows that this assumption implies $\partial v/\partial x \equiv 0$. Thus, if $\sigma \equiv \mu \equiv 0$, there is no spacial variation, and, in particular, no diffusion. Here we shall only assume that $\partial v/\partial x$ and μ are <u>initially</u> small.

The solution of the initial value problem (1.2), (6.1) generates a solution of the system (6.6) with initial data

$$v(x,0) = v_o(x), \quad \sigma(x,0) = \sigma_o(x), \quad \mu(x,0) = \mu_o(x) \quad \text{in } \mathbb{R}$$

where, for example,

$$v_o(x) = \frac{\gamma_3(x) + \frac{1}{2}\gamma_2(x)}{\gamma_1(x)+\gamma_2(x)+\gamma_3(x)} .$$

Since $\rho_j \geq 0$ and $\rho > 0$, we have

$$v \in [0, 1], \quad \sigma \in [-1, 1] \quad \text{in } \mathbb{R} \times \mathbb{R}^+. \tag{6.7}$$

Moreover, by (6.3) and (6.4), μ is bounded in every strip $\mathbb{R} \times [0,T]$.

We shall assume for simplicity that

$$\mu_o \equiv 0 \quad \text{in } \mathbb{R}. \tag{6.8}$$

In view of (6.7) the term in braces on the right-hand side of (6.6c) is bounded by $\mu^2 + \frac{5}{4}\varepsilon$, where

$$\varepsilon = |\tau_1 - \tau_2| + |\tau_3 - \tau_2|$$

is a small parameter. It follows from (6.6c) that

$$|\mu(x,t)| \leq \int_0^t \int_{\mathbb{R}} \left| \frac{\partial G}{\partial x}(x-\xi, t-\eta) \right| \{\mu(\xi,\tau)^2 + \frac{5}{4}\varepsilon\} d\xi d\eta, \quad (6.9)$$

where $G(x,t)$ denotes the fundamental solution of the equation of heat conduction. Let

$$M(t) = \sup_{\mathbb{R}\times[0,t]} |\mu(\xi,\eta)|. \quad (6.10)$$

Then (6.9) implies that

$$M(t) \leq 2\sqrt{\frac{t}{\pi}} \left\{ M(t)^2 + \frac{5}{4}\varepsilon \right\}.$$

Therefore

$$M(t) \leq \frac{1}{4}\sqrt{\frac{\pi}{t}} \left\{ 1 - \left(1 - \frac{20\varepsilon t}{\pi} \right)^{1/2} \right\},$$

provided $20t\varepsilon \leq \pi$. In particular, if $\varepsilon t \in [0, 3\pi/80]$, then $1 - (20\varepsilon t/\pi) \geq 1/4$. It follows from the mean value theorem that

$$|\mu(x,t)| \leq M(t) \leq 5\varepsilon\sqrt{\frac{t}{\pi}} \text{ in } \mathbb{R} \times \left[0, \frac{3\pi}{80\varepsilon}\right]. \quad (6.11)$$

In order to estimate the product $\mu \cdot \partial v/\partial x$ we need a suitable bound for $|\partial v/\partial x|$. For this purpose we assume that there is a constant k_1 such that

$$|v_0'(x)| \leq k_1 \varepsilon^{1/2}. \quad (6.12)$$

Here and in the sequel we denote by k, with or without subscript, a constant which does not depend on ε or r.

From (6.6a) we obtain the integral identity

$$\left.\begin{aligned}\frac{\partial v}{\partial x}(x,t) = &\int_{\mathbb{R}} G(x-\xi,\ t)v_o'(\xi)d\xi \\ &+ \int_o^t\int_{\mathbb{R}} G_x(x-\xi,\ t-\eta)\Big\{2\mu\frac{\partial v}{\partial x} + f(v) \\ &+ \frac{1}{4}[(\tau_2-\tau_1)v - (\tau_2-\tau_3)(1-v)]\sigma\Big\}d\xi d\eta.\end{aligned}\right\} \quad (6.13)$$

Let

$$m(t) = \sup_{\mathbb{R}\times[0,t]} \left|\frac{\partial v}{\partial x}(\xi,\eta)\right|.$$

In view of (1.4), we have $f(v) = O(\varepsilon)$. Thus by (6.7), (6.11), (6.12), and (6.13) we find that

$$m(t) \leq k_1\varepsilon^{1/2} + 5\varepsilon t m(t) + k_2\varepsilon t^{1/2} \quad \text{in} \quad \left[0,\ \frac{3\pi}{80\varepsilon}\right].$$

Thus

$$\left|\frac{\partial v}{\partial x}(x,t)\right| \leq m(t) \leq k\varepsilon^{1/2} \quad \text{in } \mathbb{R}\times\left[0,\ \frac{3\pi}{80\varepsilon}\right]. \qquad (6.14)$$

Moreover, for $t \in [0,\ 3\pi/80\varepsilon]$, equation (6.6b) has the form

$$\frac{\partial\sigma}{\partial t} - \frac{\partial^2\sigma}{\partial x^2} - 2\mu\frac{\partial\sigma}{\partial x} + (r-c)\sigma = g,$$

where, in view of (6.7) and (6.14), $c \leq k_3\varepsilon$ and $|g| \leq k_4\varepsilon$. If $r > 2\varepsilon k_3$, a standard comparison argument shows that

$$|\sigma(x,t)| \leq se^{-\frac{1}{2}rt} + \frac{2\varepsilon k_4}{r} \quad \text{in } \mathbb{R}\times\left[0,\ \frac{3\pi}{80\varepsilon}\right], \qquad (6.15)$$

where s is an upper bound for $|\sigma_o(x)|$. (Note that by (6.7), $s \leq 1$.)

Let u denote the solution of the equation (1.1) with

$u(x,0) = v_o(x)$. By the mean value theorem and (6.6a), the difference

$$w = v - u$$

satisfies the equation

$$\frac{\partial w}{\partial t} - \frac{\partial^2 w}{\partial x^2} - f'(\eta)w = 2\mu\,\frac{\partial v}{\partial x} + \frac{1}{4}\{(\tau_2-\tau_1)v - (\tau_2-\tau_3)(1-v)\}\sigma,$$

where $\eta(x,t)$ lies between $u(x,t)$ and $v(x,t)$ and hence between 0 and 1. We see from (6.7), (6.11), (6.14), and (6.15) that

$$\left|\frac{\partial w}{\partial t} - \frac{\partial^2 w}{\partial x^2} - f'(\eta)w\right| \leq k\varepsilon\left(se^{-\frac{1}{2}rt} + \sqrt{\varepsilon t} + \frac{\varepsilon}{r}\right)$$

$$\text{in } \mathbb{R}\times\left[0,\ \frac{3\pi}{80\varepsilon}\right].$$

According to (1.4), $f'(u) = O(\varepsilon)$ so that $tf'(\eta)$ is uniformly bounded for $t \in [0, 3\pi/80\varepsilon]$. Since $w(x,0) = 0$, a standard comparison argument shows that

$$|w(x,t)| \leq k\varepsilon t\left[\frac{s}{rt}\left(1 - e^{-\frac{1}{2}rt}\right) + \sqrt{\varepsilon t} + \frac{\varepsilon}{r}\right] \tag{6.16}$$

$$\text{in } \mathbb{R}\times\left[0,\ \frac{3\pi}{80\varepsilon}\right].$$

Therefore, the difference $|u - v|$ is very small compared to εt, provided that ε/r is sufficiently small and $\frac{1}{r} \ll t \ll \frac{1}{\varepsilon}$.

Since $f(u)/\varepsilon$ is bounded below on any closed interval which does not contain 0, α, or 1, we can expect the effect of $f(u)$ on the solution of (1.1) to be of the order εt. Thus for t which are large compared to r^{-1}, but small compared to ε^{-1}, the error made by replacing the system (1.2) by the single equation (1.1) is small compared to the effect of f.

REMARKS. 1. The estimate (6.15) shows that $\sigma = O(\varepsilon r^{-1})$ for $t \geq r^{-1} \log(r\varepsilon^{-1})$. Thus any deviation from the Hardy-Weinberg law is damped out in time $r^{-1} \log(r\varepsilon^{-1})$. For a system which has been in operation for at least this long before $t = 0$, we can assume that $s = O(\varepsilon r^{-1})$. It then follows from (6.16) that $|u-v| = t\,o(\varepsilon)$ for all $t \ll \varepsilon^{-1}$.

2. The assumption that $\mu_0 \equiv 0$ is not necessary. The inequality (6.16) is still valid if $\mu_0 = o(\varepsilon^{1/2})$.

3. By simple dimensional considerations, it follows that the propagation speed c^* associated with equation (1.1) is of order $\varepsilon^{1/2}$ when $f(u)$ is given by (1.4). Thus the time it takes a pulse to reach a particular point is of order $\varepsilon^{-1/2}$, which may be small relative to ε^{-1}.

4. Since the bounds (6.11) and (6.14) do not depend upon a or b, a simple limiting process shows that the condition (6.2) may be replaced by the condition $\gamma(x) > 0$.

BIBLIOGRAPHY

1. ARONSON, D.G. and WEINBERGER, H.F., *Asymptotic properties of nonlinear diffusion*, (in preparation).

2. COHEN, H., *Nonlinear diffusion problems*, Studies in Mathematics 7, Studies in Applied Mathematics, ed. A. Taub. Math. Assoc. of America and Prentice Hall, (Englewood Cliffs, N. J.) 1971, 27-63.

3. FRIEDMAN, A. *Partial Differential Equations of Parabolic Type*, Prentice Hall, (Englewood Cliffs, N. J.) 1964.

4. FISHER, R. A., *The advance of advantageous genes*, Ann. of Eugenics 7 (1937), 355-369.

5. GELFAND, I. M., *Some problems in the theory of quasilinear equations*, Uspekhi Mat. Nauk (N. S.) 14 (1959), 87-158; American Math. Soc. Transl. (2) 29 (1963), 295-381.

6. HODGKIN, A. L. and HUXLEY, A. F., *A quantitative description of membrane current and its application to conduction and excitation in nerve*, J. Physiol. 117 (1952), 500-544.

7. KANEL', JA. I., *The behavior of solutions of the Cauchy problem when time tends to infinity, in the case of quasilinear equations arising in the theory of combustion.* Dokl. Akad. Nauk S. S. S.R. 132 (1961), 268-271; Soviet Math. Dokl. 1, 533-536.

8. KANEL', JA. I., *Certain problems on equations in the theory of burning,* Dokl. Akad. Nauk S.S.S.R. 136 (1961), 277-280; Soviet Math. Dokl. 2, 48-51.

9. KANEL', JA. I., *Stabilization of solutions of the Cauchy problem for equations encountered in combustion theory*, Mat. Sbornik (N. S.) 59 (101) (1962), supplement, 245-288.

10. KANEL', JA. I., *On the stability of solutions of the equation of combustion theory for finite initial functions*, Mat. Sbor. (N. S.) 65 (107) (1964) 398-413.

11. KOLMOGOROFF, A., PETROVSKY, I., and PISCOUNOFF, N., *Étude de l'équation de la diffusion avec croissance de la quantité de matière et son application à un*

probleme biologique, Bull. Univ. Moskou, Ser. Internat., Sec. A, 1 (1937) # 6, 1-25.

12. NAGUMO, J., ARIMOTO, S., and YOSHIZAWA, S., *An active pulse transmission line simulating nerve axon*, Proc. Inst. of Radio Eng. 50 (1962), 2061-2070.

13. NAGUMO, J., YOSHIZAWA, S., and ARIMOTO, S., *Bistable transmission lines*, I. E. E. E. Transactions on Circuit Theory 12 (1965), 400-412.

14. PROTTER, M. H. and WEINBERGER, H. F., *Maximum Principles in Differential Equations*, Prentice Hall (Englewood Cliffs, N. J.) 1967.

15. PETROVSKI, I. G., *Ordinary Differential Equations*, Prentice-Hall (Englewood Cliffs, N. J.) 1966, Dover (New York) 1973.

A NEW METHOD IN THE STUDY OF SUBSONIC FLOWS

by

HAÏM BREZIS

Institut de Mathématiques
Université de Paris VI
75230 Paris 5e, France

We present here a preliminary report on a joint work with G. Stampacchia. We are concerned with the determination of a steady subsonic flow for a non viscous compressible fluid, past a given symmetric convex profile in the plane. Our main contribution is to observe that the corresponding free boundary value problem for the stream function (or its Legendre transform) in the hodograph plane can be easily solved by using variational inequalities.

1. THE BASIC EQUATIONS

Let P be a given profile in the plane, convex and symmetric with respect to the x-axis. We consider symmetric flows around P which are uniform at infinity with prescribed velocity q_∞ parallel to the x-axis.

Let $\vec{q} = (q_1, q_2)$ be the velocity and let $q = |\vec{q}| =$

$(q_1^2 + q_2^2)^{1/2}$. The basic equations are

(1) $$\operatorname{div} (\rho\vec{q}) = \frac{\partial}{\partial x} (\rho q_1) + \frac{\partial}{\partial y} (\rho q_2) = 0,$$

(2) $$\operatorname{curl} \vec{q} = \frac{\partial q_2}{\partial x} - \frac{\partial q_1}{\partial y} = 0,$$

where ρ is the density (ρ is not constant). Using Bernouilli's equation we get a relation relating ρ and q; let

(3) $$\rho = h(q),$$

where h is a given decreasing function depending on the physical properties of the fluid (example: for some gases $h(q) = (1 - Cq^2)^{\frac{1}{\gamma-1}}$). The *stream function* ψ is defined by

(4) $$\psi_x = \frac{\partial \psi}{\partial x} = -\rho q_2, \quad \psi_y = \frac{\partial \psi}{\partial y} = \rho q_1.$$

Hence $qh(q) = [\psi_x^2 + \psi_y^2]^{1/2}$ and thus, we can consider q as a function of ψ_x and ψ_y. Using (2) we are led to

(5) $$\left(1 - \frac{q_1^2}{a^2(q)}\right)\psi_{xx} + \left(1 - \frac{q_2^2}{a^2(q)}\right)\psi_{yy} - \frac{2q_1 q_2}{a^2(q)} \psi_{xy} = 0$$

where

$$a^2(q) = -q \frac{h(q)}{h'(q)}$$

($a(q)$ is the local speed of sound). In particular when the fluid is incompressible, (5) reduces to $\Delta\psi = 0$. The boundary condition along P is $\psi = 0$.

Equation (5) is a quasi-linear equation (the coefficients depend on ψ_x and ψ_y) of mixed type. The discriminant is

$$\delta = \frac{q_1^2 q_2^2}{a^4(q)} - \left(1 - \frac{q_1^2}{a^2(q)}\right)\left(1 - \frac{q_2^2}{a^2(q)}\right) = -1 + \frac{q^2}{a^2(q)} .$$

Thus (5) is elliptic when $\delta < 0$ i.e. $q < a(q)$ and (5) is hyperbolic when $\delta > 0$ i.e. $q > a(q)$. Let q_c be the solution of $q_c = a(q_c)$ (q_c is the speed of sound); so that (5) is elliptic in the subsonic range ($q < q_c$) and (5) is hyperbolic in the supersonic range ($q > q_c$). Existence and uniqueness results for (5) have been proved in the elliptic case by Shiffman, Bers, Finn, Gilbarg and others (see [2]). The existence part relies on Schauder fixed point theorem and is not constructive.

2. THE HODOGRAPH TRANSFORM

It is well known that if we consider ψ as a function of $\vec{q}$ instead of (x,y), equation (5) becomes linear in the new variables. More precisely in polar coordinates (i.e. $q_1 = q \cos\theta$, $q_2 = q \sin\theta$) we obtain Chaplygin's equation

$$q \frac{d}{dq}\left(\frac{1}{qh(q)}\right) \frac{\partial^2\psi}{\partial\theta^2} = \frac{\partial}{\partial q}\left(\frac{q}{h(q)} \frac{\partial\psi}{\partial q}\right),$$

which reduces to

$$q^2 \frac{\partial^2\psi}{\partial q^2} + q \frac{\partial\psi}{\partial q} + \frac{\partial^2\psi}{\partial\theta^2} = 0$$

in the incompressible case.

Equation (6) degenerates at $q = 0$, therefore it is convenient to introduce the new variable σ instead of q by

$$\sigma = \int_{q}^{q_c} h(\tau) \frac{d\tau}{\tau} \tag{7}$$

and (6) takes finally the form

$$k(\sigma) \frac{\partial^2\psi}{\partial\theta^2} + \frac{\partial^2\psi}{\partial\sigma^2} = 0 \tag{8}$$

where

$$k(\sigma) = \frac{1}{h^2(q(\sigma))} \left(1 - \frac{q^2(\sigma)}{a^2(q(\sigma))}\right)$$

and $q(\sigma)$ is the reciprocal function in (7). Thus $k(\sigma) > 0$ in the subsonic range $(\sigma > 0)$ and $k(\sigma) < 0$ in the supersonic range $(\sigma < 0)$. (When $k(\sigma)$ is replaced by a linear function near $\sigma = 0$, (8) becomes Tricomi's equation.)

3. THE BOUNDARY CONDITIONS

The main interest of the hodograph transform lies in the fact that we can deal with linear equations. However, equation (8) has to be solved on a domain which is a priori unknown (the image of the profile P under the hodograph transform is not known since we do not know the distribution of the velocities along P).

Because of the symmetry we have $\psi = 0$ along the x-axis and it is sufficient to study the problem in the upper half plane where $\psi > 0$. Assuming that the flow is *totally subsonic*, the profile P is transformed by the hodograph transform into a curve Γ contained in the region $[\sigma > 0]$. Γ is to be regarded as a free boundary; let us denote its equation by $\sigma = \ell(\theta)$.

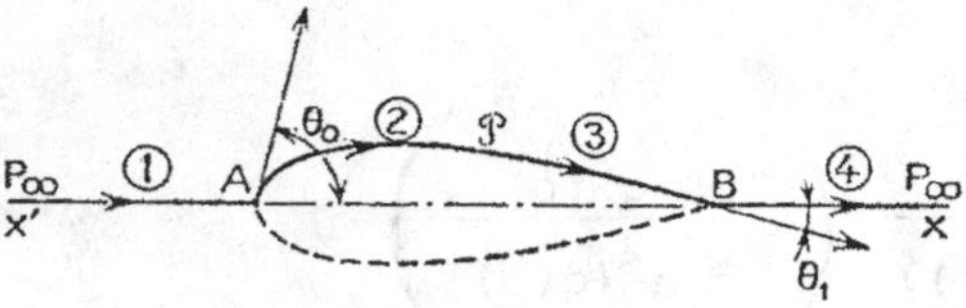

Fig. 1
The physical plane

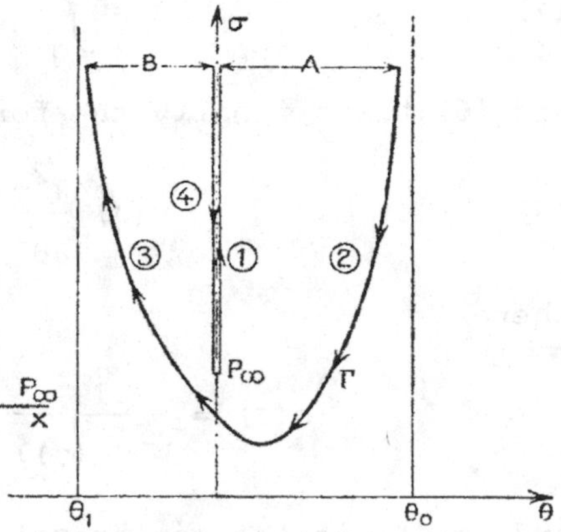

Fig. 2
The hodograph plane

The following lemma gives the boundary conditions satisfied by ψ along Γ (for a proof see e.g. [3] p. 49).

LEMMA 1. *On* Γ *we have* $\psi = 0$ *and*

$$(9) \qquad \frac{\partial\psi}{\partial\sigma} = -\frac{R(\theta)g(\sigma)}{1+k(\sigma)\left(\frac{\partial\ell}{\partial\theta}\right)^2} \quad \text{and} \quad \frac{\partial\psi}{\partial\theta} = \frac{R(\theta)g(\sigma)\frac{\partial\ell}{\partial\theta}}{1+k(\sigma)\left(\frac{\partial\ell}{\partial\theta}\right)^2}$$

where $R(\theta)$ *is the radius of curvature of* $\mathcal{P}$ *at the point* $P \in \mathcal{P}$ *where the tangent at* P *makes an angle* θ *with the x-axis; we take* $R(\theta) < 0$ *since* $\mathcal{P}$ *is convex.*

4. THE MAIN RESULT

For $\sigma > \ell(\theta)$ and $\theta_1 < \theta < \theta_0$ define

$$(10) \qquad u(\theta,\sigma) = \int_{\ell(\theta)}^{\sigma} \frac{k(s)}{q(s)} \psi(\theta,s)ds.$$

The introduction of u as new unknown is suggested by the work of Baiocchi [1] concerning some free boundary value problems in hydrodynamics.

Let $\mathcal{D}$ be defined as

$$\mathcal{D} = \{(\theta,\sigma);\ \sigma > \ell(\theta),\ \theta_1 < \theta < \theta_0,\ \theta \neq 0\} \cup$$

$$\cup \ \{(0,\sigma);\ \ell(0) < \sigma < \sigma_\infty\}$$

where $\sigma_\infty = \int_{q_\infty}^{q_c} h(s) \frac{ds}{s}$.

In the next Lemma we collect some properties of u.

LEMMA 2. *The function* u *satisfies*

$$(11) \qquad u > 0 \quad on \quad \mathcal{D}$$

$$(12) \qquad \frac{1}{q^2(\sigma)} \frac{\partial}{\partial\sigma}\left(\frac{q^2(\sigma)}{k(\sigma)} \frac{\partial u}{\partial\sigma}\right) + \frac{\partial^2 u}{\partial\theta^2} + u = -R(\theta) \quad on \quad \mathcal{D}$$

$$(13) \qquad u = 0 \quad on \quad \Gamma$$

$$(14) \qquad \frac{\partial u}{\partial\theta} = 0 \quad and \quad \frac{\partial u}{\partial\sigma} = 0 \quad on \quad \Gamma$$

$$(15) \qquad u(0,\sigma) = H \quad for \quad \sigma \geq \sigma_\infty$$

where 2H *denotes the height of the profile.*

PROOF OF LEMMA 2. (11), (13) and (14) follow directly from the definition of u and the fact that $\psi > 0$. (12) is a consequence of (8), (9) and the following relations:

$$q'(\sigma) = q_\sigma = -\frac{q(\sigma)}{h(q(\sigma))}, \tag{16}$$

$$\frac{k(\sigma)}{q(\sigma)} + \left(\frac{q_\sigma}{q^2}\right)_\sigma = 0. \tag{17}$$

Next, observe that $u(0,\sigma)$ is constant for $\sigma \geq \sigma_\infty$ since $\psi(0,s) = 0$ for $s \geq \sigma_\infty$. In order to determine the value of the constant we have to make some computation.

For a fixed θ, let $(x(s), y(s))$ be the point in the physical plane which is mapped by the hodograph transform into (θ,s) i.e.

$$\frac{1}{\rho}(\psi_y, -\psi_x) = q(\cos\theta, \sin\theta).$$

By (17) we have

$$u(\theta,\sigma) = -\int_{\ell(\theta)}^{\sigma} \psi(\theta,s)\frac{d}{ds}\left(\frac{q'}{q^2}\right) ds$$

$$= -\int_{\ell(\theta)}^{\sigma} \psi(x(s),y(s))\frac{d}{ds}\left(\frac{q'}{q^2}\right) ds$$

$$= -\frac{q'(\sigma)}{q^2(\sigma)}\psi(\theta,\sigma) + \int_{\ell(\theta)}^{\sigma} (\psi_x x' + \psi_y y')\frac{q'}{q^2} ds$$

$$= \frac{1}{q\rho}\psi + \int_{\ell(\theta)}^{\sigma} (-\sin\theta\, x' + \cos\theta\, y')\ \frac{q'\rho}{q}\, ds$$

$$= \frac{1}{q\rho}\psi + \int_{\ell(\theta)}^{\sigma} (\sin\theta\, x' - \cos\theta\, y')\, ds$$

$$= \frac{1}{q\rho}\psi + \sin\theta[x(\sigma) - x(\ell(\theta))] - \cos\theta[y(\sigma) - y(\ell(\theta))].$$

For $\theta = 0$ and $\sigma > \sigma_\infty$, we have $\psi = 0$, $y(\sigma) = 0$, $y(\ell(\theta)) = H$ and so $u(0,\sigma) = H$.

REMARK. The function

$$\Psi(\theta,\sigma) = \rho q(y(\sigma)\cos\theta - x(\sigma)\sin\theta) - \psi(\theta,\sigma)$$

is called the Legendre transform of ψ. There is a simple relation relating Ψ and u; namely

$$u(\theta,\sigma) = -\frac{1}{q\rho}\Psi - x(\ell(\theta))\sin\theta + y(\ell(\theta))\cos\theta.$$

Note that $(x(\ell(\theta)), y(\ell(\theta)))$ are the coordinates of the point $P \in \mathcal{P}$ where the tangent to $\mathcal{P}$ makes an angle θ with the x-axis.

Let $\Omega = \{(\theta,\sigma);\ \theta_1 < \theta < \theta_0,\ \sigma > 0\}$. We extend u to Ω by choosing $u(\theta,\sigma) = 0$ for $0 < \sigma \le \ell(\theta)$. Our purpose is to show that u is the solution of some variational inequality on Ω. We define first the appropriate functional space.

Let

$$V = \left\{ \begin{array}{l} v(\theta,\sigma)\,;\ q(\sigma)v(\theta,\sigma) \in L^2(\Omega),\ q(\sigma)v_\theta(\theta,\sigma) \in L^2(\Omega), \\ \dfrac{q(\sigma)}{\sqrt{k(\sigma)}}\, v_\sigma(\theta,\sigma) \in L^2(\Omega),\quad v = 0 \ \text{ on } \ \partial\Omega \end{array} \right\}$$

with the canonical norm.

Let K be the closed convex set

$$K = \{v \in V;\ v \geq 0 \ \text{ on } \ \Omega \ \text{ and } \ v(0,\sigma) = H \ \text{ for } \ \sigma \geq \sigma_\infty\}.$$

Let $a(u,v)$ be the bilinear form defined on V by

$$a(u,v) = \int_\Omega \left(\frac{1}{k(\sigma)}\, u_\sigma v_\sigma + u_\theta v_\theta - uv\right) q^2(\sigma)\, d\theta d\sigma.$$

It is clear that $a(u,v)$ is continuous on V.

LEMMA 3. a *is coercive on* K *i.e.*

$$\lim_{\substack{\|u\| \to \infty \\ u \in K}} \frac{a(u,u)}{\|u\|_V} = \infty.$$

PROOF OF LEMMA 3. We divide Ω into 3 regions

$$\Omega_1 = \{(\theta,\sigma);\ \theta_1 < \theta < 0\ ,\quad \sigma > \sigma_\infty\}$$

$$\Omega_2 = \{(\theta,\sigma);\ 0 < \theta < \theta_0\ ,\quad \sigma > \sigma_\infty\}$$

$$\Omega_3 = \{(\theta,\sigma);\ \theta_1 < \theta < \theta_0\ ,\quad 0 < \sigma \leq \sigma_\infty\}.$$

Let $v_0 \in K$ be fixed and let $v = u - v_0$ with $u \in K$, so that $v(0,\sigma) = 0$ for $\sigma \geq \sigma_\infty$. Thus, by Poincaré's inequality we have

$$\int_{\theta_1}^{0} |v_\theta|^2 d\theta \geq \frac{\pi}{|\theta_1|}\int_{\theta_1}^{0} |v|^2 d\theta \geq 2\int_{\theta_1}^{0} |v|^2 d\theta$$

(since $v = 0$ at both end points). Hence

$$\int_{\Omega_1}\left(\frac{1}{k}\, v_\sigma^2 + v_\theta^2 - v^2\right)q^2 d\theta d\sigma \geq \int_{\Omega_1}\left(\frac{1}{k}\, v_\sigma^2 + \frac{1}{3}\, v_\theta^2 + \frac{1}{3}\, v^2\right)q^2 d\theta d\sigma$$

and similarly on Ω_2.

On the other hand we have

$$\int_0^{\sigma_\infty} v^2 q^2 d\sigma \leq q_c^2 \int_0^{\sigma_\infty}\left|\int_0^{\sigma} v_\sigma ds\right|^2 d\sigma \leq \frac{1}{2}\, q_c^2 \sigma_\infty^2 \int_0^{\sigma_\infty} v_\sigma^2 d\sigma$$

$$\leq \frac{1}{C}\int_0^{\sigma_\infty} \frac{1}{k}\, v_\sigma^2 q^2 d\sigma,$$

where $C > 0$ is a constant such that

$$C\, \frac{k(\sigma)}{q^2(\sigma)} \leq \frac{2}{q^2 \sigma_\infty^2} \quad \text{for} \quad 0 \leq \sigma \leq \sigma_\infty .$$

Therefore

$$I = \int_{\Omega_3}\left(\frac{1}{k}\, v_\sigma^2 + v_\theta^2 - v^2\right)q^2 d\theta d\sigma$$

$$\geq \frac{1}{2}\int_{\Omega_3} \frac{1}{k}\, v_\sigma^2 q^2 d\theta d\sigma + \frac{C}{2}\int_{\Omega_3} v^2 q^2 d\theta d\sigma$$

$$+ \left(1 - \frac{C}{4}\right)\int_{\Omega_3} v_\theta^2 q^2 d\theta d\sigma + \frac{C}{4}\int_{\Omega_3} v_\theta^2 q^2 d\theta d\sigma - \int_{\Omega_3} v^2 q^2 d\theta d\sigma .$$

By Poincaré's inequality we have

$$\int_{\theta_1}^{\theta_0} v_\theta^2 d\theta \geq \frac{\pi}{\theta_0 - \theta_1} \int_{\theta_1}^{\theta_0} v^2 d\theta \geq \int_{\theta_1}^{\theta_0} v^2 d\theta$$

so that

$$I \geq \frac{1}{2} \int_{\Omega_3} \frac{1}{k} v_\sigma^2 q^2 d\theta d\sigma + \frac{C}{4} \int_{\Omega_3} (v_\theta^2 + v^2) q^2 d\theta d\sigma .$$

Adding these inequalities we get for $u \in K$

$$a(u-u_0 , u-u_0) \geq \alpha \|u-u_0\|_V^2$$

and therefore $a(u,u) \geq \frac{\alpha}{2} \|u\|_V^2 - \beta$ for $u \in K$.

THEOREM 1. *The function* u *defined by* (10) *is the unique solution of the variational inequality*

$$(18) \quad \begin{cases} u \in K \\ a(u, v-u) \geq \int_\Omega R(\theta)(v-u) q^2 d\theta d\sigma \quad \textit{for all} \quad v \in K. \end{cases}$$

In other words

$$\min_K \left\{ \frac{1}{2} a(v,v) - \int_\Omega R(\theta) v q^2 d\theta d\sigma \right\}$$

is achieved at u.

PROOF OF THEOREM 1. It follows from Lemma 2 that $u \in K$. On the other hand, since $u = 0$ outside $\mathcal{D}$ we have

$$a(u,v-u) = \int_{\mathcal{D}} \left[\frac{1}{k} u_\sigma (v-u)_\sigma + u_\theta (v-u)_\theta - u(v-u) \right] q^2 d\theta d\sigma$$

$$= \int_{\mathcal{D}} \left[- \left(\frac{q^2}{k} u_\sigma \right)_\sigma - q^2 u_{\theta\theta} - q^2 u \right] (v-u) d\theta d\sigma$$

(use (14) and the fact that $(v-u)(0,\sigma) = 0$ for $\sigma \geq \sigma_\infty$).
Hence by (12)

$$a(u,v-u) = \int_{\mathcal{D}} R(\theta)(v-u)q^2 d\theta d\sigma \geq \int_{\Omega} R(\theta)(v-u)q^2 d\theta d\sigma$$

since on $\Omega \backslash \mathcal{D}$, $R(\theta) \leq 0$ (by the convexity of P), $v \geq 0$ and $u = 0$.

CONCLUSION. Given the profile P (so that θ_0, θ_1, $R(\theta)$ and H are known), given the physical properties of the fluid (so that $q(\sigma)$ and $k(\sigma)$ are known), given the velocity at infinity (so that σ_∞ is known), there is a unique solution u of (18) (and there are even efficient methods for numerical computation of u). Having solved (18) let

$$\mathcal{D}' = \{(\theta,\sigma) \in \Omega;\ u(\theta,\sigma) > 0\}.$$

If $\overline{\mathcal{D}'}$ does not intersect the axis $\{\sigma = 0\}$, the curve Γ, boundary of $\mathcal{D}'$, represents the distribution of velocities along P. If $\overline{\mathcal{D}'}$ intersects the axis $\{\sigma = 0\}$ we conclude that q_∞ is too large and there exists no totally subsonic flow past P.

5. AN ESTIMATE OF q_{max}

We assume that $R_m = \min_\theta |R(\theta)| > 0$.

THEOREM 2. *Let* $q_A \geq q_\infty$ *be the solution of the equation*

(19) $$\frac{H}{R_m} - 1 = \frac{q_A}{q_\infty}\left[-1 + \frac{1}{h(q_\infty)} \int_{q_\infty}^{q_A} \frac{h(s)}{s} ds\right]$$

and suppose $q_A \le q_c$. *Then the maximum velocity of the flow is less than* q_A.

PROOF. Let $A = \int_{q_A}^{q_c} \frac{h(s)}{s} ds$ and consider the function ϕ defined for $A \le \sigma \le \sigma_\infty$ by

$$\phi(\sigma) = R_m q_A \int_A^\sigma \frac{k(s)}{q(s)} (s - A) ds .$$

Therefore $\phi \ge 0$, $\phi(A) = 0$, $\phi_\sigma(A) = 0$; and also

$$\phi(\sigma) = R_m q_A \int_A^\sigma \left(- \frac{q'}{q^2}\right)' (s-A) ds$$

$$= R_m q_A \left[- \frac{q'}{q^2} (\sigma-A) + \int_A^\sigma \frac{q'}{q^2} ds\right]$$

$$= R_m q_A \left[\frac{1}{qh(q)} (\sigma-A) - \int_{q_A}^{q} d\left(\frac{1}{q}\right)\right]$$

$$= R_m q_A \left[\frac{1}{qh(q)} \int_q^{q_A} \frac{h(s)}{s} ds - \frac{1}{q} + \frac{1}{q_A}\right] .$$

Thus $\phi(\sigma_\infty) = H$ (use (19)).

On the other hand $\phi_\sigma = R_m q_A \frac{k}{q} (\sigma-A)$ and

$$\left(\frac{q^2}{k} \phi_\sigma\right)_\sigma = R_m q_A [q + q'(\sigma-A)] .$$

Finally

$$\frac{1}{q^2} \left(\frac{q^2}{k} \phi_\sigma\right)_\sigma + \phi = R_m q_A \left[\frac{1}{q} - \frac{\sigma-A}{qh(q)}\right] + \phi = R_m .$$

We define $\phi(\sigma) = 0$ for $0 \le \sigma \le A$; by a comparison argument we are going to show that $u \le \phi$.

Therefore $\mathcal{D}' \subset [\sigma > A]$ and hence the maximum velocity of the flow is less than q_A.

Let

$$a_3(u,v) = \int_{\Omega_3} \left(\frac{1}{k} u_\sigma v_\sigma + u_\theta v_\theta - uv\right) q^2 d\theta d\sigma .$$

We check easily that

$$(20) \quad \begin{cases} a_3(u,v-u) \geq \int_{\Omega_3} R(\theta)(v-u) q^2 d\theta d\sigma & \text{for all } v \in V; \\ v \geq 0 \text{ on } \Omega_3 \text{ and } v(\theta,\sigma_\infty) = u(\theta,\sigma_\infty). \end{cases}$$

On the other hand we have also

$$(21) \quad \begin{cases} a_3(\phi,v-\phi) \geq \int_{\Omega_3} -R_m (v-\phi) q^2 d\theta d\sigma & \text{for all } v; \\ v \geq 0 \text{ on } \Omega_3,\ v(\theta,\sigma_\infty) = H,\ v(\theta,0) = 0. \end{cases}$$

We choose $v = \min\{u,\phi\} = u - (u-\phi)^+$ in (20) and $v = \max\{u,\phi\} = \phi + (u-\phi)^+$ in (21). Therefore we get

$$a_3(u,-(u-\phi)^+) \geq -\int_{\Omega_3} R(\theta)(u-\phi)^+ q^2 d\theta d\sigma$$

$$a_3(\phi,(u-\phi)^+) \geq -\int_{\Omega_3} R_m (u-\phi)^+ q^2 d\theta d\sigma$$

and by addition

$$a_3(u-\phi,\ (u-\phi)^+) \leq \int_{\Omega_3} (R(\theta)+R_m)(u-\phi)^+ q^2 d\theta d\sigma \leq 0 .$$

Hence $a_3((u-\phi)^+,\ (u-\phi)^+) \leq 0$ and $u \leq \phi$.

REMARK. In the incompressible case, equation (19) reduces to

$$\frac{H}{R_m} - 1 = \frac{q_A}{q_\infty}\left[-1 + \log \frac{q_A}{q_\infty}\right].$$

In particular when P is a sphere $H = R_m$ and

$\log \frac{q_A}{q_\infty} = 1$; we deduce that $\max q \le e q_\infty$. In this case an explicit computation shows that $\max q = 2q_\infty$.

REFERENCES

1. BAIOCCHI, C., *Su un problema di frontiera libera connesso a questioni di idraulica*, Annali di Mat. Pura ed Appl. 92 (1972), 107-127.

2. BERS, L., *Mathematical Aspects of Subsonic and Transonic Gas Dynamics*, Chapman and Hall, London, 1958.

3. FERRARI, C. and TRICOMI, F., *Transonic Aerodynamics*, Acad. Press, New York, 1968.

INTERPOLATION CLASSES FOR MONOTONE OPERATORS

by

HAÏM BREZIS
Institut de Mathématiques
Université de Paris VI
75230 Paris 5e, France

We discuss some characterizations and properties of the interpolation classes as studied by D. Brezis (see [1]).

Let H be a Hilbert space and let A be a maximal monotone operator in H with domain $D(A)$ (for properties of monotone operators see e.g. [3]). Let $J_\lambda = (I + \lambda A)^{-1}$ and let $A_\lambda = \frac{1}{\lambda}(I - J_\lambda)$ $(\lambda > 0)$.

One can build intermediate classes between $D(A)$ and $\overline{D(A)}$ by using the same approach as in the linear case (see [4]). Surprisingly many results known for linear operators are still valid in the nonlinear case.

First observe that

$$\begin{cases} \text{for } u \in D(A), & |u - J_\lambda u| \le C\lambda, \\ \text{for } u \in \overline{D(A)}, & \lim\limits_{\lambda \to 0} |u - J_\lambda u| = 0. \end{cases}$$

Roughly speaking, we are going to classify intermediate spaces between $D(A)$ and $\overline{D(A)}$ by measuring how fast $(I + \lambda A)^{-1}u \to u$ as $\lambda \to 0$. For example it is natural to consider

$$\{u \in \overline{D(A)}\ ;\ \frac{|u - J_\lambda u|}{\lambda^\alpha} \text{ remains bounded as } \lambda \to 0\}.$$

More generally for $0 < \alpha < 1$ and $1 \le p \le \infty$ we define

$$\mathcal{B}_{\alpha,p} = \{u \in \overline{D(A)}\ ;\ \frac{|u - J_\lambda u|}{\lambda^\alpha} \in L^p_*(0,1)\}$$

where

$$L^p_* = L^p\left(\frac{d\lambda}{\lambda}\right).$$

Using the fact that $|A_\lambda u|$ is a nonincreasing function of λ on $(0,\infty)$ it is easy to verify that

$$\mathcal{B}_{\alpha,p} \subset \mathcal{B}_{\beta,q} \quad \text{for all } \alpha > \beta, \text{ for all } p,q\ ,$$

$$\mathcal{B}_{\alpha,p} \subset \mathcal{B}_{\alpha,q} \quad \text{for all } \alpha \text{ and for all } p \le q.$$

1. THE TRACE, SUM AND K CHARACTERIZATIONS

THEOREM 1 (TRACE METHOD). *Let* $u_0 \in \overline{D(A)}$; *then* $u_0 \in \mathcal{B}_{\alpha,p}$ *if and only if there is a continuous function* $v(t)$ *from* $[0,1]$ *into* H *such that* v *is absolutely continuous on* $(0,1]$, $t^{1-\alpha}|\frac{dv}{dt}(t)| \in L^p_*(0,1)$, $v(t) \in D(A)$ *a.e. and* $t^{1-\alpha}|A^0 v(t)| \in L^p_*(0,1)$.

The proof of Theorem 1 relies on the following

LEMMA 1. *For every* $u_0 \in \overline{D(A)}$, $J_t u_0$ *is Lipschitz continuous on* $(0,\infty)$ *and*

$$\left|\frac{d}{dt} J_t u_0\right| \le |A_t u_0| \quad \text{a.e. on } (0,\infty)\ .$$

PROOF OF LEMMA 1. By the resolvent formula we have

$$J_{t+h}\, u_0 = J_t\left(\frac{t}{t+h}\, u_0 + \left(1 - \frac{t}{t+h}\right) J_{t+h}\, u_0\right)$$

and thus

$$\frac{|J_{t+h}\, u_0 - J_t\, u_0|}{h} \le \frac{1}{t+h}\, |u_0 - J_{t+h}\, u_0| \le |A_t\, u_0|.$$

LEMMA 2 (HARDY). *For every measurable* $\phi \ge 0$ *and* $\alpha > 0$, *one has*

$$\left\| \frac{1}{t^\alpha} \int_0^t \phi(s)\, \frac{ds}{s} \right\|_{L_*^p} \le \frac{1}{\alpha} \left\| \frac{1}{t^\alpha}\, \phi(t) \right\|_{L_*^p}.$$

PROOF OF THEOREM 1. Suppose first $u_0 \in B_{\alpha,p}$; we deduce from Lemma 1 that $v(t) = J_t\, u_0$ satisfies the conditions of Theorem 1.

Conversely we have

$$\left|\frac{d}{dt}\, J_t\, u_0\right| \le |A_t\, u_0| \le |A^0 v(t)| + \frac{1}{t}|u_0 - v(t)|$$

and therefore

$$\frac{|J_t\, u_0 - u_0|}{t^\alpha} \le \frac{1}{t^\alpha} \int_0^t (|A^0 v(s)| + \frac{1}{s}|u_0 - v(s)|)ds.$$

We apply now Lemma 2 to $\phi(t) = t|A^0 v(t)| + |u_0 - v(t)|$. Clearly $t^{-\alpha}\phi(t) \in L_*^p$ since

$$\frac{|u_0 - v(t)|}{t^\alpha} \le \frac{1}{t^\alpha} \int_0^t \left|\frac{dv}{dt}(s)\right| ds$$

(use again Lemma 2 with $t\left|\frac{dv}{dt}(t)\right|$).

Slight modifications in the proof of Theorem 1 lead to

THEOREM 2 (SUM METHOD). *Let* $u_0 \in \overline{D(A)}$; *then* $u_0 \in B_{\alpha,p}$ *if and only if there exist two functions* $v_1(t)$ *and* $v_2(t)$ *from* $(0,1)$ *to* H *such that* $u_0 = v_1(t) + v_2(t)$, $v_1(t) \in D(A)$, $t^{1-\alpha} |A^0 v_1(t)| \in L_*^p(0,1)$, $t^{-\alpha}|v_2(t)| \in L_*^p(0,1)$.

THEOREM 3 (METHOD K). *Given* $u_0 \in \overline{D(A)}$, *define*

$$K(t,u_0) = \operatorname*{Inf}_{v \in D(A)} \{|u_0 - v| + t\,|A^0 v|\}.$$

Then $u_0 \in B_{\alpha,p}$ *if and only if* $t^{-\alpha}K(t,u_0) \in L_*^p(0,1)$.

As a consequence, we deduce easily an interpolation theorem.

COROLLARY 1. *Let* A_1 *(resp.* A_2*) be a maximal monotone operator in* H_1 *(resp.* H_2*). Let* T *be a mapping from* $D(A_1)$ *into* $D(A_2)$ *and from* $\overline{D(A_1)}$ *into* $\overline{D(A_2)}$ *such that*

$$\text{for all } x,y \in \overline{D(A_1)}, \quad |Tx - Ty|_2 \le C|x-y|_1,$$

$$\text{for all } x \in D(A_1), \quad |A_2^0 Tx|_2 \le C|A_1^0 x|_1 + \omega(|x|_1)$$

where ω *is a continuous function. Then* T *maps* $B_{\alpha,p}(A_1)$ *into* $B_{\alpha,p}(A_2)$.

PROOF. Let $v_1(t) = (I+tA_1)^{-1} u_0$; we have

$$Tu_0 = Tv_1(t) + (Tu_0 - Tv_1(t))$$

with $t^{1-\alpha}|A_2^0 Tv_1(t)|_2 \le Ct^{1-\alpha}|A_1 v_1(t)|_1 + t^{1-\alpha}\omega(|v_1(t)|)$

and $t^{-\alpha}|Tu_0 - Tv_1(t)|_2 \le Ct^{-\alpha}|u_0 - v_1(t)|_1$.

It follows from Theorem 2 that $Tu_0 \in B_{\alpha,p}(A_2)$ as soon as $u_0 \in B_{\alpha,p}(A_1)$.

2. THE SEMIGROUP CHARACTERIZATION

Let $S(t)$ be the semigroup generated by $-A$.

THEOREM 4. *We have*

$$B_{\alpha,p} = \left\{ u \in \overline{D(A)}\ ;\ \frac{|u-S(t)u|}{t^{\alpha}} \in L_*^p(0,1) \right\}$$

and in addition

$$\frac{1}{3}\left\|\frac{u-S(t)u}{t^{\alpha}}\right\|_{L^p} \le \left\|\frac{u-J_t u}{t^{\alpha}}\right\|_{L_*^p} \le \frac{2}{1+\alpha}\ \left\|\frac{u-S(t)u}{t^{\alpha}}\right\|_{L_*^p} .$$

The proof of Theorem 4 relies on the next Lemma and on Hardy's inequality.

LEMMA 3. *For every* $u_0 \in \overline{D(A)}$ *we have*

$$\frac{1}{3}|u_0-S(t)u_0| \le |u_0-J_t u_0| \le \frac{2}{t}\int_0^t |u_0-S(\tau)u_0|d\tau.$$

PROOF OF LEMMA 3. For every $v \in D(A)$ we have $|u_0-S(t)u_0| \le |u_0-v| + |v-S(t)v| + |S(t)v-S(t)u_0|$ $\le 2|u_0-v| + t|A^0 v|$. Choosing $v = J_t u_0$ we obtain the first part of Lemma 3.

Next suppose $u_0 \in D(A)$, so that $u(t) = S(t)u_0 \in D(A)$ and we have

$$(Au(t) - Av,\ u(t) - v) \ge 0.$$

Hence

$$\frac{1}{2}\frac{d}{dt}|u(t)-v|^2 \le (Av,v-u(t)) = (Av,v-u_0) + (Av,u_0-u(t)).$$

After integration we obtain

$$\frac{1}{2}|u(t)-v|^2 - \frac{1}{2}|u_0-v|^2 \le t(Av,v-u_0) + |Av|\int_0^t |u_0-u(\tau)|d\tau.$$

Choosing $v = J_t u_0$ we are led to

$$\frac{1}{2}|u_0 - J_t u_0|^2 \le |u_0 - J_t u_0| \frac{1}{t}\int_0^t |u_0 - u(\tau)|d\tau ,$$

and the second inequality in Lemma 3 follows.

3. THE CASE WHERE $A = \partial\phi$.

Let $\phi : H \to (-\infty,+\infty]$ be a convex l.s.c. function such that $\phi \not\equiv +\infty$ and let $D(\phi) = \{u \in H;\ \phi(u) < +\infty\}$. For $u \in D(\phi)$ define

$$\partial\phi(u) = \{f \in H;\ \phi(v)-\phi(u) \ge (f, v-u) \quad \text{for each} \quad v \in H\}.$$

Then $A = \partial\phi$ is maximal monotone and it is known that the semigroup $S(t)$ generated by $-A$ has a smoothing action on the initial data. More precisely, for all $t > 0$, $S(t)\overline{D(A)} \subset D(A)$ and

$$(*) \quad |A^0 S(t)u_0| \le |A^0 v| + \frac{1}{t}|u_0-v| \quad \text{for all} \quad v \in D(A)$$

(for a proof of (*) see e.g. [3]).

In case $A = \partial\phi$, there are further characterizations of $\mathcal{B}_{\alpha,p}$.

THEOREM 5. *Let* $u_0 \in \overline{D(A)}$; *then* $u_0 \in B_{\alpha,p}$ *if and only if* $t^{1-\alpha}|\frac{d}{dt}S(t)u_0| = t^{1-\alpha}|A^0S(t)u_0| \in L_*^p(0,1)$.

PROOF OF THEOREM 5. Choose $v = J_tu_0$ in (*) to show that $t^{1-\alpha}|A^0S(t)u_0| \in L_*^p$ as soon as $u_0 \in B_{\alpha,p}$.

Conversely observe that $\dfrac{|u_0-S(t)u_0|}{t^\alpha} \le \dfrac{1}{t^\alpha}\displaystyle\int_0^t |\frac{d}{dt}S(\tau)u_0|d\tau$ and use Hardy's inequality.

REMARK 1. In case $A = \partial\phi$, the second inequality in Lemma 3 can be replaced by a pointwise inequality. More precisely, for every $u_0 \in \overline{D(A)}$ we have $|u_0 - J_tu_0| \le (1 + \frac{1}{\sqrt{2}})|u_0 - S(t)u_0|$. (Question: what is the best constant?) Indeed, from the inequality $\phi(v)-\phi(u) \ge (-\frac{du}{dt}, v-u)$ we get by integration

$$t\phi(v) - t\phi(u(t)) \ge \frac{1}{2}|u(t)-v|^2 - \frac{1}{2}|u_0-v|^2 .$$

Taking $v = J_tu_0$ we obtain

$$t\phi(J_tu_0) - t\phi(u(t)) \ge \frac{1}{2}|u(t) - J_tu_0|^2 - \frac{1}{2}|u_0 - J_tu_0|^2.$$

On the other hand

$$\phi(u(t)) - \phi(J_tu_0) \ge \frac{1}{t}(u_0 - J_tu_0, u(t) - J_tu_0).$$

By addition we have

$$(u_0 - J_tu_0, u(t) - J_tu_0) \le \frac{1}{2}|u_0 - J_tu_0|^2 - \frac{1}{2}|u(t) - J_tu_0|^2.$$

Therefore $\sqrt{2}\,|u(t) - J_tu_0| \le |u_0 - u(t)|$ and finally $|u_0 - J_tu_0| \le (1 + \frac{1}{\sqrt{2}})|u_0 - u(t)|$.

REMARK 2. In the linear case it is well known that even for $u_0 \in \overline{D(A)}$, $\frac{1}{t}\int_0^t S(\tau)u_0\, d\tau \in D(A)$. It is not known whether a similar result holds true in the nonlinear case. However, one can show at least that $x = \frac{1}{t}\int_0^t S(\tau)u_0\, d\tau \in B_{\frac{1}{2},\infty}$. Indeed, we have $(Av, v-u(t)) \geq (-\frac{du}{dt}(t), v-u(t))$ and therefore $(Av, v-x) \geq \frac{1}{2t}|u(t) - v|^2 - \frac{1}{2t}|u_0 - v|^2 \geq -\frac{1}{2t}|u_0 - v|^2$. Choosing $v = J_\lambda x$ we conclude that $x \in B_{\frac{1}{2},\infty}$. In case $A = \partial\phi$, one can prove further that $x \in D(\phi)$ ($= B_{\frac{1}{2},2}$ by Theorem 6). This follows from the inequality

$$\phi^*\left(\frac{u_0 - u(t)}{t}\right) + \phi\left(\frac{1}{t}\int_0^t u(s)ds\right) \leq \frac{1}{2t}|u_0|^2 - \frac{1}{2t}|u(t)|^2.$$

THEOREM 6. *We have* $D(\phi) = B_{\frac{1}{2},2}$.

PROOF OF THEOREM 6. It follows directly from the equation $\frac{du}{dt} + \partial\phi(u) \ni 0$ that

$$\int_\varepsilon^t \left|\frac{du}{dt}\right|^2 ds + \phi(u(t)) - \phi(u(\varepsilon)) = 0, \quad 0 < \varepsilon < t.$$

Therefore if $u_0 \in D(\phi)$, $\frac{du}{dt} \in L^2(0,1)$ i.e. $t^{\frac{1}{2}}\frac{du}{dt} \in L^2_*(0,1)$ and by Theorem 5, $u_0 \in B_{\frac{1}{2},2}$. Conversely, if $u_0 \in B_{\frac{1}{2},2}$, then $\frac{du}{dt} \in L^2(0,1)$ and so $\limsup_{\varepsilon\to 0} \phi(u(\varepsilon)) < \infty$; therefore $u_0 \in D(\phi)$.

Finally, one can characterize the classes $B_{\alpha,p}$ by considering the behavior of $\phi(J_t u_0)$ (resp. $\phi(S(t)u_0)$) near $t = 0$. We have to distinguish two cases.

THEOREM 7. *Suppose* $0 < \alpha < \frac{1}{2}$ *and let* $u_0 \in \overline{D(A)}$.

Then $u_0 \in B_{\alpha,p}$ *if and only if* $t^{\frac{1}{2}-\alpha}\sqrt{|\phi(J_t u_0)|} \in L^p_*(0,1)$ (*resp.* $t^{\frac{1}{2}-\alpha}\sqrt{|\phi(S(t)u_0|} \in L^p_*(0,1)$).

THEOREM 8. *Suppose* $\frac{1}{2} < \alpha < 1$ *and let* $u_0 \in D(\phi)$. *Then* $u_0 \in B_{\alpha,p}$ *if and only if* $t^{\frac{1}{2}-\alpha}\sqrt{\phi(u_0)-\phi(J_t u_0)} \in L^p_*(0,1)$ (*resp.* $t^{\frac{1}{2}-\alpha}\sqrt{\phi(u_0)-(S(t)u_0)} \in L^p_*(0,1)$).

4. AN EXAMPLE

Consider $H = L^2(\Omega)$, $Au = |u|^{q-2}u$ $(q > 2)$ with $D(A) = L^{2(q-1)}(\Omega)$.

THEOREM 9. $B_{\alpha,p} = [L^2(\Omega), L^{2(q-1)}(\Omega)]_{\theta,t} = L^{s,t}(\Omega)$ ($L^{s,t}$ *denotes the Lorentz space with*

$$\theta = \frac{\alpha(q-1)}{1+\alpha(q-2)}, \quad s = 2(1+\alpha(q-2)), \quad t = p(1+\alpha(q-2)).$$

In particular $B_{\alpha,2} = L^s(\Omega)$, $B_{\alpha,\infty} = M^s(\Omega)$ (M^s *denotes the Marcinkiewicz space*).

PROOF. By Theorem 2, $u_0 \in B_{\alpha,p}$ if and only if there exist two functions $v_1(t)$ and $v_2(t)$ such that $t^{1-\alpha}|Av_1(t)| \in L^p_*$ and $t^{-\alpha}|v_2(t)| \in L^p_*$, i.e.

$$t^{\frac{1-\alpha}{q-1}}\|v_1(t)\|_{L^{2(q-1)}} \in L^{p(q-1)}_* \quad \text{and} \quad t^{-\alpha}\|v_2(t)\|_{L^2} \in L^p_*.$$

Therefore $B_{\alpha,p}$ appears as an interpolation space (in the usual sense, see [4]) between $L^2(\Omega)$ and $L^{2(q-1)}(\Omega)$. More precisely $B_{\alpha,p} = [L^2(\Omega), L^{2(q-1)}(\Omega)]_{\theta,t}$

Further examples and applications can be found in [2].

REFERENCES

1. BREZIS, D., *Classes d'interpolation associées à un opérateur monotone*, C. R. Acad. Sc. Paris 276 (1973), 1553-1556.

2. BREZIS, D., *Perturbations singulières et problèmes d'évolution avec défaut d'ajustement*, C. R. Acad. Sci. Paris 276 (1973), 1597-1600.

3. BREZIS, H., *Opérateurs Maximaux Monotones et Semi-groupes de Contractions dans les Espaces de Hilbert*, Math. Studies 5, North Holland, 1973.

4. LIONS, J. L. and PEETRE, J., *Sur une classe d'espaces d'interpolation*, I.H.E.S. Public. Math. 19 (1964), 5-68.

SINGULAR NONLINEAR INTEGRAL EQUATIONS OF HAMMERSTEIN TYPE

by

FELIX E. BROWDER
Department of Mathematics
University of Chicago
Chicago, Illinois 60637

INTRODUCTION

Let Ω be a measure space with σ-finite measure μ. By a nonlinear integral equation of Hammerstein type, we mean an integral equation of the form

$$(1) \qquad u(x) + \int_\Omega k(x,y)f(y,u(y))\mu(dy) = h(x) \qquad (x \in \Omega).$$

Here, the unknown function u and the inhomogeneous term h lie in a given class X of measurable real valued functions on Ω, k is a given real valued measurable function on $\Omega \times \Omega$, while f is a given real valued function on $\Omega \times R$. (R denotes the real numbers.)

If we introduce the corresponding operators K and F acting on real-valued functions v and w on Ω by setting

$$(2) \qquad (Kv)(x) = \int_\Omega k(x,y)v(y)\mu(dy), \quad (x \in \Omega),$$

and

(3) $$(Fw)(x) = f(x,w(x)), \quad (x \in \Omega),$$

the integral equation (1) can be formally rewritten as

(4) $$(I + KF)(u) = h,$$

where I is the identity operator on X.

If we seek solutions u from the space X, we distinguish the regular from the singular Hammerstein equations by the criterion usually applied in the linear case, i.e. the regular equations are those in which the Hammerstein operator KF is defined on the whole space X while the singular cases are those in which KF is defined on a domain which is a proper subset of X.

In a number of recent papers ([1],[2],[3],[4],[5]), H. Brezis and the writer have made a detailed study of the regular Hammerstein equation in the spaces $L^p(\mu)$, $1 < p \leq +\infty$. A simple representative result of this investigation is the following theorem:

THEOREM ([1]). *Suppose that for the Hammerstein equation* (1), *the following hypotheses are all valid:*

(I) $f(y,r)$ *satisfies the Caratheodory condition (i.e.* $f(y,r)$ *is continuous in* r *for almost all fixed* y, *and is measurable in* y *for all fixed* r*).*

(II) $f(y,r)$ *is nondecreasing in* r *for each fixed* y.

(III) *If we set* $f_r(y) = f(y,r)$ *for each constant* r, *then* f_r *lies in* $L^1(\mu)$ *for each* r.

(IV) K *is a bounded linear mapping of* $L^1(\mu)$ *into* $L^\infty(\mu)$.

(V) *For each* u *in* $L^1(\mu)$,

$$(Ku,u) \geq 0$$

where $(w,v) = \int_{\Omega} w(y)v(y)\mu(dy)$.

Then for each h *in* $L^{\infty}(\mu)$, *there exists exactly one solution* u *in* $L^{\infty}(\mu)$ *of the integral equation* $(I + KF)(u) = h$, *and* u *depends continuously upon* h *in* $L^{\infty}(\mu)$.

In the result just stated, no growth conditions were imposed upon the behavior of $f(y,r)$ as a function of r, but on the other hand, the very strong condition was imposed upon the linear operator K with kernel $k(x,y)$ that it maps $L^1(\mu)$ into $L^{\infty}(\mu)$. (This is equivalent to assuming that k is essentially bounded in the measure $\mu \times \mu$ on the product space $\Omega \times \Omega$.) If we wish to avoid growth conditions upon f without restricting k so seriously, we must extend this result to the singular case.

It is our object in the present paper to apply variants of the methods used in the regular case to obtain the following theorem for the singular case:

THEOREM 1. *Let* (Ω,μ) *be a* σ*-finite measure space, and consider the Hammerstein equation*

$$(1) \quad u(x) + \int_{\Omega} k(x,y)f(y,u(y))\mu(dy) = h(x), \quad (x \in \Omega).$$

Suppose that:

(I) $f(y,r)$ *satisfies the Caratheodory condition.*

(II) $f(y,r)$ *is nondecreasing in* r *for fixed* y.

(III)' *If* $f_r(y) = f(y,r)$ *for a constant* r, *then each* f_r *lies in* $L^1(\mu) \cap L^{\infty}(\mu)$.

$(IV)_1$ *K is a bounded linear mapping of $L^2(\mu)$ into $L^2(\mu)$ and of $L^1(\mu)$ into $L^1(\mu)$. If $f_o(y) = f(y,0)$, Kf_o lies in $L^\infty(\mu)$.*

$(IV)_c$ *There exists a sequence $\{\Omega_j\}$ of subsets of finite measure in Ω whose union is all of Ω such that for each j, if $K_j u = Ku|_{\Omega_j}$, then K_j is a compact linear mapping of $L^1(\mu) \cap L^\infty(\mu)$ into $L^1(\mu|_{\Omega_j})$.*

(V) *For each u in $L^2(\mu)$,*

$$(Ku,u) \geq 0.$$

Then for each h in $L^1(\mu) \cap L^\infty(\mu)$, there exists u in $L^1(\mu)$ with $F(u)$ in $L^1(\mu)$ which is a solution of the integral equation

$$u + KFu = h.$$

The proof of Theorem 1 uses the following result which is interesting in its own right.

THEOREM 2. *Let (Ω,μ) be a σ-finite measure space. Suppose that*

(I) *f satisfies the Caratheodory condition.*

(II) *$f(y,r)$ is nondecreasing in r for fixed y.*

(III) *For each r, f_r lies in $L^2(\mu)$.*

$(IV)_2$ *K is a bounded linear mapping from $L^2(\mu)$ into $L^2(\mu)$.*

$(V)_2$ *There exists $c > 0$ such that for all u in $L^2(\mu)$, $(Ku,u) \geq c\,\|u\|^2_{L^2}$.*

Then for each h in $L^2(\mu)$, there exists one and only one u in $L^2(\mu)$ with $F(u)$ in $L^2(\mu)$ such that

$$u + KFu = h.$$

Moreover, the mapping $(I + KF)^{-1}$ *is continuous and bounded from* $L^2(\mu)$ *to* $L^2(\mu)$.

In Section 1, we develop some preliminary results about the Niemitskyi operator F. Section 2 contains the proof of Theorem 2. In Section 3, we use Theorem 2 to give a proof of Theorem 1.

SECTION 1

We begin with a measure theoretical discussion of basic properties of the Niemitskyi operator F.

PROPOSITION 1. *Suppose that* f *is a real valued function on* $\Omega \times R$ *which satisfies the Caratheodory condition and such that* $f(y,r)$ *is nondecreasing in* r *for each fixed* y. *Let* v *be a real valued measurable function on* Ω. *Then there exists exactly one measurable function* u *on* Ω *(up to equality a.e.) such that*

$$F(u)(y) + u(y) = v(y), \tag{5}$$

for almost all y *in* Ω.

PROOF OF PROPOSITION 1. For each y in Ω, we set

$$g_y(s) = f(y,s) + s, \quad s \in R.$$

Since $f(y,s)$ is nondecreasing in s, $g_y(s)$ is strictly increasing in s on R. The relation (5) is equivalent to

$$g_y(u(y)) = v(y), \quad (y \in \Omega).$$

Hence, there exists at most one solution $u(y)$ for each y. Moreover, if $f_0(y) = f(y,0)$, f_0 is measurable and we remark that

$$\{g_y(s) - f_0(y)\}s = s\{f(y,s) - f(y,0)\}s + s^2 \geq s^2.$$

Hence for each fixed y,

$$g_y(s)s \geq s^2 - |s||f_o(y)| \to +\infty, \quad (|s| \to +\infty).$$

Therefore, each g_y is a homeomorphism of R onto R, and for each y in Ω, there exists exactly one solution $u(y)$ of the equation

$$g_y(u(y)) = v(y).$$

We denote this solution by

$$u(y) = U(v,y).$$

We need only to prove that the function u thus defined is measurable on Ω.

Let $\Omega^{(j)} = \{y|\ |f_o(y)| \leq j\}$. Since Ω is the union of the sequence $\{\Omega^{(j)}\}$, in order to prove that u is measurable on Ω, it suffices to prove u measurable on each $\Omega^{(j)}$. If we replace Ω by $\Omega^{(j)}$, we may assume without loss of generality that $|f_o(y)| \leq M$ on all of Ω with a fixed constant M. Similarly, we may also assume that $\mu(\Omega) < +\infty$.

Let $u(y) = U(v,y)$, $u_1(y) = U(v_1, y)$ for two measurable functions v and v_1. For y in Ω,

$$\{f(y,u(y)) - f(y,u_1(y))\} + \{u(y) - u_1(y)\} = v(y) - v_1(y).$$

Since the two terms in curly brackets on the left side of this equation have the same algebraic sign, it follows that

$$u(y) - u_1(y) \leq v(y) - v_1(y).$$

By symmetry, we also have

$$u_1(y) - u(y) \leq v_1(y) - v(y).$$

Hence

$$|u(y) - u_1(y)| \leq |v(y) - v_1(y)|.$$

Suppose we replace v_1 in the last inequality by a sequence of simple functions $\{v_n\}$ converging to v almost everywhere. If $u_n(y) = U(v_n, y)$, it follows that u_n converges to u a.e.. Hence if all the u_n are measurable, so is their limit u. Thus it suffices to assume that v is itself a simple function. More precisely, since the passage from $v(y)$ to $u(y)$ does not depend on the values of v at points other than y, we may assume without loss of generality that v is a constant function on Ω, $v(y) = c$.

For all y in Ω, we then have

$$g_y(s)\, s \geq s^2 - Ms.$$

For any s for which $g_y(s) = c$, it follows that

$$s^2 \leq (|c| + M)|s|,$$

i.e.,

$$|s| \leq (|c| + M).$$

Let $k = |c| + M + 1$, and consider the closed interval $K = [-k, +k]$ in R. Since K is compact, it follows from the Caratheodory condition that for all y in the complement of a μ-null set N in Ω, g_y is uniformly continuous in y on K. Since

$$N = \bigcup_{\xi>0} \bigcap_{\gamma>0} \bigcup_{\substack{s,r \in K \\ |s-r| \leq \gamma}} \{y \mid |f(y,r) - f(y,s)| \geq \xi\},$$

we know that if we set

$$B_{\gamma,\xi} = \bigcup_{\substack{s,r \in K \\ |s-r| \le \gamma}} \{y \mid |f(r,y) - f(s,y)| \ge \xi\}$$

then

$$\mu(\bigcap_{\gamma > 0} B_{\gamma,\xi}) = 0,$$

for each $\xi > 0$. Since $\mu(\Omega) < +\infty$, $\mu(B_{\gamma,\xi}) \to 0$ as $\gamma \to 0$. Let $\beta > 0$ be given. We may choose γ_j so small that

$$\mu(B_{\gamma_j, 2^{-j}}) \le \beta 2^{-j}.$$

Let

$$B^\beta = \bigcup_{j=1}^{\infty} B_{\gamma_j, 2^{-j}}.$$

We remark that

$$\mu(B^\beta) \le \sum_{j=1}^{\infty} \mu(B_{\gamma_j, 2^{-j}}) \le \beta \sum_{j=1}^{\infty} 2^{-j} = \beta.$$

On the other hand, for y in the complement of B^β in Ω, g_y is uniformly continuous on K, uniformly in y. If we can show that u is measurable on $\Omega - B^\beta$ for each $\beta > 0$, and if we choose a sequence $\beta_k \to 0$ as $k \to \infty$, then u will be measurable on

$$\bigcup_k (\Omega - B^{\beta_k}) = \Omega - \bigcap_k B^{\beta_k} = \Omega - N_1,$$

where N_1 is a μ-null set. Hence, we may assume that Ω is replaced by $\Omega - B^\beta$ for some $\beta > 0$ and that g_y is uniformly continuous in r on K uniformly for all y in Ω.

Suppose that $\{f_k(y,r)\}$ is a sequence of nondecreasing real valued functions satisfying the Caratheodory condition converging to $f(y,r)$ uniformly on $\Omega \times K$. Let $u_k(y)$ be the unique solution of

$$f_k(y,u_k(y)) + u_k(y) = c, \quad (y \in \Omega)$$

for each k. For k sufficiently large, it follows by an analogue of the argument for f that the solutions $u_k(y)$ lie in the interval K. Suppose that

$$d_k = \sup_{y,r} |f_k(y,r) - f(y,r)|.$$

Then

$$f(y,u_k(y)) + u_k(y) = c + \{f(y,u_k(y)) - f_k(y,u_k(y))\}.$$

Since

$$f(y,u(y)) + u(y) = c,$$

it follows that

$$|u(y) - u_k(y)| \leq d_k \to 0, \quad (k \to \infty).$$

If each of the u_k are measurable, therefore, so is their limit u.

We obtain such f_k and u_k as follows: Since changing f outside the interval K will not affect $u(y)$ for any y in Ω, we set $f(y,r) = f(y,k)$ for $r \geq k$, $f(y,r) = f(y,-k)$ for $r \leq -k$. We then approximate the modified function f which is uniformly continuous in r by taking its convolution with ζ_k, where $\zeta_k(s) = k^{-1}\zeta(ks)$ and the function ζ is C^1, has compact support, is non-negative, and has

$$\int \zeta(s)ds = +1.$$

Each f_k is nondecreasing in r and satisfies a uniform Lipschitz condition

$$|f_k(y,s) - f_k(y,r)| \leq C_k|r-s|$$

for all r and s in R as well as all y in Ω. Moreover, $f_k(y,r)$ converges to $f(y,r)$ uniformly in y and r. Thus, we may assume that the given function f itself satisfies a Lipschitz condition

$$|f(y,r) - f(y,s)| \leq C|r-s|$$

on $\Omega \times R$.

The solution $s = u(y)$ of the equation $g_y(s) = c$ satisfies the equation

$$s = c - f(y,s),$$

as well as the equation

$$s = \xi\{c - f(y,s)\} + (1-\xi)s$$

for any ξ in $(0,1)$. Let

$$T_{\xi,y}(s) = \xi\{c - f(y,s)\} + (1-\xi)s.$$

Then for any r and s in R,

$$|T_{\xi,y}(s) - T_{\xi,y}(r)|^2 = (1-\xi)^2|s-r|^2 + \xi^2|f(y,s) - f(y,r)|^2$$

$$- 2\xi(1-\xi)\{f(y,s) - f(y,r)\}(s-r)$$

$$\leq (1-\xi)^2|s-r|^2 + \xi^2|f(y,s) - f(y,r)|^2$$

$$\leq \{(1-\xi)^2 + C^2\xi^2\}|s-r|^2.$$

Hence

$$|T_{\xi,y}(s) - T_{\xi,y}(r)| \leq c(\xi)|s-r|$$

where

$$c(\xi)^2 = 1 - 2\xi + (C^2+1)\xi^2 < 1$$

if

$$(C^2+1)\xi < 2.$$

For such a choice of ξ, (which is independent of y in Ω), the root $s = u(y)$ which is a fixed point of $T_{\xi,y}$ is obtainable as the limit of the iterate

$$s_{n,\xi}(y) = (T_{\xi,y})^n(0).$$

Each $s_{n,\xi}$, being a composition of functions satisfying the Caratheodory condition, is measurable. Hence u is measurable. q.e.d.

SECTION 2

From the technical results of Section 1, we derive the following important fact:

PROPOSITION 2. *Suppose that* f *satisfies the*

Caratheodory condition and that $f(y,r)$ *is nondecreasing in* r *for each fixed* y *in* Ω*. Suppose that* f_o *lies in* $L^2(\mu)$*. Let* F *be the mapping defined in* $L^2(\mu)$ *by the Niemitskyi operator with domain*

$$D(F) = \{u \mid u \in L^2(\mu),\ F(u) \in L^2(\mu)\}.$$

Then F *is a single-valued maximal monotone mapping from* $H = L^2(\mu)$ *to* 2^H.

PROOF OF PROPOSITION 2. By the standard theory of monotone operators (cf. Chap. 7 of [6]), it suffices to show that for each v in $L^2(\mu)$, the equation

$$u + KFu = v$$

has a solution u in $L^2(\mu)$. By Proposition 1, there exists a measurable function u on Ω such that a.e.,

$$u(y) + f(y,u(y)) = v(y).$$

Since $f_o(y) = f(y,0)$ and since $f(y,r)$ is nondecreasing in r, we have

$$(f(y,u(y)) - f_o(y))u(y) \geq 0.$$

On the other hand

$$u(y)^2 + \{f(y,u(y)) - f_o(y)\}u(y) = \{v(y) - f_o(y)\}u(y).$$

Hence

$$|u(y)|^2 \leq |v(y) - f_o(y)| \cdot |u(y)|,$$

and

$$|u(y)| \leq |v(y)| + |f_o(y)|.$$

Since v and f_o lies in $L^2(\mu)$, so does u. q.e.d.

The fact that F is a monotone mapping from $L^2(\mu)$ to $L^2(\mu)$ follows from the elementary calculation

$$(F(u)-F(u_1), u-u_1)$$

$$= \int_\Omega \{f(y,u(y))-f(y,u_1(y))\}\{u(y)-u_1(y))\}\mu(dy) \geq 0$$

since the integrand is non-negative.

PROOF OF THEOREM 2. By hypothesis, for all u in $L^2(\mu)$,

$$(Ku,u) \geq c\|u\|^2.$$

Hence K is an isomorphism of $L^2(\mu)$ with inverse mapping L. Let v be any element of $H = L^2(\mu)$, $u = Lv$. Then $v = Ku$, and we obtain

$$(Lv,v) = (Ku,u) \geq c\|u\|^2 \geq c\|K\|^{-2}\|v\|^2 = c_1\|v\|^2.$$

The equation

$$u + KFu = h$$

is obviously equivalent to the equation

$$Lu + Fu = Lh.$$

It is obviously sufficient to prove that $(L+F)$ is injective, has a continuous inverse, and is surjective.

Since F is monotone and L is strongly monotone, it follows trivially that $(L+F)$ is injective and that $(L+F)^{-1}$ is Lipschitzian. To show that $(L+F)$ maps onto H, we remark that by Proposition 2, F is maximal monotone in H. L is monotone and continuous. Hence, by standard monotone operator theorems, $L+F$ is maximal monotone. If $(L+F)$ is coercive, it follows from other standard monotone operator theorems ([6], Chap. 7) that $R(L+F) = H$. However, if $w_o = F(u_o)$ for some u_o in $D(F)$, we see that

$$\frac{((L+F)(u),u-u_o)}{\|u-u_o\|} \geq \frac{(Lu,u)+(w_o,u-u_o)}{\|u-u_o\|}$$

$$\geq \frac{c_1\|u\|^2 - c_2\|u-u_o\|}{\|u-u_o\|} \to +\infty$$

as $\|u-u_o\| \to +\infty$. Hence $(L+F)$ is coercive with respect to u_o and its range is all of H. q.e.d.

SECTION 3

PROOF OF THEOREM 1. We begin by a normalization of the problem. Let f_o be given as before by $f_o(y) = f(y,0)$. By hypothesis f_o lies in $L^1(\mu) \cap L^\infty(\mu)$, while Kf_o lies in $L^\infty(\mu)$. A function u is a solution in the desired sense of the equation

$$u + KFu = h$$

if and only if u is a solution of the modified equation

$$u + K\{Fu - f_o\} = h - Kf_o = h_1$$

where h_1 lies in $L^1(\mu) \cap L^\infty(\mu)$ if h does because Kf_o lies in $L^\infty(\mu) \cap L^1(\mu)$, the latter since K maps $L^1(\mu)$ into itself. If we replace $f(y,r)$ by the kernel of the modified problem $\{f(y,r) - f(y,0)\}$, we may assume that for the original problem $f_o = 0$.

For each $\xi > 0$, we may apply Theorem 2 and obtain a solution u_ξ in $L^2(\mu)$ with Fu_ξ in $L^2(\mu)$ of the approximate equation

$$u_\xi + (K + \xi I)Fu_\xi = h.$$

We choose a sequence $\{\xi_k\}$ of positive numbers converging to zero, and set $u_k = u_{\xi_k}$.

Since $f_o = 0$, and $f(y,r)$ is nondecreasing in r, it follows that for all real r,

$$f(y,r)r \geq 0.$$

It follows immediately that

$$f(y,r)r \geq |f(y,r)| \cdot |r|.$$

Since u_k is a solution of the equation

$$u_k + (K + \xi_k I)Fu_k = h,$$

we may take the inner product of both sides of the equation with Fu_k and we obtain

$$(Fu_k, u_k) + (Fu_k, (K + \xi_k I)Fu_k) = (Fu_k, h).$$

Since $(Kv,v) \geq 0$ for all v in $L^2(\mu)$ by the hypothesis, the second term on the left side of the equation is non-negative. Hence

$$(Fu_k, u_k) \leq (h, Fu_k) \leq \|h\|_{L^\infty} \|Fu_k\|_{L^1}.$$

On the other hand,

$$\begin{aligned}(Fu_k, u_k) &= \int_\Omega f(y,u_k(y))u_k(y)\mu(dy) \\ &\geq \int_\Omega |f(y,u_k(y))| \cdot |u_k(y)|\mu(dy).\end{aligned}$$

Thus

$$\int |Fu_k| \cdot |u_k| \mu(dy) \le \|h\|_{L^\infty} \|Fu_k\|_{L^1} .$$

Let $R = \|h\|_{L^\infty} + 1$. If $|u_k(y)| \le R$, then

$$f(y,-R) \le Fu_k(y) \le f(y,+R),$$

and in particular

$$|Fu_k(y)| \le |f_{-R}(y)| + |f_R(y)| .$$

We now decompose the integral

$$\int_\Omega |Fu_k(y)| \, |u_k(y)| \mu(dy)$$

$$= \left(\int_{|u_k(y)|<R} + \int_{|u_k(y)|\ge R} \right) |Fu_k(y)| \cdot |u_k(y)| \mu(dy)$$

$$= I_1 + I_2 .$$

For I_1, we have the estimate

$$I_1 \le \int_\Omega R\{|f_R(y)| + |f_{-R}(y)|\} \mu(dy) \le R\left\{\|f_R\|_{L^1} + \|f_{-R}\|_{L^1}\right\}.$$

For I_2, we have the estimate

$$I_2 \ge \int_{|u_k(y)|\ge R} R|Fu_k(y)| \mu(dy)$$

$$\ge R\|Fu_k\|_{L^1} - R \int_{|u_k(y)|<R} |Fu_k(y)| \mu(dy)$$

$$\ge R\|Fu_k\|_{L^1} - R\left\{\|f_R\|_{L^1} + \|f_{-R}\|_{L^1}\right\}.$$

Combining these estimates, we obtain the complete estimate

$$R\|Fu_k\|_{L^1} - Rc(R) \le \|h\|_{L^\infty}\|Fu_k\|_{L^1}.$$

Since $R = \|h\|_{L^\infty} + 1$, this inequality becomes

$$\|Fu_k\|_{L^1} \le Rc(R),$$

where the latter constant is independent of the index k. If we introduce this estimate into the earlier estimates, we obtain

$$\int |Fu_k(y)| \cdot |u_k(y)| \mu(dy) \le M$$

with a constant M independent of k.

Let $E_{k,n} = \{y \mid |u_k(y)| < n\}$, and let $p_{k,n}$ be the characteristic function of the set $E_{k,n}$. Then $p_{k,n}u_k$ and $(1-p_{k,n})u_k$ have disjoint supports. It follows from the definition of the Niemitskyi operator F that

$$Fu_k = w_{k,n} + z_{k,n}$$

where

$$w_{k,n} = F(p_{k,n}u_k), \quad z_{k,n} = F((1-p_{k,n})u_k).$$

Since

$$|p_{k,n}u_k| \le n,$$

it follows from the hypothesis of Theorem 1 that

$$f(y,-n) \le w_{k,n}(y) \le f(y,n).$$

Since f_{-n} and f_n both lie in $L^1(\mu) \cap L^\infty(\mu)$ by hypothesis, so does $w_{k,n}$, and their norms in both spaces are bounded as $k \to \infty$ for fixed n. (Indeed, the sequence $\{w_{k,n}\}$ for fixed n lies in a weakly compact subset of $L^1(\mu)$.)

By the condition $(IV)_c$ of the hypothesis, Ω is the union of the sequence of subsets $\{\Omega_j\}$ such that for each j, the mapping of u into $K(u)|_{\Omega_j}$ is a compact mapping of $L^1(\mu) \cap L^\infty(\mu)$ into $L^1(\mu|_{\Omega_j})$. Fix one such index j. Then the sequence $\{Kw_{k,n}|_{\Omega_j}\}$ for fixed j and n is a relatively compact subset of $L^1(\mu|_{\Omega_j})$. For $z_{k,n}$, on the other hand, we have

$$n\|z_{k,n}\|_{L^1} \leq \int_{|u_k(y)|\geq R} |Fu_k(y)| \cdot |u_k(y)|\mu(dy) \leq M,$$

i.e.

$$\|z_{k,n}\|_{L^1(\mu|_{\Omega_j})} \leq \|z_{k,n}\|_{L^1(\mu)} \leq \frac{M}{n} .$$

What we have shown above is that for each $\xi_n = \frac{M}{n}$, there exists a strongly relatively compact subset of $L^1(\mu|_{\Omega_j})$ such that the sequence $\{Fu_k\}$ lies at distance at most ξ_n from that subset. It follows that the sequence $\{Fu_k|_{\Omega_j}\}$ for each fixed j is a relatively strongly compact subsequence in $L^1(\mu|_{\Omega_j})$. If we use this fact together with the standard result that from a strongly convergent sequence in an L^1 space, we may extract an infinite subsequence which converges almost everywhere, we may apply the diagonal procedure to obtain an infinite subsequence of the sequence $\{Fu_k\}$

(which we will continue to denote as $\{Fu_k\}$) such that for each j, $Fu_k(y)$ converges almost everywhere on Ω_j. Using the fact that Ω is the union of the sequence Ω_j, it follows that $Fu_k(y)$ converges almost everywhere in Ω.

We assert that because of this almost everywhere convergence and the uniform estimates

$$\int |Fu_k(y)| \cdot |u_k(y)| \mu(dy) \leq M$$

it follows that Fu_k converges as $k \to \infty$ in $L^1(\mu)$. To show this we apply the Vitali theorem which asserts that strong convergence in $L^1(\mu)$ is equivalent to the uniform absolute continuity of the set functions

$$\delta_k(E) = \int_E |Fu_k(y)| \mu(dy)$$

together with the fact that for any $\xi > 0$, we can find a set E_ξ of finite measure such that $|\delta_k|(\Omega - E_\xi) < \xi$ for all k. To check the uniform absolute continuity of the sequence $\{\delta_k\}$, we note that

$$\delta_k(E) \leq \left\{ \int_{E\cap\{y \mid |u_k(y)|<n\}} + \int_{E\cap\{y \mid |u_k(y)|\geq n\}} \right\} |Fu_k(y)| \mu(dy)$$

$$\leq \{ \|f_n\|_{L^\infty} + \|f_{-n}\|_{L^\infty}) \mu(E) + \frac{M}{n} .$$

If $\gamma > 0$ is given, we can first choose n so large that $Mn^{-1} < \gamma/2$, and then (E) so small that $\{\|f_n\|_{L^\infty} + \|f_{-n}\|_{L^\infty}\}\mu(E) < \gamma/2$. A similar proof yields the existence of the set E_ξ using the fact that f_n and f_{-n} lie in $L^1(\mu)$.

Since Fu_k converges strongly in $L^1(\mu)$ to an element w of $L^1(\mu)$ as $k \to \infty$ and since K is continuous from $L^1(\mu)$ to $L^1(\mu)$, it follows that since

$$u_k = h - KFu_k - \xi_k Fu_k ,$$

u_k converges in $L^1(\mu)$ to u in $L^1(\mu)$ with $u = h - Kw$. Passing once more to an infinite subsequence, we may assume that $u_k(y)$ converges almost everywhere in Ω to $u(y)$. Since $f(y,r)$ is continuous in r for almost y in Ω, it follows that

$$Fu_k(y) = f(y,u_k(y))$$

converges almost everywhere in Ω to $F(u)$. On the other hand, $Fu_k(y)$ must converge almost everywhere to $w(y)$. Hence $Fu = w$. Finally, we see that

$$u = h - Kw = h - KFu,$$

i.e. u is the desired solution of the equation

$$u + KFu = h.$$

q.e.d.

BIBLIOGRAPHY

[1] BREZIS, H. and BROWDER, F. E., *Some new results about Hammerstein equations*, Bull. Amer. Math. Soc. 80 (1974), 568-572.

[2] BREZIS, H. and BROWDER, F. E., *Maximal monotone operators in nonreflexive Banach spaces and nonlinear integral equations of Hammerstein type*, Bull. Amer. Math. Soc. 80 (1974).

[3] BREZIS, H. and BROWDER, F. E., *Existence theorems for nonlinear integral equations of Hammerstein type*, Bull. Amer. Math. Soc. 80 (1974).

[4] BREZIS, H. and BROWDER, F. E., *Equations integrales nonlineaires du type Hammerstein,* C. R. Acad. Sci. Paris 279 (1974), 1-2.

[5] BREZIS, H. and BROWDER, F. E., *Nonlinear integral equations and systems of Hammerstein type,* (to appear in Advances in Mathematics).

[6] BROWDER, F. E., *Nonlinear operators and nonlinear equations of evolution in Banach spaces,* Proc. of Symposia in Pure Math., vol. 18, Part II, American Mathematical Society, Providence, 1974.

[7] BROWDER, F. E., *Nonlinear functional analysis and nonlinear integral equations of Hammerstein and Urysohn type,* Contributions to Nonlinear Functional Analysis, Academic Press 1971, 425-500.

THE LEFSCHETZ FIXED POINT THEOREM AND ASYMPTOTIC FIXED POINT THEOREMS

by

FELIX E. BROWDER

Department of Mathematics
University of Chicago
Chicago, Illinois 60637

INTRODUCTION

Let X be a topological space, f a continuous mapping of X into X. The mapping f is said to be locally compact if each point x_o of X has a neighborhood U such that $f(U)$ is relatively compact in X. The mapping f is said to have a compact attractor if there exists a compact subset A of X such that for each x in X, the orbit

$$O_f(x) = \bigcup_{j=0}^{\infty} \{f^j(x)\}$$

has a point of A in its closure.

In a recent paper ([4]), the writer has established the following theorem:

THEOREM 1. *Let* X *be a locally convex topological vector space,* f *a locally compact, continuous self-mapping of* X *with a compact attractor. Then* f *has a fixed point in* X.

The proof of Theorem 1 as given in [4] depends in a very essential way upon the Lefschetz fixed point theorem. In a lecture given by the writer upon this result in the Conference on Fixed Point Theory and its Applications at the University of Montreal in 1973, the writer remarked that this result could clearly be generalized to yield an extension of the Lefschetz fixed point theorem for locally compact mappings with compact attractors, and that this generalization would have Theorem 1 as a consequence. It is our purpose in the present paper to put forward a detailed verification of this remark.

We begin with some remarks upon the formulation of the Lefschetz fixed point theorem. If X is a topological space, f a continuous self-mapping of X, and if we are given a homology theory with coefficients in a given field F which applies to a category of spaces and mappings which includes X and f, respectively, then for each non-negative integer n, f induces a homology endomorphism

$$f_{*n} : H_n(X) \to H_n(X).$$

If $H_n(X)$ is of finite dimension, or more generally, if f_{*n} has finite rank, we can form the trace of the endomorphism f_{*n} and denote this trace by $tr(f_{*n})$. The classical Lefschetz number of the mapping f is given

$$L(f) = \sum_{n=0}^{\infty} (-1)^n \, tr(f_{*n}),$$

and is defined whenever all the f_{*n} are of finite rank and all but a finite number of the summands vanish.

In 1959, Leray [7] defined a generalized Lefschetz

number applying to a wider class of mappings by extending the notion of trace to the following notion of a generalized trace: An endomorphism h of a vector space V over a field F has a generalized trace (denoted by $tr_{gen}(h)$) if there exists a finite dimensional subspace V_o of V such that $h(V_o) \subset V_o$ and if the endomorphism $\tilde{h}$ of V/V_o induced by h is pointwise nilpotent (i.e. for each w, there exists $n(w)$ such that $(\tilde{h})^{n(w)} w = 0$). The generalized trace $tr_{gen}(h)$ is then defined to be the classical trace of $h|_{V_o}$ and is obviously independent of the choice of V_o. The generalized Lefschetz number of a mapping f is then defined to be

$$L_{gen}(f) = \sum_{n=0}^{\infty} (-1)^n \, tr_{gen}(f_{*n})$$

if each f_{*n} has a generalized trace in the above sense and if all but a finite number of the numbers $tr_{gen}(f_{*n})$ vanish.

In terms of this set of ideas, we can now formulate our basic result: (We use singular homology theory):

THEOREM 2. *Let* X *be a topological space which can be imbedded as a closed subset of a locally convex topological vector space* E *such that* $X \subset G$, *where* G *is an open subset of* E *which can be retracted on* X. *Let* f *be a continuous mapping of* X *into* X *with* f *locally compact and having a compact attractor.*

Then the generalized Lefschetz number of f, $L_{gen}(f)$, *is well defined and the Lefschetz fixed point theorem is valid for* f *in the sense that if* $L_{gen}(f) \neq 0$, *then* f *has a fixed point.*

As a specialization of Theorem 2, we obtain:

THEOREM 3. *Suppose that under the hypotheses of Theorem 2, there exists a positive integer* s *and a subset* B *of* X *which is contractible to a point in* X *such that* $f^s(X) \subset B$. *Then* $L_{gen}(f) = 1$ *and* f *has a fixed point in* X.

Another specialization and extension of Theorem 2 is the following:

THEOREM 4. *Let* f *be a locally compact, continuous self-mapping of a space* X *of the type described in Theorem 2 such that* f *has a compact attractor. Consider homology over the field* Q *of rational numbers. Suppose that for a given prime* p, $L_{gen}(f^p)$ *is not a multiple of* p. *Then* $L_{gen}(f) \neq 0$, *and* f *has a fixed point in* X. *(Indeed, we have* $L_{gen}(f) \equiv L_{gen}(f^p)$ mod p.*)*

We begin our discussion in Section 1 with a detailed discussion of the topological properties of locally compact mappings with compact attractors. In Section 2, we develop basic properties of the generalized trace and the generalized Lefschetz number. In Section 3, we give the proofs of Theorems 2, 3, and 4.

SECTION 1

We begin with the following basic result:

PROPOSITION 1. *Let* X *be a Hausdorff space,* f *a*

continuous mapping of X *into* X, *with* f *locally compact and having a compact attractor* A. *Let* K_o *be a compact subset of* X. *Then there exists a compact subset* K *of* X *with* $A \cup K_o \subset K$ *such that* K *has the following properties:*

(a) $f(K) \subset K$.

(b) *For each point* x *of* X, *there exists a neighborhood* $U(x)$ *of* x *in* X *and an integer* $n(x) \geq 1$ *such that*

$$f^n(U(x)) \subset K$$

for $n \geq n(x)$.

(c) *For each compact subset* K_1 *of* X, *there exists an integer* $n(K_1)$ *and an open neighborhood* U_1 *of* K_1 *in* X *such that*

$$f^n(U_1) \subset K$$

for $n \geq n(K_1)$.

PROOF OF PROPOSITION 1. Since $A \cup K_o$ is a compact attractor if A is, we may assume that K_o is empty. Since A is compact, there exists an open neighborhood U_o of A such that $cl(f(U_o))$ is compact since f is a locally compact mapping. (We use cl to denote the closure in X.) Let $S = A \cup cl(f(U_o))$. Then S is compact. Since A is an attractor (i.e. contains a point of the closure of each orbit under f), for each point x in X there exists an integer $k(x)$ such that $f^{k(x)}(x)$ lies in U_o. By the continuity of f, there exists an open neighborhood $W(x)$ of x in X such that

$$f^{k(x)}(W(x)) \subset U_o.$$

By the compactness of S, we may cover S by a finite number of these neighborhoods $\{W(x_j): 1 \le j \le s\}$. Let $r = \max\{k(x_j): 1 \le j \le s\}$. We now set

$$K = \bigcup_{j=0}^{r} f^j(S).$$

K as thus defined is compact. We assert that f maps K into K. Since $f(f^j(S)) = f^{j+1}(S)$, it suffices to show that $f^{r+1}(S)$ is contained in K. Let x be a point of S. Then x lies in $W(x_j)$ for some j. Hence

$$f^{k(x_j)}(x) \in U_o.$$

On the other hand,

$$f^{r+1}(x)$$

$$= f^{r-k(x_j)}(f(f^{k(x_j)}(x))) \in f^{r-k(x_j)}(f(U_o)) \subset f^{r-k(x_j)}(S) \subset K.$$

Hence $f(K) \subset K$.

For each point x of X, there exists an integer $n(x) \ge 2$ such that

$$f^{n(x)-1}(x) \in U_o,$$

since U_o is a neighborhood of the attractor A. Since each iterate of f is a continuous mapping, there exists a neighborhood $U(x)$ of x in X such that

$$f^{n(x)-1}(U(x)) \subset U_o.$$

Therefore

$$f^{n(x)}(U(x)) \subset f(U_o) \subset K.$$

Since K is invariant under f,

$$f^n(U(x)) \subset K$$

for $n \geq n(x)$.

If K_1 is a compact subset of X, we may cover K_1 by a finite family of such neighborhoods $U(x_j)$. If we let U_1 be the union of these neighborhoods and let $n(K_1)$ be the maximum of the corresponding $n(x_j)$, we see that $f^n(U_1) \subset K$ for $n \geq n(K_1)$. q.e.d.

PROPOSITION 2. *Let* X *be a topological space which is regular (in the sense of the separation axioms). Let* f *be a continuous mapping of* X *into* X *with* f *locally compact and having a compact attractor* A. *Let* K *be a compact subset of* X *which contains* A *with* K *invariant under* f *and with the property that for each compact subset* K_1 *of* X, *there is an integer* $n(K_1)$ *and an open neighborhood* U_1 *such that* $f^m(U_1) \subset K$ *for* $m \geq n(K_1)$. *Then there exists a neighborhood* W *of* K *such that* f *is a compact mapping of* W *into* W *(i.e.* $cl(f(W))$ *is a compact subset of* W*).*

PROOF OF PROPOSITION 2. We may choose a neighborhood U' of K such that $cl(f(U'))$ is compact and such that $f^m(U') \subset K$ for some integer m. We set

$$U = \bigcap_{j=0}^{m} f^{-j}(U').$$

Then

$$f(U) = \bigcap_{j=1}^{m} f^{-(j-1)}(U') \subset U$$

since $f^{-m}(U')$ contains U'. Hence, $f(U) \subset U$.

Since K is compact and the space X is regular, we may find a sequence of open sets

$$K \subset U_m \subset U_{m-1} \subset \ldots \subset U_1 \subset U_0 = U$$

such that

$$cl(U_j) \subset U_{j-1}.$$

Let

$$W = \bigcap_{j=0}^{m} f^{-j}(U_j).$$

W is an open subset of U which contains K, and since W is contained in U', $cl(f(W))$ is a compact subset of X. Moreover

$$cl(f(W)) = cl\left(\bigcap_{j=1}^{m} f^{-(j-1)}(U_j)\right) \subset \bigcap_{j=1}^{m} f^{-(j-1)}(cl(U_j))$$

$$\subset \bigcap_{j=1}^{m} f^{-(j-1)}(U_{j-1}) \subset W$$

since $f^{-m}(U_m)$ contains all of $U_0 = U$. Hence $cl(f(w))$ is a compact subset of W, and f is a compact mapping of W into W. q.e.d.

SECTION 2

We turn now to the discussion of the generalized trace and the generalized Lefschetz number.

PROPOSITION 3. *Let* V *and* V_1 *be two vector spaces over a given field* F, h *a linear mapping of* V *into*

V_1, h_1 *a linear mapping of* V_1 *into* V. *Suppose that the linear mapping* $h_1 h$ *of* V *into* V *has a generalized trace.*

Then the mapping hh_1 *of* V_1 *into* V_1 *has a generalized trace and*

$$\mathrm{tr}_{gen}(h_1 h) = \mathrm{tr}_{gen}(hh_1).$$

PROOF OF PROPOSITION 3. Since $h_1 h$ is assumed to have a generalized trace, there exists a subspace V_o of V with $\dim(V_o) < +\infty$ and with $(h_1 h)(V_o) \subset V_o$ such that if p is the linear map of V/V_o into V/V_o induced by $h_1 h$, then for each w in V/V_o, $p^{n(w)} w = 0$ for some integer $n(w)$. Let $V'_o = h(V_o)$. Then V'_o is a subspace of V_1 with $\dim(V'_0) < +\infty$, and since $h_1 h$ maps V_o into V_o, h_1 must map V'_o into V_o. Hence hh_1 maps V'_o into $h(V_o) = V'_o$.

Let $\tilde{h}$ be the mapping of V/V_o into V_1/V'_o induced by h, and let $\tilde{h}_1$ be the mapping of V_1/V'_o into V/V_o induced by h_1. Then $p = \tilde{h}_1 \tilde{h}$. For any w in V/V_o, therefore, there exists an integer $n(w)$ such that $(\tilde{h}_1 \tilde{h})^{n(w)} w = 0$. If we apply the mapping $\tilde{h}$ to both sides of this last equation, we obtain

$$(\tilde{h}\tilde{h}_1)^{n(w)} (\tilde{h} w) = 0.$$

Let w_1 be any element of V_1/V'_o, and set $w = \tilde{h}_1 w_1$. Then

$$(\tilde{h}\tilde{h}_1)^{n(w)+1} (w_1) = (\tilde{h}\tilde{h}_1)^{n(w)} (\tilde{h} w) = 0.$$

Hence the mapping $(\tilde{h}\tilde{h}_1)$ of V_1/V'_o into V_1/V'_o induced by hh_1 is pointwise nilpotent. It follows from the definition that hh_1 has a generalized trace.

By the definition of generalized trace, $tr_{gen}(h_1h)$ is equal to the trace of h_1h as a linear map of the finite dimensional space V_o. Similarly, $tr_{gen}(hh_1)$ is equal to the trace of hh_1 as a linear map of the finite dimensional space V'_o. Since h maps V_o into V'_o and h_1 maps V'_o into V_o, the conclusion of Proposition 3 follows from the fact that it holds for mappings of finite dimensional spaces. q.e.d.

PROPOSITION 4. *Let* V *and* V_1 *be two vector spaces over the field* F, *let* k *be a linear mapping of* V_1 *into* V, h *a linear mapping of* V *into* V, h_1 *a linear mapping of* V_1 *into* V_1. *Suppose that there exists a closed subspace* W *of* V_1 *such that* k *is injective on* W, *and such that the following two conditions hold:*

(1) *For each* v_1 *in* V_1, *there exists a positive integer* $n(v_1)$ *such that* $h_1^{n(v_1)}(v_1)$ *lies in* W.

(2) *For each* v *in* V, *there exists a positive integer* $m(v)$ *such that* $h^{m(v)}(v)$ *lies in* $k(W)$.

Suppose further that $hk = kh_1$.

Then if h_1 *has a generalized trace on* V_1, h *will have a generalized trace on* V *and*

$$tr_{gen}(h) = tr_{gen}(h_1).$$

PROOF OF PROPOSITION 4. Since h_1 is assumed to have a generalized trace, there exists a subspace V_o of finite dimension in V_1 such that $h_1(V_o) \subset V_o$ while the induced mapping h_1 of V_1/V_0 into V_1/V_0 is pointwise nilpotent. We note that since (aside from the condition of finite dimensionality) W satisfies the same conditions, so does $W \cap V_0$. Therefore we may assume that $V_0 \subset W$. It follows immediately that $h_1|_W = h_2$ has a generalized trace, and that $tr_{gen}(h_1) = tr(h_1|_{V_0}) = tr_{gen}(h_2)$.

Let $W' = k(W)$. Since $hk = kh_1$ and since W is invariant under h_1, W' is invariant under h. By assumption, $k|_W$ is an isomorphism of W with W' and the isomorphism k carries h_2 into the mapping $h' = h|_{W'}$. Hence, h' has a generalized trace, and $tr_{gen}(h') = tr_{gen}(h_2) = tr_{gen}(h_1)$.

Finally, let V'_0 be a finite dimensional subset of W' such that V'_0 is invariant under h' and such that the induced mapping of W'/V'_0 into itself is pointwise nilpotent. We assert that this latter property also holds for the mapping $\tilde{h}$ of V/V'_0 into V/V'_0 induced by h. Indeed, for any v in V/V'_0, by the assumption (2) there exists an integer $m(v)$ such that $(\tilde{h})^{m(v)}v$ lies in W'/V'_0. By the pointwise nilpotency of $\tilde{h}$ on W/V'_0, there is another integer k depending upon v such that $\tilde{h}^k(\tilde{h}^{m(v)}v) = 0$ in V/V'_0.

Thus, h has a generalized trace and $tr_{gen}(h) = tr(h'|_{V'_0}) = tr_{gen}(h') = tr_{gen}(h_1)$. q.e.d.

PROPOSITION 5. *Let* V *be a vector space of finite dimension over the field* Q *of rational numbers, and suppose that there exists a basis* $\{v_1,\ldots,v_n\}$ *for* V *in terms of which a given linear mapping* h *of* V *into*

V *has a matrix with integer coefficients.*

Then: $tr(h) - tr(h^p) \equiv 0 \pmod p$ *for any prime integer* p.

PROOF OF PROPOSITION 5. Let A be the matrix of the linear mapping h with respect to the given basis. Then $Det(\zeta I - A) = 0$ is a polynomial equation with integer coefficients and with highest power ζ^n. Let $\{\gamma_1, \ldots, \gamma_n\}$ be the complex characteristic roots of A given with their multiplicity. Then

$$tr(h) = \sum_{j=1}^{n} \gamma_j ,$$

$$tr(h^p) = \sum_{j=1}^{n} \gamma_j^p .$$

Since $tr(h)$ is an integer, we know that

$$(tr(h))^p \equiv tr(h), \quad (\text{mod } p).$$

Hence

$$tr(h) \equiv \left(\sum_{j=1}^{n} \gamma_j\right)^p, \quad (\text{mod } p)$$

where

$$\left(\sum_{j=1}^{n} \gamma_j\right)^p = \sum_{j=1}^{n} \gamma_j^p + R,$$

and

$$R = \sum_{\substack{|\beta| = p \\ \beta \text{ mixed}}} \binom{p}{\beta} \gamma^{\beta}$$

where $\beta = (\beta_1, \ldots, \beta_n)$ is a n-tuple of non-negative integers which is said to be mixed if at least two terms are different from zero, $|\beta| = \sum_{j=1}^{n} \beta_j$,

$$\gamma^\beta = \gamma_1^{\beta_1} \cdots \gamma_n^{\beta_n},$$

$$\binom{p}{\beta} = \frac{p!}{(\beta_1)!\ldots(\beta_n)!}.$$

For any mixed β with $|\beta| = p$, consider the class $C(\beta)$ of those multi-indices ξ obtainable from β by a permutation of the integers from 1 to n. Let

$$q_\beta(x) = \sum_{\xi \in C(\beta)} x^\xi.$$

Then each q_β is a symmetric function of $\{x_1,\ldots,x_n\}$ with integer coefficients, and therefore $q_\beta(\gamma)$ is polynomial with integer coefficients in the coefficients of the polynomial equation $\mathrm{Det}(\zeta I - A)$ and hence an integer. However,

$$R = \sum_{C(\beta)} \binom{p}{\beta} q_\beta(\gamma)$$

where the sum is taken over the distinct classes $C(\beta)$. Since p is a prime, $\binom{p}{\beta}$ is divisible by p. Since each $q_\beta(\gamma)$ is an integer, it follows that R is divisible by p.

Therefore, $\mathrm{tr}(h) \equiv \sum_{j=1}^{p} \gamma_j^p = \mathrm{tr}(h^p)$, (mod p).

q.e.d.

SECTION 3

Before we turn to the detailed proof of Theorem 2, let us note an important characteristic of singular homology theory which we propose to apply in that proof. Given an element h of the homology group $H_n(X)$, there exists a compact subset K_0 of the space X such that h lies in the image of the homomorphism of $H_n(K_0)$ into $H_n(X)$ induced by the injection of K_0 into X. Similarly, if V is an open subset of X, h a homology class in $H_n(V)$ such that h is homologous to zero in X, then there exists a compact subset K_1 of X such that h is mapped into 0 by the homomorphism of $H_n(V)$ into $H_n(V \cup K_1)$ induced by the injection.

As a preliminary to the Proof of Theorem 2, we give the following elementary explicit proof of the Lefschetz fixed point theorem for compact mappings of open subsets of locally convex topological vector spaces.

PROPOSITION 6. *Let* E *be a locally convex topological vector space,* U *an open subset of* E, f *a compact mapping of* U *into* U *(i.e.* f *is continuous and* $f(U)$ *has compact closure in* U*). Then* $L_{gen}(f)$ *exists and if* $L_{gen}(f) \neq 0$, f *has a fixed point in* U.

PROOF OF PROPOSITION 6. Let V be a symmetric convex open neighborhood of 0 in E which is sufficiently small. By hypothesis, $C = cl(f(U))$ is a compact subset of U. Hence, there exists a finite covering of C by sets of the form $x + V$ where x is a point of C and $(x + V)$ lies in U. Corresponding to this covering, we choose a partition of unity on C, i.e. if $\alpha = \{x_j + V: 1 \le j \le n\}$ is the covering, we have a family $\{\xi_j : 1 \le j \le n\}$ of continuous functions on E with $0 \le \xi_j(x) \le 1$ for all x, the support of ξ_j a closed subset of $x_j + V$, and satisfying the condition

$$\sum_{j=1}^{n} \xi_j(u) = 1, \qquad (u \in C).$$

We now form an approximate mapping f_V for f by setting

$$f_V(x) = \sum_{j=1}^{n} \xi_j(f(x))x_j.$$

The image $f_V(x)$ lies in the nerve P_V of the covering α, which is a finite polytope whose vertices are $\{x_1,\ldots,x_n\}$. A subcollection of these vertices $\{x_{j_1},\ldots,x_{j_r}\}$ span a simplex of P_V if and only if

$$\bigcap_{k=1}^{r} (x_{j_k} + V) \ne \emptyset.$$

If for a given x, $\{j_1,\ldots,j_r\}$ are the indices such that $\xi_{j_k}(f(x)) \ne 0$, then $f(x)$ lies in $(x_{j_k} + V)$ for each such k and the corresponding intersection is non-empty. The mapping f_V is obviously continuous since all the functions ξ_j are continuous and so is the mapping f. A point u of P_V lies in U since it can always be written in the form

$$u = \sum_{k=1}^{r} \xi_{j_k} x_{j_k}, \quad \left(0 \le \xi_{j_k},\ \sum_{k=1}^{r} \xi_{j_k} = 1\right)$$

where for k_1, k_2 among the indices of summation

$$x_{k_2} - x_{k_1} \in V.$$

Hence

$$u - x_{j_1} = \sum_{k=1}^{r} \xi_{j_k}(x_{j_k} - x_{j_1}) \in V,$$

i.e. $u \in x_{j_1} + V$, where the latter does not intersect $E - U$.

We assert that f is homotopic to f_V as mappings of U into U with the linear homotopy given by

$$g_t(x) = (1-t)f(x) + tf_V(x).$$

Since $g_t(x)$ is continuous in both x and t, we need only verify that $g_t(x)$ always lies in U. However,

$$g_t(x) = \sum_{j=1}^{n} \xi_j(f(x))\{(1-t)f(x) + tx_j\}.$$

A given term in this sum is different from 0 only if $f(x)$ lies in $x_j + V$, and in that case $\{(1-t)f(x) + tx_j\}$ also lies in $x_j + V$ for any t in $[0,1]$. Indeed, $g_t(x)$ always lies in the line joining $f(x)$ to the point $f_V(x)$ in P_V. Both points lie in $x_{j_1} + V$ for any vertex of the simplex in P_V which contains $f_V(x)$, and by assumption $x_{j_1} + V$ is contained in U.

Since f and f_V are homotopic, they induce the same homology endomorphisms. The mapping f_V can be factored into the composition of the mapping g_V of U into P_V induced by f_V with the injection mapping of P_V into X. Hence, $(f_V)_{*n}$ is of finite rank for each n and non-zero only for a finite number of integers n. It follows immediately that $tr_{gen}(f_{*n})$ is well-defined and is non-null only for finitely many n. Hence

$$L_{gen}(f) = \sum_{n=0}^{\infty} (-1)^n tr_{gen}(f_{*n})$$

is well-defined.

Since $f_V = j_V g_V$, with j_V the injection map of P_V into U, it follows from Proposition 3 that for each n,

$$tr_{gen}((f_V)_{*n}) = tr((g_V)_{*n}(j_V)_{*n}) = tr((g_V j_V)_{*n})$$

$$= tr((f_V|_{P_V})_{*n}).$$

Since

$$tr_{gen}((f_V)_{*n}) = tr(f_{*n})$$

for all n, we see that

$$L_{gen}(f) = L((f_V|_{P_V})).$$

Suppose that $L_{gen}(f) \neq 0$. Then by the Lefschetz fixed point theorem for self-mappings of finite polytopes, f_V must have a fixed point x_V in P_V. For this point x_V, we have $f(x_V) - x_V = f(x_V) - f_V(x_V) \in V$. Suppose that f has no fixed points. For each point u of C, we can find a convex symmetric open neighborhood V_u of the origin in E such that $(u + 2V_u) \cap f(u + 2V_u) = \emptyset$. We may cover the compact set by a finite number of neighborhoods of the form $\{u + V_u\}$. Let $V = \bigcap_{j=1}^{s} V_{u_j}$ where $\{u_j + V_{u_j} : 1 \leq j \leq s\}$ is the given covering. For the point x_V constructed above, $f(x_V)$, being a point of C, must lie in $u_j + V_{u_j}$ while x_V, which lies in $f(x_V) + V$, must lie in $u_j + 2V_{u_j}$. This is a contradiction, proving that f must have a fixed point. q.e.d.

PROPOSITION 7. *Let* E *be a locally convex topological vector space,* U *an open subset of* E, f *a compact mapping of* U *into* U. *Then:*

$$L_{gen}(f) \equiv L_{gen}(f^p), \quad (\text{mod } p)$$

for any prime p.

PROOF OF PROPOSITION 7. By the argument of the proof of Proposition 6, it suffices to prove the conclusion of Proposition 7 for the mapping f_V whose image is contained in the finite polytope P_V contained in U. Since $f^p = j_V g_V^p$, where g_V is the induced mapping of P_V into P_V, it follows immediately that

$$L_{gen}(f) = L((f_V|_{P_V}))$$

$$L_{gen}(f^p) = L((f_V|_{P_V})^p).$$

Hence, it suffices to prove the conclusion of Proposition 7 for a self-mapping f of a finite polytope P. For the latter, however, it follows from Proposition 4 and the fact that the Lefschetz number for a finite polytope is the same as the sum of the alternating trace of the chain mappings. Each of the chain mappings for $C_n(P)$ is the composition of a chain mapping induced by a simplicial mapping from an iterated barycentric decomposition to the original simplicial decomposition with the natural chain mapping from chains of the given simplicial decomposition to the chain groups of the barycentric decomposition. In their natural bases consisting of elementary chains, these mappings have integer coefficients. q.e.d.

PROPOSITION 8. *The conclusions of Theorems* 2, 3, *and* 4 *will hold under their given assumptions if we can establish them in the case in which* X *is itself an open subset of a locally convex topological vector space.*

PROOF OF PROPOSITION 8. Our basic hypothesis is that X is a closed subset of E and a retract of the open subset G of E which contains X. Let r be the retraction map of G on X, j the injection map of X into G. Then $rj = id_X$, the identity map of X.

If f is a continuous mapping of X into X, we form the map g of G into G by setting

$$g = jfr.$$

For any positive integer s, we then have

$$g^s = jf^s r.$$

If f is locally compact, so is g. A compact attractor A for f in X is also a compact attractor for g in G. Moreover, the fixed point sets of f and g coincide. Indeed, if $f(x) = x$, then $g(j(x)) = jfrj(x) = jf(x) = j(x)$, while if $g(u) = u$, then $jfr(u) = u$, so that u lies in X, $r(u) = u$, and $f(u) = u$.

We apply Proposition 3 and note that since $g_{*n} = j_{*n}f_{*n}r_{*n}$, g_{*n} will have a generalized trace if and only if $f_{*n}r_{*n}j_{*n} = f_{*n}$ does, and then

$$tr_{gen}(g_{*n}) = tr_{gen}(f_{*n}).$$

Similarly,

$$tr_{gen}(g^s_{*n}) = tr_{gen}(f^s_{*n}).$$

Hence, assuming either Lefschetz number exists,

$$L_{gen}(f) = L_{gen}(g)$$

and

$$L_{gen}(f^s) = L_{gen}(g^s).$$

For any of the Theorems concerned, if the hypotheses are true for f, they are true for g. If the conclusions are true for g, they are true for f. Hence, we may replace X by G and assume that X is itself an open subset of the locally convex topological vector space E. q.e.d.

PROOF OF THEOREM 2. By Proposition 8, we may assume that X is an open subset of the locally convex topological vector space E. Let A be a compact attractor for f. By Proposition 1, there exists a compact subset K of X containing A which is invariant under f and such that for any compact subset K_1 of X, there exists an open neighborhood U_1 of K_1 in X and a positive integer $n(K_1)$ such that for $n \geq n(K_1)$, $f^n(U_1) \subset K$.

By Proposition 2, there exists an open neighborhood U_o of K in X such that f is a compact mapping of U_o into U_o. Let $f_o = f|_{U_o}$. By Proposition 6 and its proof, for each n, $(f_o)_{*n}$ is of finite rank and is null for all but a finite number of n. In particular, f_o has a generalized Lefschetz number and if $L_{gen}(f_o) \neq 0$, then f has a fixed point in U_o. For each n, we let $W_n = (f_o)_{*n}(H_n(U_o)) \subset H_n(U_o)$. If j_o is the injection map of U_o into X, we let Y_n be the linear subspace of W_n which lies in the kernel of $(j_o)_{*n}$. For each n, let $\{y_n^{(1)},\ldots,y_n^{(r_n)}\}$ be a base for Y_n. By our remark about singular homology theory, there exists a compact subset $K_n^{(j)}$ such that $y_n^{(j)}$ is homologous to zero in $U_o \cup K_n^{(j)}$. Let

$$K_o = \bigcup_{n=0}^{\infty} \bigcup_{j=1}^{r_n} K_n^{(j)} .$$

Since the set of indices over which the union is taken is really finite and each $K_n^{(j)}$ is compact, K_o is a compact subset of X. Moreover, if j' is the injection mapping of U_o into $U_o \cup K_o$, then $(j')_{*n}$ annihilates the subspace Y_n.

If we apply Propositions 1 and 2, once more, we may find a compact set K_1 containing $K \cup K_o$ such that $f(K_1) \subset K_1$ and an open neighborhood U_1 of K_1 such that $f_1 = f|_{U_1}$ is a compact mapping of U_1 into U_1. We may assume without loss of generality that U_1 contains U_o since $cl(f(U_1 \cup U_o)) = cl(f(U_o)) \cup cl(f(U_1))$ is a compact subset of $U_o \cup U_1$.

By Proposition 6, f_1 has a generalized Lefschetz

number $L_{gen}(f_1)$ and the Lefschetz fixed point theorem is valid in the sense that if $L_{gen}(f_1) \neq 0$, then f_1 has a fixed point in U_1.

We now apply Proposition 4 with the following specifications:

$V = H_n(X)$

$V_1 = H_n(U_1)$

$h = f_{*n}$

$h_1 = (f_1)_{*n}$

k = the homology homomorphism of $H_n(U_1)$ into $H_n(X)$ induced by the injection mapping of U_1 into X.

W = image of W_n in $H_n(U_1)$ under the homology homomorphism ψ_n of $H_n(U_o)$ into $H_n(U_1)$ induced by the injection map of U_o into U_1.

We note that since U_1 contains $U_o \cup K_o$, each element of the subspace Y_n of W_n of elements homologous to zero in X is already homologous to zero in U_1. Hence W does not intersect the kernel of k, i.e. k is injective on W. Furthermore,

$$hk = kh_1$$

by the functorial properties of homology homomorphisms. Finally, any homology class c_1 in $H_n(U_1)$ is obtained from the injection of $H_n(K_2)$ for some compact subset K_2 of U_1. For some integer m, $f^m(K_2) \subset K \subset U_o$, i.e. $(f_1)^m_{*n}(c_1) = \psi(c_o)$ for some element c_o in $H_n(U_o)$. Hence

$$(f_1)_{*n}^{m+1}(c_1) = (f_1)_{*n}\psi(c_o) = \psi((f_o)_{*n}(c_o)) \in W.$$

Similarly, suppose c is an element of $H_n(X)$. Then c is obtained from the homomorphism of $H_n(K_3)$ into $H_n(X)$ induced by the injection of K_3 into X for some compact subset K_3 of X. There exists a positive integer r such that $f^r(K_3) \subset K$. Hence

$$(f_{*n})^r(c) = k\psi(c_o)$$

for some element c_o of $H_n(U_o)$. It follows that

$$(f_{*n})^{r+1}(c) = f_{*n}k\psi(c_o) = k\psi(f_{*n}(c_o)) \in k\psi(W_n) = k(W).$$

Thus all the hypotheses of Proposition 4 are satisfied and it follows from that Proposition that f_{*n} has a generalized trace and

$$tr_{gen}(f_{*n}) = tr_{gen}((f_1)_{*n}).$$

Since f_1 has a generalized Lefschetz number, it follows that f has a generalized Lefschetz number and

$$L_{gen}(f) = L_{gen}(f_1).$$

Furthermore, if $L_{gen}(f) \neq 0$, then $L_{gen}(f_1) \neq 0$ and it follows from the Lefschetz fixed point theorem as proved for f_1 that f_1 and hence f itself has a fixed point in U_1. q.e.d.

PROOF OF THEOREM 3. It follows from the previous proof that

$$L_{gen}(f) = L_{gen}(f_1), \quad L_{gen}(f_1^s) = L_{gen}(f^s)$$

for any positive integer s. If $f^s(X)$ is contained

in a contractible subset B of X, it follows that

$$(f_{*n})^s = (f^s)_{*n} = 0$$

for n positive, and $f^s_{*,0}$ has a single eigenvalue +1. Since $f_{*,0}$ has at least one eigenvalue +1, it follows that

$$\mathrm{tr}_{gen}(f_{*n}) = 0, \quad (n \geq 1)$$

and

$$\mathrm{tr}_{gen}(f_{*,0}) = +1.$$

Hence,

$$L_{gen}(f) = +1.$$

q.e.d.

PROOF OF THEOREM 4. We know that $L_{gen}(f) = L_{gen}(f_1)$ and $L_{gen}(f^p) = L_{gen}(f_1^p)$ by Theorem 2 and its proof.

By Proposition 7,

$$L_{gen}(f_1) \equiv L_{gen}(f_1^p), \quad (\mathrm{mod}\ p).$$

Hence

$$L_{gen}(f) \equiv L_{gen}(f^p), \quad (\mathrm{mod}\ p).$$

q.e.d.

We remark that Theorem 4 is an analogue of a result proved for the local fixed point index for compact mappings by Zabreiko and Krasnoselski [10] as well as by Steinlein [9] by geometrical arguments of a rather high order of complication.

APPENDIX

After reading the manuscript of the present paper, Arunas Liulevicius has commented that the algebraic result of Proposition 5 should be available in the mathematical literature. This does not seem to be the case, as the writer's discussions of the matter with Irving Kaplansky and I. N. Herstein would seem to indicate. There is some literature on the extension of Fermat's theorem to matrices, but no specific reference to the particular result that we apply. Both Kaplansky and Herstein have suggested the translation of the Proposition into a theorem about matrices over fields of characteristic p. Let us note an argument of this type which contains another version of the proof of Proposition 5.

PROPOSITION 5'. *Let* F *be the field of integers modulo* p, V *a finite-dimensional vector space over* F, h *a linear mapping of* V *into* V. *Let* $s_j(h)$ *be the* j*-th coefficient of the polynomial* $\mathrm{Det}(\xi I - h)$ *in* ξ. *Then:*

$$s_j(h^p) = s_j(h).$$

PROOF. Let $\gamma(u) = u^p$. Let F_1 be a finite extension of F which contains all the characteristic roots $\{\xi_1,\dots,\xi_n\}$ of h. Since F_1 is a field of characteristic p, γ is a homomorphism of F_1 into F_1. The characteristic roots of h^p are $\{\xi_1^p,\dots,\xi_n^p\}$. We know that

$$s_j(h) = p_j(\xi)$$

for a given polynomial p_j. Similarly

$$s_j(h^p) = s_j(\gamma(h)) = p_j(\gamma(\xi)).$$

Since γ is a homomorphism of F_1,

$$p_j(\gamma(\xi)) = \gamma(p_j(\xi)).$$

However, $p_j(\xi)$ lies in F, and by Fermat's theorem, $\gamma(u) = u$ for each u in F. Therefore

$$s_j(h^p) = \gamma(p_j(\xi)) = p_j(\xi) = s_j(h).$$

q.e.d.

BIBLIOGRAPHY

[1] BROWDER, F. E., *On a generalization of the Schauder fixed point theorem*, Duke Math. Jour., 26 (1959), 291-303.

[2] BROWDER, F. E., *Fixed point theorems on infinite dimensional manifolds*, Trans. Amer. Math. Soc., 119 (1965), 179-194.

[3] BROWDER, F. E., *Asymptotic fixed point theorems*, Math. Annalen, 185 (1970), 38-60.

[4] BROWDER, F. E., *Some new asymptotic fixed point theorems*, Proc. Nat. Acad. Sci., 71 (1974), 2734-2735.

[5] BROWDER, F. E., *Nonlinear operators and nonlinear equations of evolution in Banach spaces*, Proc. of Symposia in Pure. Math., vol. 18, Part II, Amer. Math. Society, Providence, 1974.

[6] GRANAS, A., *The Leray-Schauder index and the fixed point theory for arbitrary* ANR's, Bull. Soc. Math. France, 100 (1972), 209-228.

[7] LERAY, J., *Theorie des points fixes: Indice total et nombre de Lefschetz*, Bull. Soc. Mat. France, 87 (1959), 221-233.

[8] NUSSBAUM, R., *Asymptotic fixed point theorems for local condensing mappings*, Math. Annalen, 191 (1971), 181-195.

[9] STEINLEIN, H., *Über die verallgemeinerten Fixpunktindizes von Iterierten verdichtender Abbildungen*, Manuscripta Math., 8 (1972), 251-266.

[10] ZABREIKO, P. P. and KRASNOSELSKI, M. A., *Iterations of operators and fixed points*, (in Russian), Dokladi Akad. Nauk USSR, (1971), 1006-1009 (Soviet Math. Dokladi, 12 (1971), 294-298).

L^p *DECAY RATES, p BIG (≤∞), AND ENERGY DECAY IN NONBICHARACTERISTIC CONES FOR FIRST ORDER HYPERBOLIC SYSTEMS*

by

DAVID G. COSTA
Department of Mathematics
Federal University of Rio de Janeiro
Rio de Janeiro, GB, Brasil

0. INTRODUCTION

The purpose of these notes is to study the uniform (also L^p, p big) asymptotic behavior and the behavior along nonbicharacteristic rays (and cones) of solutions of a first order hyperbolic system of k equations,

$$(*)\qquad \frac{\partial u}{\partial t} + A\cdot\nabla u = 0 \quad \left(A\cdot\nabla = \sum_{j=1}^{n} A_j \frac{\partial}{\partial x_j} \right) ;$$

i.e., $A = (A_1, \dots, A_n)$ is an n-tuple of $k \times k$ matrices with constant complex coefficients such that the eigenvalues $\lambda_1(\xi), \dots, \lambda_k(\xi)$ of $A\cdot\xi = \sum_{j=1}^{n} \xi_j A_j$ are real for each $\xi = (\xi_1, \dots, \xi_n) \in \mathbb{R}^n \setminus \{0\}$.

We shall be concerned with smoothly diagonalizable systems. In other words, there exists a smooth (C^∞) matrix $P(\omega)$, $\omega \in S^{n-1}$, such that

$$(H_1) \qquad P(\omega)A\cdot\omega P(\omega)^{-1} = \begin{pmatrix} \lambda_1(\omega) & & O \\ & \ddots & \\ O & & \lambda_k(\omega) \end{pmatrix}.$$

$$(H_2) \qquad \dim(\ker A\cdot\omega) = \text{const.} \ \textit{for all} \ \ \omega \in S^{n-1}.$$

We remark that (H_1) is always true for strictly hyperbolic systems.

If we consider the single equation $\frac{\partial^2 u}{\partial t^2} - \Delta u = 0$, the wave equation, it is a known result that solutions subject to smooth Cauchy data of compact support decay uniformly (in all of space) as $t \to +\infty$ at the rate $O(t^{-(n-1)/2})$, where n is the space dimension. Since any second order homogeneous hyperbolic equation can be reduced to the form $(*)$ (with the A_j's actually Hermitian symmetric), it is natural to ask whether solutions of $(*)$ decay uniformly and, in affirmative case, at what rate. The answer is yes, with the same decay rate $O(t^{-(n-1)/2})$ for a large class of systems. More precisely, in Section 2, we assume that $(*)$ is smoothly diagonalizable and each of the λ_ℓ-connected components of the "normal surface" is a strictly convex surface. Then, for n odd, $n \geq 3$, we show that solutions subject to "nonstationary" Cauchy data of compact support decay uniformly (in all of space) as $t \to +\infty$ at the rate $O(t^{-(n-1)/2})$. Also obtained are L^p estimates for p big. From our estimates, it follows that the same uniform decay rate is valid for more general (nonstationary) Cauchy data in the Sobolev space $W^{(n+1)/2,1}$. Slightly nonoptimal results are obtained in the case of n even. Proofs of some lemmas and propositions in

Sections 2 and 3 will appear elsewhere.

Previous results on decay for first order hyperbolic systems were obtained by other people (see [4], [5] and [10], for example). Those are of a local nature, in contrast to our uniform type of result. Quantitative results of our type for the wave (and Klein-Gordon) equation can be found in [7] and [9].

Our method uses as a primary tool the Radon transform, that is, the plane wave decomposition of solutions [3]. This approach proves to be very useful in having a closer look at the "Riemann matrix".

Section 3 is concerned with decay along nonbicharacteristic rays and energy decay in cones. The results in this section were obtained in joint work with Professor C. Bardos.

In Section 4, we discuss in detail L^p decay rates for the wave equation by the method of Sections 1 and 2.

We shall start by defining and reviewing a few facts about the Radon transform, and deriving a solution formula for the Cauchy problem for (*) in Section 1.

1. THE RADON TRANSFORM AND THE CAUCHY PROBLEM

For a function $f \in \mathcal{S}(\mathbb{R}^n)$, the space of rapidly decreasing functions, the Radon transform of f, $\hat{f}$ (or $\mathcal{R}f$), is defined by the formula

$$\hat{f}(s,\omega) = \int_{x\cdot\omega=s} f(x)\,dx, \quad s \in \mathbb{R}, \quad \omega \in S^{n-1}.$$

One can show that $\hat{f} \in \mathcal{S}(\mathbb{R} \times S^{n-1})$ if $f \in \mathcal{S}(\mathbb{R}^n)$ (see [2], [6]). Moreover, the following properties are immediate from the definition:

(1.1) (i) $\hat{f}$ is an even function.

(ii) For each integer $k \geq 0$, $\int_{-\infty}^{\infty} s^k \hat{f}(s,\omega)\,ds$ can be written as a polynomial in ω of degree k.

(iii) $R\left(\frac{\partial}{\partial x_j} f\right) = \omega_j \frac{\partial}{\partial s}(Rf)$.

In fact, R establishes a 1-1 mapping of $S(\mathbb{R}^n)$ onto the subspace of $S(\mathbb{R} \times S^{n-1})$ consisting of all functions $\phi(s,\omega)$ satisfying (1.1) (i) - (ii).

The following theorem (see [2], pg. 163) gives the inversion formula for the Radon transform:

THEOREM 1.1. *Let* $f \in S(\mathbb{R}^n)$, $n \geq 3$.

(a) If n *is odd,*

$$f = c_1 \Delta^{\frac{n-1}{2}} ((\hat{f})^{\vee}),$$

where c_1 *is a constant independent of* f *and* $\vee$ *is defined, for a function* $\phi \in S(\mathbb{R} \times S^{n-1})$, *by* $\overset{\vee}{\phi}(x) = \int_{S^{n-1}} \phi(x\cdot\omega,\omega)\,d\omega$.

(b) If n *is even,*

$$f = c_2 J((\hat{f})^{\vee}),$$

where c_2 *is a constant independent of* f *and the operator* J *is defined by analytic continuation,*

$$Jf(x) = \underset{\alpha=1-2n}{\text{anal. cont.}} \int_{\mathbb{R}^n} |x-y|^{\alpha} f(y)\,dy.$$

Ludwig ([6]) also derives the same formulae in a different form. For example, if n is odd, he shows that

$$f = c\left(\left(\frac{\partial}{\partial s}\right)^{n-1} \hat{f}\right)^{\vee}, \quad f \in S(\mathbb{R}^n), \tag{1.2}$$

where c is a constant independent of f.

We now derive an explicit formula for the solution of the Cauchy problem for (*) with initial datum $f \in C_0^\infty(\mathbb{R}^n, \mathbb{C}^k)$.

Let $\hat{u}(s,\omega;t)$ denote the Radon transform of $u(x,t)$ (in the x variable). Then, as u satisfies $\frac{\partial u}{\partial t} + A \cdot \nabla u = 0$ and in view of (1.1) (iii), we see that $\hat{u}$ is the solution of

$$(1.3) \qquad \frac{\partial \hat{u}}{\partial t} + A \cdot \omega \frac{\partial \hat{u}}{\partial s} = 0, \quad \hat{u}(s,\omega;0) = \hat{f}(s,\omega).$$

Since $A \cdot \omega$ is assumed to be smoothly diagonalizable, we let $\hat{v}(s,\omega;t) = P(\omega)\hat{u}(s,\omega;t)$, $P(\omega) = [p_{ij}(\omega)]$, $P(\omega)^{-1} = [q_{ij}(\omega)]$ (see (H_1)), to obtain, for each $\ell = 1, \dots, k$,

$$\frac{\partial \hat{v}_\ell}{\partial t} + \lambda_\ell(\omega) \frac{\partial \hat{v}_\ell}{\partial s} = 0, \quad \hat{v}_\ell(s,\omega;0) = \sum_{j=1}^{k} p_{\ell j}(\omega)\hat{f}_j(s,\omega).$$

This 1-dimensional Cauchy problem is clearly solved by

$$\hat{v}_\ell(s,\omega;t) = \sum_{j=1}^{k} p_{\ell j}(\omega)\hat{f}_j(s-\lambda_\ell(\omega)t,\omega),$$

and hence,

$$\hat{u}_i(s,\omega;t) = \sum_{\ell,j=1}^{k} q_{i\ell}(\omega)p_{\ell j}(\omega)\hat{f}_j(s-\lambda_\ell(\omega)t,\omega).$$

Finally, we arrive at

$$(1.4) \qquad \hat{u}(s,\omega;t) = \sum_{\ell=1}^{k} G_\ell(\omega)\hat{f}(s-\lambda_\ell(\omega)t,\omega),$$

where $G_\ell(\omega) = \left[g_{ij}^{\ell}(\omega)\right] = \left[q_{i\ell}(\omega)p_{\ell j}(\omega)\right]$, which relates the Radon transform of $u(\cdot,t)$ and that of the initial datum f. In order to get $u(\cdot,t)$ explicitly, we use Theorem 1.1. This will be done in the following section.

2. L^p DECAY RATES, p BIG ($\le\infty$)

Here, until otherwise stated, the space dimension n will be assumed odd with $n \geq 3$. In the previous section, we derived formula (1.4) relating $\hat{u}$ and $\hat{f}$. Using Theorem 1.1 (a), we obtain

$$u(x,t) = \sum_{\ell=1}^{k} \Delta^{\frac{n-1}{2}} \left[c_1 \int_{S^{n-1}} G_\ell(\omega) \left(\int_{y\cdot\omega = x\cdot\omega - \lambda_\ell(\omega)t} f(y)\,dy \right) d\omega \right]. \tag{2.1}$$

Let $\delta(\omega,\lambda_\ell(\omega),\cdot)$ denote the distribution

$$f \longmapsto \int_{y\cdot\omega=\lambda_\ell(\omega)} f(y)\,dy$$

and put

$$\langle I_\ell\, , f\rangle = c_1 \langle \int_{S^{n-1}} G_\ell(\omega)\, \delta\,(\omega,\lambda_\ell(\omega),\cdot)\,d\omega, f\rangle.$$

Then, (2.1) yields the following

LEMMA 2.1.

$$u(\cdot,t) = \Delta^{\frac{n-1}{2}} \left[\sum_{\lambda_\ell=0} I_\ell * f + \sum_{\lambda_\ell\neq 0} \frac{1}{t}\, I_\ell\left(\frac{\cdot}{t}\right) * f \right], \quad t > 0.$$

We remark that hypothesis (H_2) in the introduction implies that, for each $\ell = 1,\ldots,k$, $\lambda_\ell(\omega)$ is either identically zero or else never vanishes on S^{n-1}, a fact that justifies our summation notation above.

So, for each $\lambda_\ell \neq 0$, let us denote by $\partial\Lambda_\ell$ the smooth surface $\{\lambda_\ell(\omega)^{-1}\omega \mid \omega \in S^{n-1}\}$, by Λ_ℓ the interior of $\partial\Lambda_\ell$, and by ∂H_ℓ the polar of $\partial\Lambda_\ell$, i.e., the hull of the hyperplanes $\{x \mid x\cdot\omega = \lambda_\ell(\omega)\}$, $\omega \in S^{n-1}$ (see [1]). From now on we shall assume

(H_3) $\partial\Lambda_\ell$ *is strictly convex for each* $\lambda_\ell \neq 0$.

Then, for any $y \notin H_\ell$ (interior of ∂H_ℓ), we will denote by $M_\ell^{n-2}(y)$ the convex $(n - 2)$ smooth manifold

$$M_\ell^{n-2}(y) = \left\{\frac{\omega}{\lambda_\ell(\omega)} \;\middle|\; y\cdot\omega = \lambda_\ell(\omega)\right\}$$

and by $\nu_\ell(\omega)$ the outward normal to $\partial\Lambda_\ell = \left\{\frac{\omega}{\lambda_\ell(\omega)} \;\middle|\; \omega \in S^{n-1}\right\}$. We obtain the following propositions, whose proofs will appear elsewhere.

PROPOSITION 2.2. *Assume* (H_3). *Then, for any* ℓ *such that* $\lambda_\ell \neq 0$, *the distribution* I_ℓ *is the function given by*

$$I_\ell(y) = \begin{cases} 0 & , \; y \in H_\ell \\ \dfrac{1}{|y|}\displaystyle\int_{M_\ell^{n-2}(y)} \frac{G_\ell(\sigma)h_\ell(\sigma)}{\left(1 - \left(\nu_\ell(\sigma)\cdot\frac{y}{|y|}\right)^2\right)^{1/2}}\, d\sigma, & y \notin H_\ell \end{cases}$$

where $h_\ell(\sigma)$ *is smooth.*

Still assuming (H_3), we have the following regularity result on the I_ℓ's (corresponding to $\lambda_\ell \neq 0$):

PROPOSITION 2.3. I_ℓ *is bounded together with all of its derivatives up to order* $(n-3)/2$, *i.e.,* $I_\ell \in W^{(n-3)/2,\infty}(\mathbb{R}^n, \mathbb{C}^{k^2})$; *in fact, we have the estimates*

$$|D^\alpha I_\ell(y)| \leq \frac{C_\alpha}{|y|^{1+|\alpha|}}\left(\frac{s_\ell(y)-1}{|y|}\right)^{\frac{n-3}{2}-|\alpha|}$$

for any multiindex α *and any* $y \notin H_\ell = \{\xi \mid S_\ell(\xi) \le 1\}$, *where* $S_\ell(\xi) = \sup_{x\in\Lambda_\ell} \xi \cdot x$ *is the support function of* Λ_ℓ.*

Now, in view of Proposition 2.2, we can rewrite $u(\cdot,t)$ in Lemma 2.1 as

$$u(x,t) = \Delta^{\frac{n-1}{2}} \left[\sum_{\lambda_\ell=0} (I_\ell * f)(x) + \frac{1}{t}\int_{\mathbb{R}^n} \sum_{\lambda_\ell\neq 0} I_\ell\left(\frac{y}{t}\right) f(x-y)\,dy \right]. \tag{2.2}$$

Since the first term in the bracket above does not depend on t, we can only expect decay for solutions whose initial data f satisfy $\sum_{\lambda_\ell=0} I_\ell * f = 0$. Such f's are said to be nonstationary.

THEOREM 2.4. *Assume* (H_1) - (H_3). *Then, the solution* $u(\cdot,t)$ *of* $(*)$ *with nonstationary* $f \in C_0^\infty(\mathbb{R}^n,\mathbb{C}^k)$ *satisfies*

$$\begin{cases} \|u(\cdot,t)\|_r \le C_q t^{-(n-1)/2+n/q} \|f\|_{(n+1)/2,p} \\ \text{for all } t > 0^{**} \end{cases} \tag{2.3}$$

where $\frac{1}{r} = \frac{1}{p} + \frac{1}{q} - 1 \ge 0$, $\infty \ge q > \frac{2n}{n-1}$ *and* $C_q > 0$ *does not depend on* f. *In particular,*

---

* Recall also that $I_\ell(y) \equiv 0$ for $y \in H_\ell$, by Proposition 2.2.

** We denote by $\|\cdot\|_{m,p}$ the usual norm in the Sobolev space $W^{m,p}(\mathbb{R}^n, \mathbb{C}^k)$; $\|\cdot\|_r = \|\cdot\|_{0,r}$ denotes the norm in $L^r(\mathbb{R}^n, \mathbb{C}^k)$.

$$(2.4) \qquad \begin{cases} \|u(\cdot,t)\|_\infty \le C_\infty\, t^{-(n-1)/2} \|f\|_{(n+1)/2,1} \\ \textit{for all} \quad t > 0. \end{cases}$$

PROOF. By (2.2), we have

$$(2.5) \quad u(x,t) = \frac{1}{t}\,\Delta^{\frac{n-1}{2}} \sum_{\lambda_\ell \neq 0} \int_{\mathbb{R}^n} I_\ell\left(\frac{y}{t}\right) f(x-y)\,dy.$$

On the other hand, for $|\alpha| = (n-3)/2$ and each ℓ, Proposition 2.3 yields $D^\alpha I_\ell(y) = 0$ for $y \in H_\ell$ and $|D^\alpha I_\ell(y)| \le \text{const.}|y|^{-(n-1)/2}$ for $y \notin H_\ell$ and hence, each $D^\alpha I_\ell$ is in L^q provided $\infty \ge q > \frac{2n}{n-1}$.

Now,

$$(2.6) \quad \begin{aligned} &\Delta^{(n-1)/2}\left[I_\ell\left(\frac{\cdot}{t}\right) * f\right] \\ &= t^{-(n-3)/2}\left[\left(\Delta^{(n-3)/4} I_\ell\right)\left(\frac{\cdot}{t}\right) * \Delta^{(n+1)/4} f\right], \end{aligned}$$

if n is of the form $4N + 3$, and equals

$$(2.7) \quad t^{-(n-3)/2} \sum_{j=1}^{n} \left[\left(\Delta^{(n-5)/4} \frac{\partial I_\ell}{\partial y_j}\right)\left(\frac{\cdot}{t}\right) * \Delta^{(n-1)/4} \frac{\partial f}{\partial y_j}\right],$$

if n is of the form $4N + 5$, $N = 0, 1, \ldots$.

We have just seen that each of the functions $\left(\Delta^{(n-3)/2} I_\ell\right)\left(\frac{\cdot}{t}\right)$, $\left(\Delta^{(n-5)/4} \frac{\partial I_\ell}{\partial y_j}\right)\left(\frac{\cdot}{t}\right)$ is in L^q for $\infty \ge q > \frac{2n}{n-1}$. Denoting by $J(\cdot,t)$ an arbitrary one, we clearly have

$$\|J(\cdot,t)\|_q \le C_q\, t^{n/q}, \quad \frac{2n}{n-1} < q \le \infty.$$

Therefore, applying the Hausdorff-Young Inequality to (2.6) (or (2.7)) gives

$$\left\|\Delta^{(n-1)/2}\left[I_\ell\left(\frac{\cdot}{t}\right) * f\right]\right\|_r \le C_q\, t^{-(n-3)/2+n/q} \sum_{|\beta|=(n+1)/2} \|D^\beta f\|_p ,$$

with $\frac{1}{r} = \frac{1}{p} + \frac{1}{q} - 1 \geq 0$, $\infty \geq q > \frac{2n}{n-1}$. This completes the proof in view of (2.5).

As a corollary, we obtain the following existence and uniqueness theorem, whose proof we omit here.

THEOREM 2.5. *The Cauchy problem for* (*) *with nonstationary* initial datum* $f \in W^{(n+1)/2,1}(\mathbb{R}^n,\mathbb{C}^k)$ *has a unique (weak) solution satisfying* (2.4).

Before closing this section, a few words on the even-dimensional case are in order. The same explicit formula given by Lemma 2.1 holds true (using Theorem 1.1 (b)), with $\Delta^{(n-1)/2}$ being now interpreted as a Riesz potential (see [8]). The estimates of Proposition 2.3 imply that the $D^{\alpha}I_{\ell}$'s, with $|\alpha| = (n-4)/2$, are Hölder continuous of exponent 1/2. Using a result of Stein ([8]), we are then able to obtain a slightly nonoptimal analogue of Theorem 2.4, namely, that for each $\varepsilon > 0$, $\|u(\cdot,t)\|_{\infty} = O(t^{-(n-1)/2+\varepsilon})$. We shall not go into the details here.

3. ENERGY DECAY IN NONBICHARACTERISTIC CONES

With the same notations as in the previous sections, we now set $\partial H = \bigcup_{\lambda_{\ell}\neq 0} \partial H_{\ell}$ and investigate the behavior

* For a distribution $f \in S'(\mathbb{R}^n,\mathbb{C}^k)$, we say that f is nonstationary if $\langle f,\phi\rangle = 0$ for all $\phi \in S(\mathbb{R}^n,\mathbb{C}^k)$ such that $A^* \cdot \nabla\phi = 0$. It is easy to check that the definition given previously is equivalent to this one.

along <u>nonbicharacteristic rays</u> $\{(t\xi,t) \mid t > 0\}$, $\xi \notin \partial H$. By $d(\xi,\partial H)$ we denote the distance from ξ to the set ∂H, and we shall assume that the space dimension n is odd and $n \geq 3$ throughout this section.

THEOREM 3.1. *Let $\xi \in \mathbb{R}^n$ be such that $\xi \notin \partial H$ and let $f \in L^1(\mathbb{R}^n,\mathbb{C}^k)$ be nonstationary with support in the ball of radius* R. *Then, for* $t > 2R/d(\xi,\partial H)$, *we have*

$$(3.1) \qquad |u(t\xi,t)| \leq Ct^{-n} \|f\|_1 .$$

PROOF. We first prove (3.1) assuming that $f \in C_0^\infty(\mathbb{R}^n,\mathbb{C}^k)$. The result is then obtained by passing to the limit.

We have:

$$\int_{\mathbb{R}^n} I_\ell\left(\frac{y}{t}\right) f(x-y)\,dy = t^n \int_{\mathbb{R}^n} I_\ell(z) f(x-tz)\,dz.$$

Let $0 < \eta < 1/2\, d(\xi,\partial H) - \frac{R}{t}$. For $|x - t\xi| < t\eta$, the support of $f(x - t\cdot)$ is contained in the ball of center ξ and radius $1/2\, d(\xi,\partial H)$. But, in this ball, I_ℓ is smooth and therefore we have

$$|\Delta^{(n-1)/2} \int_{\mathbb{R}^n} I_\ell\left(\frac{y}{t}\right) f(x-y)\,dy|$$

$$= |t^n \int_{\mathbb{R}^n} I_\ell(z)\Delta_x^{(n-1)/2} f(x-tz)\,dz|$$

$$\leq t^n t^{-(n-1)} \int_{|z-\xi|\leq\frac{1}{2} d(\xi,\partial H)} |\Delta_z^{(n-1)/2} I_\ell(z)|\,|f(x-tz)|\,dz$$

$$\leq c(\ell) t^{-(n-1)} t^n \int_{\mathbb{R}^n} |f(x-tz)|\,dz,$$

provided $|x - t\xi| < t\eta$. In particular,

$$|u(t\xi,t)| \le Ct^{-n}\int_{\mathbb{R}^n} t^n|f(t\xi-tz)|dz$$
$$= Ct^{-n}\,\|f\|_1 \;,$$

and the proof of Theorem 3.1 is complete.

From this theorem, we can derive the following estimate on the energy decay in a <u>nonbicharacteristic cone</u>:

COROLLARY 3.2. *Let* $f \in L^1(\mathbb{R}^n, \mathbb{C}^k)$ *be nonstationary with support in the ball of radius* R. *Let* Σ *be a closed measurable set contained in* $\mathbb{R}^n \backslash \partial H$ *and denote by* $d(\Sigma,\partial H)$ *the distance between* Σ *and* ∂H. *Then, for* $t > 2R/d(\Sigma,\partial H)$, *we have*

$$\int_{t\Sigma} |u(x,t)|^2\,dx \le Ct^{-n}\,\|f\|_1^2 \;,$$

where C *is a constant depending only on* Σ *and* $A\cdot\nabla$.

A sharper version of the above corollary is obtained when we restrict ourselves to the Hermitian, spherically symmetric (<u>isotropic</u>) case. So, we shall assume now that the matrices A_j, $j = 1,\dots,n$, are Hermitian. Then, the principle of conservation of energy holds, namely, if $f \in L^2(\mathbb{R}^n, \mathbb{C}^k)$,

$$\int_{\mathbb{R}^n} |u(x,t)|^2\,dx = \int_{\mathbb{R}^n} |f(x)|^2\,dx \quad \text{for all } t.$$

The assumption of spherical symmetry means that the eigenvalues $\lambda_\ell(\omega)$ are independent of $\omega \in S^{n-1}$. Since $A\cdot\omega$ is an odd function of ω, for any λ_ℓ, $-\lambda_\ell$ is also an eigenvalue. Therefore, we will write $\mu_j = |\lambda_\ell|$ and have the inequality

$$0 \le \mu_1 < \mu_2 < \ldots < \mu_s ,$$

where $s(\le k/2)$ depends on the multiplicity of each eigenvalue.

THEOREM 3.3. *Let* $0 < \mu_j < \mu_{j+1}$ *be two successive strictly positive absolute values of eigenvalues of* $A \cdot \omega$. *Let* $f \in L^1(\mathbb{R}^n, \mathbb{C}^k)$ *be nonstationary with support in the ball of radius* R. *Then, if* $\rho > 2R$, $\mu_j T > 2R$, $T > (\rho + R)/(\mu_{j+1} - \mu_j)$, *we have*

$$\int_{\Sigma(T,\rho)} |u(x,T)|^2 \, dx \le \frac{C}{\rho^n} \|f\|_1^2 ,$$

where $\Sigma(T,\rho) = \{x \mid \rho + \mu_j T < |x| < \mu_{j+1} T - R\}$.

For the proof of this theorem, we shall use (without proof) the following lemma on the representation of the distribution I_ℓ, when $\lambda_\ell = 0$ (see Section 2 for notations).

LEMMA 3.4. *The distribution* $h \longmapsto \int_{S^{n-1}} G_\ell(\omega) \hat{h}(0,\omega) d\omega$ *is given by the function* $\frac{1}{|x|} \widetilde{G}_\ell\left(\frac{x}{|x|}\right)$, *where* $\widetilde{G}_\ell$ *is defined by* $\widetilde{G}_\ell(\omega) = \int_{S^{n-1} \cap \{\sigma \mid \sigma \cdot \omega = 0\}} G_\ell(\sigma) d\sigma$.

PROOF OF THEOREM 3.3. We write

$$u(x,t) = c_1 \Delta^{\frac{n-1}{2}} \left[\sum_{\lambda_\ell \neq 0} \int_{S^{n-1}} G_\ell(\omega) \hat{f}(x \cdot \omega - \lambda_\ell t, \omega) d\omega \right].$$

Since the support of f is contained in the ball of radius R, the support of $\hat{f}$ is contained in the set $[-R,R] \times S^{n-1}$. On the other hand, for $|\lambda_\ell| \ge \mu_{j+1}$ and $x \in \Sigma(t,\rho)$ we have

$$|x\cdot\omega-\lambda_\ell t| \ge |\lambda_\ell|t - |x| > R.$$

Therefore, for $x \in \Sigma(t,\rho)$, we obtain

$$u(x,t) = \sum_{|\lambda_\ell|\le\mu_j} c_1 \Delta^{\frac{n-1}{2}} \int_{S^{n-1}} G_\ell(\omega)\hat{f}(x\cdot\omega-\lambda_\ell t,\omega)\,d\omega.$$

Henceforth, we will denote by λ_ℓ any eigenvalue such that $0 < |\lambda_\ell| \le \mu_j$ and we set

$$(3.2) \qquad u_\ell(x,t) = \Delta^{\frac{n-1}{2}} \int_{S^{n-1}} G_\ell(\omega)\hat{f}(x\cdot\omega-\lambda_\ell t,\omega)\,d\omega.$$

Using the relation $(\Delta f)^\wedge = \frac{\partial^2}{\partial s^2}\hat{f}$ and differentiating twice with respect to t on the right hand side of (3.2) we see that u_ℓ is a solution of the wave equation $\frac{\partial^2 u_\ell}{\partial t^2} - \lambda_\ell^2 \Delta u_\ell = 0$. Therefore, if we put

$$v_\ell(x) = \frac{-1}{\lambda_\ell}\int_{S^{n-1}} \Delta^{\frac{n-1}{2}-1} G_\ell(\omega)\,\frac{\partial\hat{f}}{\partial s}(x\cdot\omega,\omega)\,d\omega$$

and define

$$\omega_\ell(x,t) = \int_0^t u_\ell(x,\sigma)\,d\sigma + v_\ell(x),$$

$\omega_\ell(x,t)$ is the solution of the wave equation $\frac{\partial^2\omega_\ell}{\partial t^2} - \lambda_\ell^2\Delta\omega_\ell = 0$ in $\mathbb{R}^n \times \mathbb{R}_t$, with initial data $\omega_\ell(x,0) = v_\ell(x)$, $\frac{\partial\omega_\ell}{\partial t}(x,0) = u_\ell(x,0)$.

Now, by Lemma 3.4, we have

$$u_\ell(x,0) = \Delta^{\frac{n-1}{2}}\int_{\mathbb{R}^n} \tilde{G}_\ell\left(\frac{y}{|y|}\right)\cdot\frac{1}{|y|}\,f(x-y)\,dy.$$

Since the function $y \longmapsto \tilde{G}_\ell\left(\frac{y}{|y|}\right)$ is smooth and homogeneous of degree zero, it is not hard to see that, for every multiindex α, we have

$$\left| D^{\alpha} \tilde{G}_{\ell}\left(\frac{y}{|y|}\right) \right| \le \frac{C_{\alpha}}{|y|^{|\alpha|}} \quad , \; y \neq 0,$$

where C_{α} is a constant independent of y. Therefore, using Leibniz' formula, we obtain

$$\left| \Delta^{\frac{n-1}{2}} \frac{1}{|y|} \tilde{G}_{\ell}\left(\frac{y}{|y|}\right) \right| \le C'_{\ell} \; |y|^{-n} \; ,$$

so that, for $|x| > 2R$,

$$\begin{aligned} |u_{\ell}(x,0)| &\le C'_{\ell} \int_{|y| > \frac{|x|}{2}} |y|^{-n} |f(x-y)| dy \\ &\le C_{\ell} |x|^{-n} \|f\|_{1} . \end{aligned} \tag{3.3}$$

Similarly, we obtain the following estimate on ∇v_{ℓ} for $|x| > 2R$:

$$|\nabla v_{\ell}(x)| \le D_{\ell} |x|^{-n} \|f\|_{1} . ^{*} \tag{3.4}$$

Hence, the function $\omega_{\ell}(x,t)$ is of finite energy and we obtain, using the finite speed of propagation property and recalling that $|\lambda_{\ell}| \le \mu_{j}$:

$$\begin{aligned} \int_{|x| \ge \rho + \mu_j T} |u_{\ell}(x,T)|^2 dx &= \int_{|x| \ge \rho + \mu_j T} \left| \frac{\partial \omega_{\ell}}{\partial t}(x,T) \right|^2 dx \\ &\le \int_{|x| \ge \rho + |\lambda_{\ell}| T} \left(\left| \frac{\partial \omega_{\ell}}{\partial t}(x,T) \right|^2 + \lambda_{\ell}^2 |\nabla \omega_{\ell}(x,T)|^2 \right) dx \\ &\le \int_{|x| \ge \rho} (|u_{\ell}(x,0)|^2 + \lambda_{\ell}^2 |\nabla v_{\ell}(x)|^2) dx \\ &\le C\rho^{-n} \|f\|_{1}^{2} \quad \text{(by (3.3) and (3.4))}. \end{aligned}$$

---

* The constants C_{ℓ} and D_{ℓ} appearing in (3.3) and (3.4) are independent of f and x.

The proof is complete.

Theorem 3.3 shows that the energy of the solution concentrates along characteristics (= bicharacteristics, in the spherically symmetric case), namely, that for ρ large enough the energy in the sets

$$\Sigma_j(T) = S(\mu_j, \mu_{j+1}, \rho) \cap \{(x,T)\}^*$$

is arbitrarily small.

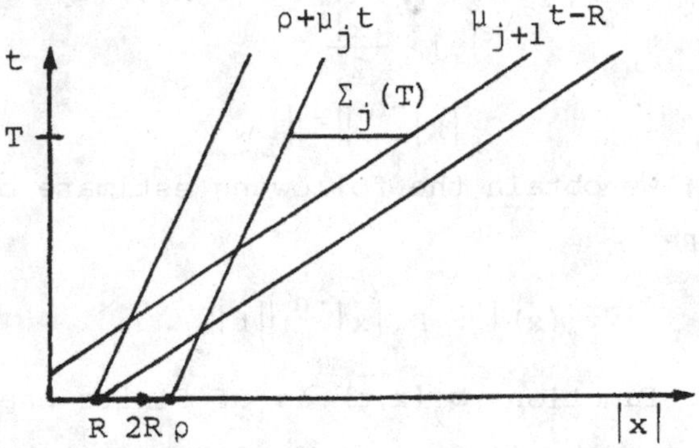

Combining this result with the conservation of energy (Hermitian case), we easily obtain the following result which shows the optimality of the uniform decay rate given by Theorem 2.4.

THEOREM 3.5. *Assume that the* A_j's *are Hermitian and that the eigenvalues of* $A \cdot \omega$ *are independent of* ω. *Let* $f \in C_0^\infty(\mathbb{R}^n, \mathbb{C}^k)$ *be a nonstationary function (not identically zero). Then, there exists a constant* $C > 0$, *independent of* t, *such that*

$$\sup_{x \in \mathbb{R}^n} |u(x,t)| \geq Ct^{-(n-1)/2} .$$

---

* $S(\mu_j, \mu_{j+1}, \rho)$ is defined as $\{(x,t) \mid t > (\rho + R)/(\mu_{j+1} - \mu_j),\ \mu_j t + \rho \leq |x| \leq \mu_{j+1} t - R\}$.

4. THE WAVE EQUATION

In this section, we apply in detail the method of Sections 1 and 2 to rediscover known results on the L^p-behavior (p big, $\leq \infty$) of the wave equation in $\mathbb{R}^n$ (n odd, ≥ 3):

$$(4.1)\qquad \begin{cases} \dfrac{\partial^2 u}{\partial t^2} - \Delta u = 0, \quad u(\cdot,0) = f \in \mathcal{S}(\mathbb{R}^n), \\ \dfrac{\partial u}{\partial t}(\cdot,0) = g \in \mathcal{S}(\mathbb{R}^n). \end{cases}$$

Without loss of generality, we may assume $g = 0$. Then, the Radon transform of u, $\hat{u}$, is the solution of the one-dimensional wave equation $\dfrac{\partial^2 \hat{u}}{\partial t^2} - \dfrac{\partial^2 \hat{u}}{\partial s^2} = 0$* with initial data $\hat{u}(s,\omega;0) = \hat{f}(s,\omega)$, $\dfrac{\partial \hat{u}}{\partial t}(s,\omega;0) = 0$. Clearly, $\hat{u}$ is given by

$$\hat{u}(s,\omega;t) = \frac{1}{2}\Big[\hat{f}(s-t,\omega) + \hat{f}(s+t,\omega)\Big],$$

so that the Radon inversion formula (Theorem 1.1 (a)) yields

$$\begin{aligned} u(x,t) &= c_1 \Delta^{\frac{n-1}{2}} \int_{S^{n-1}} \hat{u}(x\cdot\omega,\omega;t)\,d\omega \\ (4.2)\qquad &= c_1 \Delta^{\frac{n-1}{2}} \int_{S^{n-1}} \frac{1}{2}\Big[\hat{f}(x\cdot\omega - t,\omega) + \hat{f}(x\cdot\omega + t,\omega)\Big]d\omega \\ &= c_1 \Delta^{\frac{n-1}{2}} \int_{S^{n-1}} \hat{f}(x\cdot\omega - t,\omega)\,d\omega\,, \end{aligned}$$

---

* Recall that (1.1) (iii) of Section 1 implies $(\Delta u)^{\wedge} = \left(\dfrac{\partial}{\partial s}\right)^2 \hat{u}$.

the last equality holding because $\hat{f}(-s,-\omega) = \hat{f}(s,\omega)$ (recall (1.1) (i) of Section 1). Now, letting I denote the temperate distribution

$$h \longmapsto \langle I,h\rangle = c_1 \int_{S^{n-1}} \hat{h}(1,\omega)\,d\omega, \quad h \in \mathcal{S}(\mathbb{R}^n),$$

we get

LEMMA 4.1. $u(\cdot,t) = \frac{1}{t}\,\Delta^{\frac{n-1}{2}}\left[I\left(\frac{\cdot}{t}\right) * f\right], \quad t > 0.$

Indeed, by definition, $\hat{f}(x\cdot\omega - t,\omega) = \hat{h}(t,\omega)$, where $h = \tau_x f^-$, τ_x denotes "translation by x", $\tau_x f(y) = f(y-x)$, and $f^-(z) = f(-z)$. On the other hand, a change of variables gives

$$c_1 \int_{S^{n-1}} \hat{h}(t,\omega)\,d\omega = c_1 \int_{S^{n-1}} \left(\int_{y\cdot\omega=t} h(y)\,dy\right) d\omega$$

$$= c_1 t^{n-1} \int_{S^{n-1}} \left(\int_{z\cdot\omega=1} h(tz)\,dz\right) d\omega$$

$$= t^{n-1}\langle I, h(t\cdot)\rangle.$$

Therefore, as $\langle I\left(\frac{\cdot}{t}\right), h\rangle = t^n \langle I, h(t\cdot)\rangle$ by definition*, we can rewrite (4.2) as

$$u(x,t) = \Delta^{\frac{n-1}{2}}\left[t^{n-1}\cdot t^{-n}\langle I\left(\frac{\cdot}{t}\right), h\rangle\right]$$

$$= \Delta^{\frac{n-1}{2}}\left[t^{-1}\langle I\left(\frac{\cdot}{t}\right), \tau_x f^-\rangle\right]$$

$$= \frac{1}{t}\,\Delta^{\frac{n-1}{2}}\left[I\left(\frac{\cdot}{t}\right) * f\right](x).$$

The next result gives a better description of I.

---

* If $a > 0$ and T is a distribution, the distribution $T(a\cdot)$ is defined by $\langle T(a\cdot),\phi\rangle = a^{-n}\langle T,\phi(a^{-1}\cdot)\rangle$.

LEMMA 4.2. *The distribution* $h \longmapsto \langle I,h\rangle$ *is given by the function*

$$I(y) = c_1 \Omega_{n-2} \begin{cases} 0 & , \ |y| \le 1 \\ |y|^{-1}(1-|y|^{-2})^{\frac{n-3}{2}} & , \ |y| > 1 \end{cases}$$

(Ω_{n-2} = *surface measure of* S^{n-2}).

PROOF. For each $t > 0$, let us define the function I_t by

$$(4.3) \qquad I_t(y) = \frac{d}{dt}\int_{S^{n-1}} H(t-y\cdot\omega)\,d\omega,$$

where $H(\sigma) = \begin{cases} 0 \ , \ \sigma < 0 \\ 1 \ , \ \sigma \ge 0 \end{cases}$ is the Heaviside function.

Then, using Fubini's theorem and observing that $\hat{h}(t,\omega) = \frac{d}{dt}\int_{y\cdot\omega\le t} h(y)\,dy$, we get that

$$(4.4) \qquad \int_{S^{n-1}} \hat{h}(t,\omega)\,d\omega = \int_{\mathbb{R}^n} h(y)\,I_t(y)\,dy.$$

Clearly, (4.3) gives $I_t(y) = 0$ if $|y| \le t$. On the other hand, writing $\int_{S^{n-1}} H(t-y\cdot\omega)\,d\omega$ as an iterated integral yields

$$\int_{S^{n-1}} H(t-y\cdot\omega)\,d\omega = \Omega_{n-2}\int_{-1}^{1} H(t-\sigma|y|)(1-\sigma^2)^{\frac{n-3}{2}}\,d\sigma$$

$$= \Omega_{n-2}\int_{-1}^{t/|y|}(1-\sigma^2)^{\frac{n-3}{2}}\,d\sigma, \quad \text{if} \quad |y| > t,$$

and hence,

$$(4.5) \qquad I_t(y) = \Omega_{n-2}\,\frac{1}{|y|}\left(1-\frac{t^2}{|y|^2}\right)^{\frac{n-3}{2}} \quad \text{if} \quad |y| > t.$$

In particular, for $t = 1$, (4.4) and (4.5) give the desired result.

Now, we can see that I as well as all of its derivatives up to order $\frac{n-3}{2}$ are bounded, i.e.,

$$I \in W^{(n-3)/2,\infty}(\mathbb{R}^n).$$

In fact, an easy induction shows the estimates

$$|D^\alpha I(y)| \le \frac{C_\alpha}{|y|^{1+|\alpha|}}\left(1-\frac{1}{|y|^2}\right)^{\frac{n-3}{2}-|\alpha|}, \quad |y|>1, \text{ any } \alpha.$$

Therefore, recalling that $I(y) = 0$ for $|y| \le 1$, we obtain the following L^q-estimates on the functions $D^\alpha\left[I\left(\frac{\cdot}{t}\right)\right]$, with $|\alpha| = \frac{n-3}{2}$:

$$\left\|D^\alpha\left[I\left(\frac{\cdot}{t}\right)\right]\right\|_q \le C_q t^{\frac{n}{q}-\frac{n-3}{2}}, \quad \frac{2n}{n-1} < q \le \infty.$$

Finally, if in Lemma 4.1 we move $\frac{n-3}{2}$ derivatives over $I\left(\frac{\cdot}{t}\right)$, the remaining $\frac{n+1}{2}$ derivatives over f, and use the Hausdorff-Young Inequality, we get

$$\|u(\cdot,t)\|_r \le C_q t^{\frac{n}{q}-\frac{n-1}{2}} \|f\|_{\frac{n+1}{2},p} \quad \text{for all } t > 0.$$

with $\frac{1}{r} = \frac{1}{p} + \frac{1}{q} - 1 \ge 0$, $\frac{2n}{n-1} < q \le \infty$. In particular, for $p = 1$, $r = q = \infty$, this gives the uniform behavior:

$$\|u(\cdot,t)\|_\infty \le C_\infty t^{-\frac{n-1}{2}} \|f\|_{\frac{n+1}{2},1} \quad \text{for all } t > 0.$$

BIBLIOGRAPHY

1. BUSEMAN, H., *Convex Surfaces*, Interscience, New York, 1958.

2. HELGASON, S. *The Radon transform on Euclidean spaces...*, Acta Math. 113 (1965), 93-106.

3. JOHN, F. *Plane Waves and Spherical Means Applied to Partial Differential Equations*, Interscience, New York, 1955.

4. LAX, P. D. and PHILLIPS, R. S., *Scattering theory*, Rocky Mountain J. of Math.1 (1971), 173-223.

5. LUDWIG, D., *Examples of the behavior of hyperbolic equations for large times*, J. Math. Mech. 12 (1963), 557-566.

6. LUDWIG, D., *The Radon transform on Euclidean spaces*, Comm. Pure Appl. Math. 19 (1966), 49-81.

7. SEGAL, I., *Quantization and dispersion for non-linear relativistic equations*, in *Mathematical Theory of Elementary Particles*, M. I. T. Press, Cambridge, Mass. (1966), 79-108.

8. STEIN, E. M., *Singular Integrals and Differentrability Properties of Functions*, Princeton University Press, Princeton, 1970.

9. WAHL, W. v., *L^p-decay rates for homogeneous wave equations*, Math. Z. 120 (1971), 93-106.

10. WILCOX, C. H., *Wave operators and asymptotic solutions of wave propagation problems of classical physics*, Arch. Rat. Mech. Anal. 22 (1966), 37-78.

THE DIRICHLET PROBLEM FOR NONLINEAR ELLIPTIC EQUATIONS: A HILBERT SPACE APPROACH

by

DJAIRO GUEDES DE FIGUEIREDO*

Department of Mathematics
Universidade de Brasília
Brasília, D. F., Brasil

1. INTRODUCTION

Let Ω be a bounded domain in R^N, and

$$(1) \qquad Lu = \sum_{\substack{|\alpha|\le m \\ |\beta|\le m}} (-1)^{|\beta|} D^{\beta}(a_{\alpha\beta}(x) D^{\alpha} u)$$

be a uniformly strongly elliptic operator acting on functions defined in Ω. Besides the existence theory of generalized solutions for the Dirichlet problem associated with L, we shall also use the regularity theory of solutions. For that matter, we will assume some regularity conditions on the coefficients of L and on the boundary $\partial\Omega$ of Ω. Let $F(x,\eta)$ be a real-valued function defined on $\Omega \times R^M$, where M is the number of all derivatives D^{α}, for $0 \le |\alpha| \le 2m - 1$.

---

* The material presented here corresponds roughly to the lectures given by the author at Tulane, in a program sponsored by the Ford Foundation.

In this paper we study a Hilbert space theory for the Dirichlet problem for the equation

(2) $$Lu = F(x, Du(x)), \text{ in } \Omega,$$

where $Du(x)$ stands for the M-vector whose components are the derivatives $D^{\alpha}u(x)$ of u, for $0 \le |\alpha| \le 2m-1$. As usual the Dirichlet boundary conditions are $\frac{\partial^j u}{\partial \nu^j} = 0$ on the boundary $\partial\Omega$, for all $0 \le j \le m-1$, where $\partial/\partial\nu$ stands for the outward normal derivative at a point on $\partial\Omega$. We will consider the generalized Dirichlet problem

(3) $$a[u,\phi] = (F(\cdot,Du), \phi), \text{ for all } \phi \in C_o^{\infty}(\Omega),$$

where $a[\cdot,\cdot]$ is the Dirichlet form associated to L, and $(\cdot,\cdot)$ stands for the L_2-inner product. A solution of (3) will be a function in some appropriate Sobolev space, namely $H^{2m}(\Omega) \cap H_o^m(\Omega)$.

In Section 2 we collect some known results on the theory of the Dirichlet problem for linear elliptic equations, stating them in the form that they are needed here. The short Section 3 contains a simple proof that the Niemytski mapping $Nu = F(x, Du)$ is continuous from H^{2m-1} to L^2.

In Section 4 we consider the Dirichlet problem when the nonlinearity $F(x, Du)$ grows at most linearly in u, and in a certain sense is bounded away from the eigenvalues of L. For that reason, we say that, in this case, there is no resonance. The main theorem of Section 4 has been proved by the author in [2], and generalizes results of Landesman-Lazer [8], Adams [1] and Schultz [12]. The proof of Theorem 2 presented here is simpler than the one given in [2], and follows a suggestion of F. E. Browder.

In Section 5 we consider nonlinearities $F(x, Du)$ that grow like powers k $(0 \leq k < 1)$ of u and its derivatives of order $\leq 2m - 1$. In this case we do not assume that $F(x, Du)$ stays away from the eigenvalues of L. For that reason we say that there is resonance. The result presented in this section generalizes a result of [2], and uses the regularity theory, as suggested by H. Brezis and L. Nirenberg. Theorem 3 generalizes also previous results of Nečas [10], Fučik-Kučera-Nečas [5], Hess [6], Landesman-Lazer [8], Nirenberg [11], Williams [14] and the author [3].

2. LINEAR ELLIPTIC EQUATIONS

Let us consider a linear differential operator

$$(2.1) \qquad Lu = \sum_{|\alpha|,|\beta|\leq m} (-1)^{|\beta|} D^{\beta}(a_{\alpha\beta}(x) D^{\alpha} u)$$

acting on real-valued functions $u(x)$ defined on a bounded domain Ω in R^N, whose boundary $\partial\Omega$ is assumed to be of class C^{2m}. No regularity of the boundary would be necessary if we were to use only the existence of H_0^m-solutions for the Dirichlet problem, as we did in [2]. The regularity of $\partial\Omega$ will be required to improve the class of differentiability of these solutions through the so-called regularity theory. The following assumptions will be made on L throughout this paper.

(L - 1) *The coefficients* $a_{\alpha\beta}$ *are real-valued functions defined in* $\bar{\Omega}$ *and* $a_{\alpha\beta} \in C^{|\beta|}(\bar{\Omega})$, *for all*

$0 \leq |\alpha|, |\beta| \leq m.$

(L - 2) L *is uniformly strongly elliptic, i.e., there exists a constant* $c > 0$ *such that*

$$(2.2) \qquad \sum_{|\alpha|=|\beta|=m} a_{\alpha\beta}(x)\xi^\alpha\xi^\beta \geq c|\xi|^{2m}$$

for all $\xi \in R^N$ *and all* $x \in \bar{\Omega}$.

The Dirichlet from associated with L is given by

$$(2.3) \quad a[\psi,\phi] = \sum_{|\alpha|,|\beta|\leq m} \int_\Omega a_{\alpha\beta}(x)D^\alpha\psi(x)D^\beta\phi(x)dx,$$

which, in view of (L-1), is a bounded bilinear form in the space $C_o^\infty(\Omega)$ of the infinitely differentiable functions with compact support in Ω, endowed with the m-norm. If k is a non-negative integer and ϕ is a continuously differentiable function up to order k, the k-norm is defined by

$$(2.4) \qquad ||\phi||_k = \left[\sum_{|\alpha|\leq k} \int_\Omega |D^\alpha\phi|^2\right]^{1/2}.$$

Let us denote by H_o^m the space obtained by completion of $C_o^\infty(\Omega)$ with respect to the m-norm. H_o^m is a Hilbert space, $H_o^m \subset L^2$, and $H_o^o = L^2$. The Dirichlet form (2.3) can be extended by H_o^m.

Using (L - 1) and (L - 2) we can prove Gårding's inequality: there exist real constants $c_o > 0$ and k_o such that, for all $\phi \in H_o^m$, one has

$$(2.5) \qquad a[\phi,\phi] \geq c_o||\phi||_m^2 - k||\phi||_o^2.$$

We remark that the whole power of (L - 1) is not used in the derivation of Gårding's inequality. Namely (2.5) holds under the following weaker assumptions on the coefficients: (i) $a_{\alpha\beta} \in L^{\infty}(\Omega)$, for $0 \leq |\alpha|, |\beta| \leq m$, (ii) $a_{\alpha\beta}$, for $|\alpha| = |\beta| = m$, are uniformly continuous in Ω. See Friedman [4] for the proof of Gårding's inequality and other results stated in this section.

The classical Dirichlet problem is the following. Given a real-valued function $f: \Omega \to \mathbb{R}$, to find $u \in C^{2m}(\Omega) \cap C^{m-1}(\overline{\Omega})$ such that

$$Lu = f, \text{ in } \Omega$$

$$\frac{\partial^j u}{\partial \nu^j} = 0, \text{ on } \partial\Omega, \text{ for } j = 0,1,\ldots,m-1.$$

It is known that this problem does not always have a solution. Indeed there are continuous functions f such that $\Delta u = f$, in Ω, and $u = 0$, on $\partial\Omega$, has no solution $u \in C^2(\Omega) \cap C^o(\overline{\Omega})$. However, the existence of solution is guaranteed whenever f is Hölder-continuous. It is possible to deal directly with the classical Dirichlet problem, using Schauder estimates and working in spaces of Hölder continuous functions. It became clear in the 50's that a better approach to treat the Dirichlet problem is through a Hilbert space formulation of the problem. In this way, a generalized Dirichlet problem is set and solved using essentially two tools, Gårding's inequality and the Lax-Milgram lemma. The solution so obtained is in the Sobolev space H^m_o. Then comes the much harder part, called the regularity theory, which consists in proving that this solution is actually in a Sobolev space H^k, for some $k > m$. Once this done, one can obtain information about continuity and differentiability of this solution, via the Sobolev imbedding theorem.

The generalized Dirichlet problem is the following. Given a real valued function $f \in L^2(\Omega)$, to find $u \in H_o^m$ such that

$$(2.6) \qquad a[u,\phi] = (f,\phi), \quad \text{for all} \quad \phi \in H_o^m ,$$

where $(\ ,\ )$ stands for the L^2-inner product.

The following result is a complete answer to the existence and uniqueness questions for the generalized Dirichlet problem.

PROPOSITION 1. (Fredholm Alternative) *Assume* (L-1) *and* (L-2) *above. Then, one and only one of the following possibilities holds true:*

(i) There exists a unique $u \in H_o^m$, *for each given* $f \in L^2$, *such that* (2.6) *holds. In particular* $a[u,\phi] = 0$, *for all* $\phi \in H_o^m$, *implies* $u = 0$.

(ii) There exists $0 \neq u \in H_o^m$ *such that*

$$(2.7) \qquad a[u,\phi] = 0, \quad \textit{for all} \quad \phi \in H_o^m .$$

Furthermore, when (i) holds, the solution operator $A: L^2 \to H_o^m$, *defined by* $Af = u$, *is linear and bounded. In the case when (ii) holds, the set of solutions of* (2.7) *is a finite dimensional subspace of* H_o^m, *whose dimension is the same as the dimension of*

$$N^* = \{v \in H_o^m : \ a[\phi,v] = 0, \ \textit{for all} \ \ \phi \in H_o^m\}.$$

Moreover, still in case (ii) equation (2.6) *has a solution if and only if* $(f,v) = 0$, *for all* $v \in N^*$.

The eigenvalue problem associated with the generalized

Dirichlet problem consists in the following. To find λ, such that for some $0 \neq u \in H_o^m$ one has

$$(2.8) \qquad a[u,\phi] = \lambda(u,\phi), \quad \text{for all} \quad \phi \in H_o^m.$$

λ is an eigenvalue, u is an eigenfunction corresponding to the eigenvalue λ, and the set of all eigenfunctions corresponding to λ form the eigenspace corresponding to the eigenvalue λ.

The differential operator L is said to be symmetric if $a_{\alpha\beta} = a_{\alpha\beta}$, for all $0 \leq |\alpha|, |\beta| \leq m$. In this case $a[\phi,\psi] = a[\psi,\phi]$. Now we state the main facts about the eigenvalue problem for a symmetric operator L.

PROPOSITION 2. *Assume* (L-1) *and* (L-2) *above, and that* L *is symmetric. Then:*

(a) The eigenvalues of (2.8) *are all real, they form a sequence* (λ_n) *such that* $\lambda_n \to +\infty$, *and the corresponding eigenspaces are finite dimensional.*

(b) If part (i) of the alternative holds, then the operator A *thought as an operator from* L^2 *to* L^2 *is symmetric, bounded and its norm is given by*

$$(2.9) \quad \|A\|_{L^2 \to L^2} = \max\{|\lambda|^{-1} : \lambda \text{ an eigenvalue of } (2.8)\}.$$

Now we state the regularity result needed for the present paper. Let us first introduce the Sobolev spaces H^k, where k is some fixed non-negative integer. Let $\hat{C}^k(\Omega)$ denote the space of all functions $u: \Omega \to \mathbb{R}$, which are of class C^k and $\|u\|_k < \infty$, where this norm is defined in (2.4). Now define H^k as the completion of $\hat{C}^k(\Omega)$ with respect to this norm. When $\partial\Omega$ is of class C^k, this definition is equivalent to taking the completion, in the norm (2.4), of the space $C^k(\overline{\Omega})$ of all functions $u: \overline{\Omega} \to \mathbb{R}$, which are continuously differentiable up to order k in $\overline{\Omega}$.

PROPOSITION 3. *Assume* (L-1) *and* (L-2) *above, and that* $\partial\Omega$ *is of class* C^{2m}. *Then, if* $u \in H_o^m$ *is a solution of* (2.6), *it follows that* $u \in H^{2m}$ *and*

(2.10) $$\|u\|_{2m} \le c_1(\|f\|_o + \|u\|_o).$$

Thus, when part (i) of the Alternative holds the solution operator A maps L^2 onto $H_o^m \cap H^{2m}$. Moreover, A <u>viewed as an operator from</u> L^2 <u>to</u> H^{2m} <u>is bounded</u>. Indeed, it follows from (2.10) that

$$\|Af\|_{2m} \le c_1(\|f\|_o + \|Af\|_o),$$

and, from Proposition 1, $\|Af\|_o \le \|Af\|_m \le \text{const}\|f\|_o$. Consequently $\|Af\|_{2m} \le \text{const}\|f\|_o$ for all $f \in L^2$.

3. NIEMYTSKI MAPPINGS

We shall consider perturbations of the linear elliptic equations studied in Section 2 by a nonlinear term of the form $F(x, Du)$ where Du stands for the M-vector whose components are $D^\alpha u$, for all $0 \le |\alpha| \le 2m - 1$. The following basic assumption is made on F:

(F-1) *The function* $F: \Omega \times R^M \to R$ *satisfies Carathéodory's conditions, i.e., (i) for each fixed* $\eta \in R^M$, *the function* $x \longmapsto F(x,\eta)$ *is measurable in* Ω, *(ii) for fixed* $x \in \Omega$ *(a.e.), the function* $\eta \longmapsto F(x,\eta)$ *is continuous in* R^M.

The above assumption (F-1) implies that if the mapping $H: \Omega \to R^M$ is measurable, then the function $x \mapsto F(x, H(x))$ is also measurable, see Krasnoselski [7] or Vainberg [13]. We shall need the following result, whose proof can also be seen in the afore-mentioned books. However, we will include here a proof, which seems to be of some interest.

PROPOSITION 4. *Suppose that a function* $F: \Omega \times R^M \to R$ *satisfies Carathéodory's conditions* (F-1), *and moreover, that there exist constants* $b_\alpha \geq 0$ *and a function* $b \in L^2$ *such that*

$$|F(x,\eta)| \leq \sum_{|\alpha| \leq 2m-1} b_\alpha |\eta_\alpha| + b(x) \tag{3.1}$$

for all $\eta \in R^M$ *and* $x \in \Omega$ *(a.e.). Then the Niemytski mapping* N *given by* $(Nu)(x) = F(x, Du(x))$, *where* $Du(x)$ *is the* M-*vector whose components are* $D^\alpha u(x)$, *is bounded and continuous from* H^{2m-1} *to* L^2.

PROOF. It follows from Minkowski inequality that

$$\|Nu\|_o \leq \sum_{|\alpha| \leq 2m-1} b_\alpha \|D^\alpha u\|_o + \|b\|_o$$

which implies, via the Cauchy-Schwarz inequality, that

$$\|Nu\|_o \leq B\|u\|_{2m-1} + \|b\|_o , \tag{3.2}$$

where B is a constant depending only on the b_α's. Thus (3.2) implies that N is a mapping from H^{2m-1} into L^2, and also that such a mapping is bounded. To prove the continuity of N we use the following characterization of L^1-convergence: "Let $f_n: \Omega \to R$ be a sequence of measurable functions which converges almost everywhere to a function f. Then f_n converges to f in L^p, $p \geq 1$, if and only if, given any subsequence $(f_{n'})$ of (f_n), there exists a further subsequence $(f_{n''})$ of $(f_{n'})$ and an L^p-function g (depending on this last subsequence) such that $|f_{n''}(x)| \leq g(x)$, for all n'' and $x \in \Omega$ (a.e.)." Now let $u_n \in H^{2m-1}$ be a sequence converging to u in the H^{2m-1}-norm, and we claim that $F_n(x) = F(x, Du_n(x))$ converges in L^2 to $F(x) = F(x, Du(x))$. To apply the result stated above, we take a subsequence $(F_{n'})$ of (F_n). Then, using the above result, we take a subsequence $(u_{n''})$ of $(u_{n'})$ such that $|D^\alpha u_{n''}(x)| \leq g(x)$, for $0 \leq |\alpha| \leq 2m-1$, where $g \in L^2$. Using now inequality (3.1) we get

$$|F_{n''}(x)| = |F(x, Du_{n''}(x))| \leq \left(\sum_{|\alpha| \leq 2m-1} b_\alpha \right) g(x) + b(x),$$

where the last term of the inequality is an L^2-function. So we apply the result above once more to get the convergence of F_n to F in the L^2-sense.

We recall that a mapping $N: X \to Y$ from a Banach space X into another Banach space Y is <u>compact</u> if it is continuous and takes bounded sets of X into relatively compact sets of Y.

COROLLARY. *Same hypothesis as in Proposition* 4. *Then the Niemytski mapping* N *is a compact mapping from* H^{2m} *to* L^2.

PROOF. This is a direct consequence of Proposition 4 and the fact that H^{2m} is compactly imbedded in H^{2m-1}.

The generalized Dirichlet problem for

$$\text{(3.3)} \qquad Lu = F(x, Du)$$

where F satisfies (F-1) and an inequality of type (3.1) above, consists in finding $u \in H^{2m} \cap H_o^m$ such that

$$\text{(3.4)} \qquad a[u,\phi] = (Nu,\phi), \quad \text{for all } \phi \in H_o^m.$$

4. THE DIRICHLET PROBLEM WHEN THERE IS NO RESONANCE

First we consider the generalized Dirichlet problem (3.4) in the case when there is no resonance. This terminology is justified by the fact that in the differential equation (3.3) the perturbation $F(x, Du)$ "stays away" from the eigenvalues of L, in a sense that will be made precise below.

THEOREM 1. *Suppose that the differential operator* L *is symmetric and satisfies assumptions* (L-1) *and* (L-2) *above. Let* $\lambda_1 < \lambda_2$ *be two consecutive eigenvalues of the eigenvalue problem* (2.8). *Let* λ *and* γ *be such that* $\lambda_1 < \lambda < \lambda_2$ *and* $0 \le \lambda < \min\{\lambda-\lambda_1, \lambda_2-\lambda\}$. *Assume that the function* F *satisfies* (F-1) *and the following assumption:*

(F-2) *There exists constants* $0 \le \kappa < 1$, $c_\alpha \ge 0$, *for* $|\alpha| \le 2m-1$, *and an* L^2*-function* $b(x)$ *such that*

$$\text{(4.1)} \quad |F(x,\eta) - \lambda\eta_o| \le \gamma|\eta_o| + \sum_{|\alpha|\le 2m-1} c_\alpha |\eta_\alpha|^\kappa + b(x),$$

for all $\eta \in R^M$ *and* $x \in \Omega$ *(a.e.). Here* η_o *stands for the first component of the vector* η, *which corresponds to* u *in the vector* Du.

Then there exists $u \in H^{2m} \cap H_o^m$ *such that*

(4.2) $$a[u,\phi] = (Nu,\phi), \quad \textit{for all} \quad \phi \in H_o^m.$$

PROOF. First of all we observe that in view of the fact that Ω is bounded, it follows from (4.1) that the function $F(x,\eta)$ satisfies an inequality like (3.1). So Proposition 4 and its corollary apply. Now since (4.2) is equivalent to

(4.3) $$a[u,\phi] - \lambda(u,\phi) = (Nu - \lambda u,\phi), \quad \text{for all} \quad \phi \in H_o^m,$$

the present theorem will follow as a consequence of Theorem 2, below, and the following remarks:

(i) The problem

(4.4) $$a[u,\phi] - \lambda(u,\phi) = (f,\phi), \quad \text{for all} \quad \phi \in H_o^m$$

is uniquely solvable, for each $f \in L^2$; see Propositions 1 and 2, Section 2.

(ii) The mapping $A: L^2 \to H^{2m}$ defined by $Af = u$, where u is the solution of (4.4), is linear and continuous, and, moreover, its range $R(A) \subset H_o^m$, and its norm as mapping from L^2 to L^2 can be estimated as follows

$$\|A\|_{L^2 \to L^2} \leq \max\left\{\frac{1}{\lambda-\lambda_1}, \frac{1}{\lambda_2-\lambda}\right\} < \frac{1}{\gamma}.$$

The next result will be stated separately because it holds without the symmetry of L.

THEOREM 2. *Suppose that the differential operator* L *satisfies assumptions* (L-1) *and* (L-2), *and that part (i) of the Alternative in Proposition* 1 *holds. Assume that the function* F *satisfies* (F-1) *and*

(F-3) *There exist constants* $0 \le c' < \|A\|^{-1}_{L^2 \to L^2}$, $0 \le \kappa < 1$, $c_\alpha \ge 0$, *for* $|\alpha| \le 2m-1$, *and an* L^2*-function* $b(x)$, *such that*

$$(4.5) \qquad |F(x,\eta)| \le c'|\eta_o| + \sum_{|\alpha| \le 2m-1} c_\alpha |\eta_\alpha|^\kappa + b(x)$$

for all $\eta \in R^M$ *and* $x \in \Omega$ *(a.e.)*.

Then there exists $u \in H^m_o \cap H^{2m}$ *such that*

$$(4.6) \qquad a[u,\phi] = (Nu,\phi), \quad \textit{for all } \phi \in H^m_o.$$

PROOF. In view of Proposition 1, the problem of solving (4.6) is equivalent to finding $u \in H^{2m}$ such that

$$(4.7) \qquad u = ANu.$$

Observe that (4.7) says that $u \in \mathcal{R}(A)$, and so any solution of (4.7) is also in H^m_o. Now Propositions 3 and 4 imply that $AN: H^{2m} \to H^{2m}$ is compact. So the theorem will be proved if we show that AN has a fixed point. First we observe that (4.7) has a solution whenever the equation

$$(4.8) \qquad v = NAv$$

has a solution $v \in L^2$. Indeed, if v is a solution of (4.8) then $u = Av$ is in H^{2m} and u is a solution of (4.7). To prove that (4.8) has a solution we use Schauder fixed point theorem. So we show that there exists a ball of radius R about the origin in L^2 which is mapped into itself by NA. Using (4.5) and Hölder's inequality we get

$$(4.9) \qquad \|Nu\|_o \leq c' \|u\|_o + c'' \|u\|_{2m}^{\kappa} + \|b\|_o ,$$

for all $u \in H^{2m}$, where c'' is a constant depending on the c_α's and on κ. On the other hand it follows from (2.10) that

$$(4.10) \quad \|Av\|_{2m} \leq c_1(\|v\|_o + \|Av\|_o) \leq c_1(1+\|A\|_{L^2 \to L^2})\|v\|_o$$

for all $v \in L^2$. Now using (4.9) we can obtain the following estimate:

$$(4.11) \quad \|NAv\|_o \leq c' \|Av\|_o + c'' \|Av\|_{2m}^{\kappa} + \|b\|_o$$

for all $v \in L^2$. Using (4.10) in (4.11) we conclude that there are constants c_2 and c_3 such that

$$\|NAv\|_o \leq c' \|A\|_{L^2 \to L^2} \|v\|_o + c_2 \|v\|_o^{\kappa} + c_3$$

for all $v \in L^2$. Since $c' \|A\|_{L^2 \to L^2} < 1$ and $0 \leq \kappa < 1$, we see that, for R sufficiently large

$$\|v\|_o \leq R \text{ implies } \|NAv\|_o \leq R.$$

Thus the proof of Theorem 2 is complete.

5. THE DIRICHLET PROBLEM WHEN THERE IS RESONANCE

Now we consider the case when the perturbation $F(x, Du)$ does not stay properly between the eigenvalues of L. In fact it could touch at most one eigenvalue.

More precisely, we assume

(F-4) *Let* λ_1 *be an eigenvalue of the eigenvalue problem* (2.8), *where* L *is assumed to be symmetric. There exist constants* $c_\alpha \geq 0$, $0 \leq \kappa < 1$ *and an* L^2-*function* $d(x)$ *such that*

$$(5.1)\quad |F(x,\eta) - \lambda_1\eta_o| \leq \sum_{|\alpha|\leq 2m-1} c_\alpha|\eta_\alpha|^\kappa + d(x)$$

for all $\eta \in R^M$ *and* $x \in \Omega$ *(a.e.).*

Compare this condition with (F-2): Observe that if λ_1 and λ_2 are consecutive eigenvalues of the eigenvalue problem (2.8), and we take λ, such that $\lambda_1 < \lambda < (\lambda_1 + \lambda_2)/2$, then inequality (5.1) implies

$$(5.2)\quad |F(x,\eta) - \lambda\eta_o| \leq (\lambda-\lambda_1)|\eta_o| + \sum_{|\alpha|\leq 2m-1} c_\alpha|\eta_\alpha|^\kappa + d(x).$$

So, it is $\lambda - \lambda_1$ that corresponds to the constant γ of (4.1), and, unlike γ, it cannot be taken strictly smaller than $\min\{\lambda - \lambda_1, \lambda_2 - \lambda\}$, no matter how we choose λ.

THEOREM 5-3. *Suppose that the operator* L *is symmetric and satisfies conditions* (L-1) *and* (L-2). *Assume that* F *satisfies* (F-1), (F-4) *and the two additional conditions below:*

(F-5) *For each* $\eta \in S_1 = \{\eta \in R^M : \sum_{|\alpha|\leq 2m-1} |\eta_\alpha|^2 = 1\}$, *and* $x \in \Omega$ *(a.e.), the limit below exists*

$$(5.3) \qquad \lim_{n\to+\infty} \frac{F\left(x, r_n\eta^{(n)}\right) - \lambda_1 r_n \eta_0^{(n)}}{r_n^{\kappa}} = h(x, \eta),$$

for all sequences (r_n), $r_n \in R$ *and* $(\eta^{(n)})$, $\eta^{(n)} \in S_1$, *such that* $r_n \to +\infty$ *and* $\eta^{(n)} \to \eta \in S_1$.

(F-6) *For all* $v \in W = \{v \in H_0^m : a[v,\phi] = \lambda_1(v,\phi),$ *for all* $\phi \in H_0^m\}$, *with* $\|v\|_{2m} = 1$, *one has*

$$(5.4) \qquad \int_\Omega h\left(x, \frac{Dv(x)}{|Dv(x)|}\right)|Dv(x)|^{\kappa} v(x)\,dx > 0,$$

where $h(x,\eta)$ *is a real measurable function defined in* $\Omega \times S_1$, *and* $|h(x,\eta)| \leq g(x)$, *for* $x \in \Omega$ *(a.e.) and* $\eta \in S_1$, *with a* $g \in L^{2/(1-\kappa)}(\Omega)$.

Then there exists $u \in H_0^m \cap H^{2m}$ *such that*

$$(5.5) \qquad a[u,\phi] = (Nu,\phi), \quad \textit{for all} \quad \phi \in H_0^m.$$

REMARKS. 1) If F depends only on x and u, conditions (F-4), (F-5) and (F-6) take the following simpler forms.

$(F\text{-}4)_0$ *There exist constants* $c > 0$, $0 \leq \kappa < 1$ *and an* L^2*-function* $d(x)$ *such that*

$$|F(x,s) - \lambda_1 s| \leq c|s|^{\kappa} + d(x)$$

for all $s \in R$ *and* $x \in \Omega$ *(a.e.).*

$(F\text{-}5)_0$ *There exist functions* $h_+ \in L^{2/(1-\kappa)}$ *and* $h_- \in L^{2/(1-\kappa)}$ *such that*

$$\lim_{s \to \pm\infty} \frac{F(x,s) - \lambda_1 s}{|s|^{\kappa}} = h_{\pm}(x)$$

for $x \in \Omega$ *(a.e.).*

$(F\text{-}6)_o$ *For all* $v \in W = \{v \in H_o^m : a[v,\phi] = \lambda_1(v,\phi),$ *for all* $\phi \in H_o^m\}$, *with* $\|v\|_{2m} = 1$ *one has*

$$\int_{v>0} h_+|v|^{1+\kappa} - \int_{v<0} h_-|v|^{1+\kappa} > 0.$$

These are the conditions we have given in [2]. We remark that assumption (F-4) of [2] is superfluous.

2) If F depends only on x and u, and $\kappa = 0$, conditions $(F\text{-}4)_o$, $(F\text{-}5)_o$ and $(F\text{-}6)_o$ coincide with the ones given by Nirenberg [11], although he considers more general linear operators L.

3) If F depends only on u and $\kappa = 0$, conditions $(F\text{-}4)_o$, $(F\text{-}5)_o$ and $(F\text{-}6)_o$ coincide with the ones given by Landesman-Lazer [9], for second order symmetric operators L. The condition $\kappa = 0$ implies that the function $F(u)$ is bounded and continuous for $-\infty < u < \infty$. This condition has been assumed also in the papers of Hess [6], Nečas [10] and Williams [14].

4) Nonlinearities F depending on derivatives of u have been considered before in [11], Fučik-Kučera-Nečas [5], and the author [3]. The conditions in [11] seem hard to check.

PROOF. Consider the approximate equations, for each positive integer n,

(5.6) $a[u,\phi] - \frac{1}{n}(u,\phi) = (Nu,\phi)$, for all $\phi \in H_o^m$.

Let λ_2 be first eigenvalue of (2.8) larger than λ_1. So $\lambda_1 - \frac{1}{n}$ and $\lambda_2 - \frac{1}{n}$ are consecutive eigenvalues of the eigenvalue problem

$$a[u,\phi] - \frac{1}{n}(u,\phi) = \mu(u,\phi), \quad \text{for all} \quad \phi \in H_o^m.$$

For n sufficiently large, $\lambda_1 - \frac{1}{n} < \lambda_1 < \lambda_2 - \frac{1}{n}$. So we see, using (F-4), that for equation (5.6) condition (F-2) holds with $\lambda = \lambda_1$ and $\gamma = 0$. Thus Theorem 1 applies to (5.6) and we obtain $u_n \in H_o^m \cap H^{2m}$ such that

$$(5.7) \quad a[u_n, \phi] - \frac{1}{n}(u_n, \phi) = (Nu_n, \phi), \quad \text{for all} \quad \phi \in H_o^m.$$

Now we claim that $\|u_n\|_{2m}$ is bounded. Assuming for the moment that this has been established, the proof of the theorem is completed as follows. Selecting subsequences, if necessary, we can assume that $u_n \to u$ in H^{2m}, $u_n \to u$ in H_o^m, $Nu_n \to g$ in L^2. So passing to the limit in (5.7) we get

$$(5.8) \qquad a[u,\phi] = (g,\phi), \quad \text{for all} \quad \phi \in H_o^m.$$

From (5.7) and (5.8) we get

$$(5.9) \quad a[u_n - u, \phi] = (Nu_n - g + \frac{1}{n}u_n, \phi), \quad \text{for all} \quad \phi \in H_o^m.$$

Using (2.10) we obtain from (5.9)

$$\|u_n - u\|_{2m} \le c_1 (\|Nu_n - g + \frac{1}{n} u_n\|_o + \|u_n - u\|_o)$$

which implies that $u_n \to u$ in H^{2m}. So $Nu_n \to Nu$ in L^2, and then $g = Nu$. Thus (5.8) gives

$$a[u,\phi] = (Nu,\phi), \quad \text{for all } \phi \in H_o^m,$$

and then the theorem would be proved. So all we have to do is to prove that $(\|u_n\|_{2m})$ is a bounded sequence. Let us assume by contradiction that $\|u_n\|_{2m} \to \infty$. Let $v_n = u_n / \|u_n\|_{2m}$. Taking subsequences, if necessary, we may assume that there exist $v \in H_o^m \cap H^{2m}$ such that

$$v_n \rightharpoonup v \text{ in } H^{2m}, \quad v_n \to v \text{ in } H_o^m,$$

$$D^\alpha v_n \to D^\alpha v \text{ in } \Omega \text{ (a.e.)}$$

for all $0 \le |\alpha| \le 2m - 1$. From (5.7) we have

$$\text{(5.10)} \quad a[v_n, \phi] - \lambda_1 (v_n, \phi) - \frac{1}{n}(v_n, \phi) = \left(\frac{Nu_n - \lambda_1 u_n}{\|u_n\|_{2m}}, \phi \right),$$

for all $\phi \in H_o^m$. On the other hand, (F-4) and Hölder's inequality implies that there is a constant c_4, such that

$$\text{(5.11)} \quad \|Nu_n - \lambda_1 u_n\|_o \le c_4 \|u\|_{2m-1}^{\kappa} + \|d\|_o .$$

Now passing to the limit in (5.10) and using (5.11) we have

$$\text{(5.12)} \quad a[v,\phi] - \lambda_1 (v,\phi) = 0, \quad \text{for all } \phi \in H_o^m,$$

i.e. $v \in W$. Now we proceed to prove that $v \ne 0$, which is a consequence of the fact that $v_n \to v$ in H^{2m}. And this last fact is readily proved in the following way. It follows from (5.10) and (5.12) that

$$a[v_n - v, \phi] = \left(\frac{1}{n} v_n + \frac{Nu_n - \lambda_1 u_n}{\|u_n\|_{2m}} + \lambda_1 (v_n - v), \phi \right)$$

which implies, in view of (2.10), that

$$\|v_n - v\|_{2m} \le c_1\left(\left\|\frac{1}{n}v_n + \frac{Nu_n - \lambda_1 u_n}{\|u_n\|_{2m}} + \lambda_1(v_n - v)\right\|_o + \|v_n - v\|_o\right).$$

Using (5.11) and the fact that $v_n \to v$ in L^2, we then obtain that $v_n \to v$ in H^{2m}. Now taking $\phi = v$ in (5.10) and using the symmetry of L we obtain

$$-\frac{1}{n}(v_n, v) = \left(\frac{Nu_n - \lambda_1 u_n}{\|u_n\|_{2m}}, v\right).$$

Thus, for large n, we have

$$(Nu_n - \lambda_1 u_n, v) < 0$$

or, equivalently,

$$\text{(5.13)} \qquad \int_\Omega \frac{F(x, Du_n(x)) - \lambda_1 u_n(x)}{\|u_n\|_{2m}^{\kappa}} v(x)ds < 0.$$

Let $\Omega_o = \{x \in \Omega: v(x) \ne 0\}$. So, for each $x \in \Omega_o$,

$$|Dv(x)| = \left[\sum_{|\alpha| \le 2m-1} |D^\alpha v(x)|^2\right]^{1/2} > 0.$$

Thus, for each fixed $x \in \Omega_o$ (a.e.), there is an n_o (depending on x) such that, for $n \ge n_o$:

$$Du_n(x) = \|u_n\|_{2m} Dv_n(x) = \|u_n\|_{2m}|Dv_n(x)| \frac{Dv_n(x)}{|Dv_n(x)|}.$$

Now since $\|u_n\|_{2m}|Dv_n(x)| \to +\infty$, for $x \in \Omega_o$ (a.e.),

and $\dfrac{Dv_n(x)}{|Dv_n(x)|} \longrightarrow \dfrac{Dv(x)}{|Dv(x)|}$, for each fixed $x \in \Omega_o$

(a.e.), we use (F-5) to conclude that

$$(5.14) \qquad \frac{F(x, Du_n(x)) - \lambda_1 u_n(x)}{\|u_n\|_{2m}^{\kappa}} \longrightarrow |Dv(x)|^{\kappa} h\left(x, \frac{Dv(x)}{|Dv(x)|}\right) .$$

On the other hand, using (F-4), we have

$$(5.15) \qquad \left| \frac{F(x, Du_n(x)) - \lambda_1 u_n(x)}{\|u_n\|_{2m}^{\kappa}} \right|$$

$$\leq \frac{1}{\|u_n\|_{2m}^{\kappa}} \sum_{|\alpha| \leq 2m-1} c_\alpha |Du_n(x)|^{\kappa} + d(x) .$$

Now using Hölder's inequality, we see that the right side of (5.15) is an L^2-function. So we can apply the Lebesgue dominated convergence theorem to (5.13) to get, in view of (5.14),

$$\int_\Omega h\left(x, \frac{Dv(x)}{|Dv(x)|}\right) |Dv(x)|^{\kappa} v(x) dx \leq 0 ,$$

and this is a contradiction to assumption (F-6).

REFERENCES

1. ADAMS, R.A., *A quasi-linear elliptic boundary value problem*, Canad. J. Math. 18 (1966), 1105-1112.

2. DE FIGUEIREDO, D. G., *Some remarks on the Dirichlet Problem for semilinear elliptic equations*, Anais Acad. Brasil. Ci., to appear.

3. DE FIGUEIREDO, D. G., *On the range of nonlinear operators with linear asymptotes which are not invertible*, Comm. Math. Univ. Carolinae, to appear.

4. FRIEDMAN, A., *Partial Differential Equations*, Holt, Rinehart and Winston, New York, 1969.

5. FUČIK, KUČERA, NEČAS, *Ranges of nonlinear asymptotically linear operators*, to appear.

6. HESS, P., *On a theorem by Landesman and Lazer*, Indiana U. Math. J. 23 (1974), 827-829.

7. KRASNOSELSKII, M. A., *Topological Methods in the Theory of Nonlinear Integral Equations*, Pergamon Press, New York, 1964.

8. LANDESMAN, E. M., and LAZER, A. C., *Linear eigenvalues and a nonlinear boundary value problem*, Pacific J. Math. 33 (1970), 311-328.

9. LANDESMAN, E. M., and LAZER, A. C., *Nonlinear perturbations of linear elliptic boundary value problems at resonance*, J. Math. Mech. 19 (1970), 609-623.

10. NEČAS, J., *On the range of nonlinear operators with linear asymptotes which are not invertible*, Comm. Math. Univ. Carolinae 14 (1973), 63-72.

11. NIRENBERG, L., *An application of generalized degree to a class of nonlinear problems*, Proc. Symp. Fctl. Anal. Liège, Sept, 1971.

12. SHULTZ, M. H., *Quasi-linear elliptic boundary value problems*, J. Math. Mech. 18 (1968), 21-25.

13. VAINBERG, M., *Variational Methods for the Study of Nonlinear Operators*, Holden Day, San Francisco, 1964.

14. WILLIAMS, S. A., *A sharp sufficient condition for solution of a nonlinear elliptic boundary value problem*, J. Diff. Eq. 8 (1970), 580-586.

*EXACT CONTROLLABILITY OF LINEAR SYSTEMS IN INFINITE DIMENSIONAL SPACES**

by

H. O. FATTORINI

Department of Mathematics
University of California
Los Angeles, California 90024

and

Department of Mathematics
University of Buenos Aires
Buenos Aires, Argentina

1. INTRODUCTION

We deal here with the control system

$$u'(t) = Au(t) + Bf(t), \quad t \geq 0. \tag{1.1}$$

The linear operator A is the infinitesimal generator of a semigroup $S(t)$, $t \geq 0$ in the Banach space E, whereas the linear bounded operator B maps another Banach space F into E.

We shall be concerned with the case where F is n-dimensional Euclidean space. Then there exist $b_1, \ldots, b_n \in E$ such that, if $x = (x_1, \ldots, x_n) \in R^n$,

---

*This research was partially supported by the National Science Foundation through contract No. GP-27973.

$$Bx = x_1 b_1 + \dots + x_n b_n . \tag{1.2}$$

The control function $f(\cdot)$, defined in $t \geq 0$ and taking values in R^n will be assumed measurable (which means that each of its n components is measurable) and locally integrable in $t \geq 0$, that is

$$\int_0^T |f(t)| dt < \infty \tag{1.3}$$

for any $T > 0$, where $|\cdot|$ denotes any of the usual norms in R^n. (Note that f is locally integrable if and only if the same is true of each of its n components). We shall denote by F the class of all such functions.

If $f(\cdot)$ is, say, continuously differentiable and $u_0 \in D(A)$, the domain of A, then

$$\begin{aligned} u(t) &= S(t)u_0 + \int_0^T S(t-s)Bf(s)ds \\ &= S(t)u_0 + \sum_{k=1}^{n} \int_0^T S(t-s)b_k f_k(t)dt \quad (t \geq 0) \end{aligned} \tag{1.4}$$

is a strong or genuine solution of (1.1) with initial value

$$u(0) = u_0. \tag{1.5}$$

On the other hand, if f is an arbitrary element of F and $u_0 \in E$ we <u>define</u> $u(t)$, given by the "variation-of-constants" expression (1.4), to be the solution of the initial value problem (1.1), (1,5), the integrals in (1.4) understood in the sense of Bochner [7]. It is easy to see that $u(\cdot)$ is continuous for any $f \in F$.

Define, as in [3],

$$K_T = \{u \in E;\ u = \int_0^T S(t)Bf(t)dt,\ \ f \in F\}.$$

Clearly, K_T consists of the values (for $t = T$) of all the solutions of (1.1) with $u(0) = 0$ and $f \in \mathcal{F}$ arbitrary. Define also

$$K_\infty = \bigcup_{T>0} K_T .$$

We say that (1.1) is <u>exactly reachable in time</u> T if, given $u \in E$ there exists $f \in \mathcal{F}$ such that the solution $u(\cdot)$ of (1.1) with $u(0) = 0$ satisfies

$$u(T) = u.$$

Clearly, this is equivalent to

$$K_T = E. \tag{1.6}$$

The system is <u>exactly reachable</u> if

$$K_\infty = E. \tag{1.7}$$

On the other hand, (1.1) is <u>exactly controllable in time</u> T if, given $u \in E$ there exists $f \in \mathcal{F}$ such that the solution of (1.1) with $u(0) = u$ satisfies

$$u(T) = 0.$$

Clearly, this is equivalent to

$$S(T)E \subseteq K_T .$$

Finally, we say that (1.1) is <u>exactly controllable</u> if the conditions for exact controllability in time T are satisfied but with T depending in general on u. (The terminology employed here is similar to that in [5]. Other names are also used; see, for instance [6] where "strong" is written in place of "exact".)

We shall also consider a different space of controls, namely $\mathcal{M}$, the set of all n-ples $(\mu_1, \ldots, \mu_n)$ of

σ-additive measures with support in $t \geq 0$ and such that

$$\|\mu_k\|_T = |\mu_k|([0,T]) < \infty \qquad (1 \leq k \leq n).$$

The trajectory $u(\cdot)$ corresponding to a $\mu \in m$ starting at $u_0 \in E$ for $t = 0$ is by definition

$$u(t) = S(t)u_0 + \sum_{k=1}^{n} \int_0^t S(t-s)b_k\mu_k(ds) \qquad (t \geq 0).$$

By analogy with formula (1.4) and comments preceding it, we may say that $u(\cdot)$ is a solution of the "differential equation"

$$du(t) = Au(t)dt + \sum_{k=1}^{n} b_k\mu_k(dt). \tag{1.9}$$

Note that, in contrast with the case $f \in F$, $u(\cdot)$ need not be continuous. However, all the definitions previously made in that case can be immediately generalized. (When confusion could arise we shall write $K_T(m)$, F-exactly controllable, ... etc.)

We examine in the sequel the question: How large can K_T (or K_∞) be under the conditions outlined above? (That is, with $\dim F < \infty$.) Some insight into the problem can be gained from some simple examples. Let H be separable Hilbert space, $\{\phi_j;\ -\infty < j < \infty\}$ a complete orthonormal system in H,

$$A\phi_j = ij\phi_j \quad (j = \ldots -1, 0, 1, \ldots),$$
$$n = 1,\ b_1 = b = \sum_{j=-\infty}^{\infty} b_j\phi_j. \tag{1.10}$$

Then, if

$$u = \sum_{j=-\infty}^{\infty} u_j \phi_j \in H ,$$

$$S(t)u = \sum_{j=-\infty}^{\infty} e^{ijt} u_j \phi_j .$$

Let now $f \in F$ (that is, let f be a locally integrable function). Then,

$$\int_0^T S(t)bf(t)dt = \sum_{j=-\infty}^{\infty} (\int_0^T e^{ijt} f(t)dt)b_j\phi_j$$
$$= \sum_{j=-\infty}^{\infty} c_j\phi_j . \tag{1.11}$$

It is clear that, once we have chosen b and T, (1.11) will not represent every element of H; in fact

$$|c_j| \le C|b_j| \quad (j = \ldots -1,0,1, \ldots).$$

Accordingly, $K_T(F) \neq E$ for any $T > 0$. A similar argument shows that $K_T(m) \neq E$ as well and then, in view of the forthcoming Lemma 3.1 $K_\infty(m) \neq E$, that is, the system is not m-reachable. On this basis, we may guess that

$$K_\infty(m) \neq E \tag{1.12}$$

when $\dim E = \infty$, for any infinitesimal generator A, for any positive integer n and any $b_1, \ldots, b_n \in E$. That this is in fact the case has been proved by Kuperman and Repin [9] who proved that (1.12) holds (although they state and sketch the proof for the case A <u>bounded</u> it applies without change to the present case). Their proof consists in observing that the operator

$$\mathcal{B}(\mu_1, \ldots, \mu_n) = \sum_{k=1}^{n} \int_0^T S(s) b_k \mu_k(ds) \qquad (1.13)$$

from $m_T \times \ldots \times m_T$ is <u>compact</u> (m_T is the space of σ-additive finite measures in $[0,T]$ endowed with the norm $\|\cdot\|_T$); in fact, since $K_T(m)$ is the range of the operator $\mathcal{B}$, if $\mathcal{B}$ is compact it will never coincide with all of E unless E is finite dimensional. (Compactness of $\mathcal{B}$ can be easily proved as follows: since $S(\cdot)$ is strongly continuous, the sets

$$U_k = \{S(t)b_k;\ 0 \le t \le T\} \qquad (0 \le k \le n)$$

are compact. But then, if, say, $\|\mu_1\|_T, \ldots, \|\mu_n\|_T \le 1$, $\mathcal{B}(\mu_1, \ldots, \mu_n)$ is contained in $\tilde{U} = \tilde{U}_1 + \ldots + \tilde{U}_n$, where $\tilde{U}_k$ consists of the closure of the set of all elements of the form

$$\Sigma_m \alpha_m S(t_m),$$

$\{\alpha_m\}$ a finite sequence of real numbers such that $\Sigma|\alpha_m| = 1$, $\{t_m\}$ a sequence in $[0,T]$. Now, it is easy to see that compactness of U_k implies compactness of $\tilde{U}_k$ and then of $\tilde{U}$. (Note that closure is understood here in the weak topology although, as the sets involved are convex it may be also taken in the strong topology.)

Going back to the example, we see that if we set, say

$$b_1 = \phi_0 + \sum_{|j|>0} \frac{1}{j} \phi_j \qquad (1.14)$$

and if we take $T \ge 2\pi$, it follows from the Riesz-Fischer theorem that if $u = \Sigma u_j \phi_j \in D(A)$ (that is, if $\Sigma j^2 |u_j|^2 < \infty$) we can choose $f \in L^2(0,2\pi)$ such that

$$c_j = u_j \qquad (-\infty < j < \infty)$$

(c_n the coefficients in (1.11)). This of course means that

$$K_T(F) \supseteq D(A) \quad \text{if} \quad T \geq 2\pi. \tag{1.15}$$

(This corresponds to controllability of the wave equation; see [1] for details and [10] for a far-reaching generalization.) We obtain a sort of converse of this as a particular case of Theorem 3.2. In fact, if $K_\infty(F)$ or $K_\infty(m) \supseteq D(A^m)$ for some $m \geq 1$ or even $D(A^\infty) = \cap_m D(A^m)$, then A must be, at least from the spectral point of view, somewhat similar to (1.10); in particular, $\sigma(A)$ must consist of isolated points at which $R(\lambda;A) = (\lambda I - A)^{-1}$ has poles and must satisfy a growth condition on the poles and their order. (See the statement of Theorem 3.2 for a precise formulation.)

Consider now a second example,

$$A\phi_j = -j^2\phi_j \quad (j = 1,2,\ldots)$$

where $\{\phi_j; j \geq 1\}$ is a complete orthonormal system in H. Let, further $n = 1$ and let b be given by (1.14). Then, if $S(\cdot)$ is the semigroup generated by A_1,

$$\int_0^T S(t)bf(t)dt = \sum_{j=1}^\infty \left(\int_0^T e^{-j^2t} f(t)dt \right) b_j\phi_j = \sum_{j=1}^\infty c_j\phi_j .$$

It can be seen that, unlike in the previous example, $D(A^\infty) \not\subseteq K_\infty(m)$. However, it follows as a very particular case of the results in [5] that if $u = \Sigma u_j\phi_j \in S(\xi)E$ for some $\xi > 0$ (that is, if $\Sigma e^{-2j^2\xi}|u_j|^2 < \infty$) then for any $T > 0$ we may find $f \in L^2(0,T)$ with

$$b_j \int_0^T e^{-j^2t} f(t)dt = u_j \quad (1 \leq j < \infty).$$

Note that since we may take $\xi = T$ this implies exact controllability in any time $T > 0$. Also with this we obtain from Theorem 3.2 a sort of converse which states roughly that if $K_\infty(m) \supseteq S(\xi)E$ for some $\xi > 0$ then A enjoys properties similar to that in the previous example. These results have some interesting consequences in the design of control systems which are discussed in Remark 3.4.

The results in Section 2 are obtained with the help of a functional calculus somewhat similar to (but simpler than) the one developed in [3] and [4]. Because of this, and to avoid repetition, some reference to [3] will be necessary.

2. A FUNCTIONAL CALCULUS

Given $T > 0$, let E_T denote the class of all entire functions $f(\lambda)$ of the form

$$f(\lambda) = \hat{\mu}(\lambda) = \int_0^T e^{\lambda t} \mu(dt) \quad (\mu \in m). \tag{2.1}$$

For $f \in E_T$ we define

$$f(A) = \int_0^T S(t)\mu(dt), \tag{2.2}$$

$S(\cdot)$ the semigroup generated by A (Strictly speaking, (2.2) means that $f(A)$ is the unique bounded operator that satisfies

$$\langle u^*, f(A)u\rangle = \int_0^T \langle u^*, S(t)u\rangle \mu(dt)$$

for all $u \in E$ and all $u^* \in E^*$, the dual space of E, $\langle u^*, u\rangle$ indicating the value of u^* at u.) Since μ is uniquely determined by f, $f(A)$ is well defined by

(2.2).

The connection between this functional calculus and the control problem in Section 1 is obvious: $u \in K_T(m)$ of the control system

$$du(t) = Au(t)dt + \sum_{k=1}^{n} b_k \mu_k(dt) \tag{2.3}$$

if and only if there exist $f_1, \ldots, f_n \in \mathcal{E}_T$ such that

$$u = \sum_{k=1}^{n} f_k(A) b_k. \tag{2.4}$$

(The reason for taking $f = \hat{\mu}$ rather than μ itself as a basis for the calculus is that some computations later in this section become simpler.)

If we set $\mathcal{E} = \bigcup_{T>0} \mathcal{E}_T$, then $\mathcal{E}$ is an algebra with respect to the ordinary (pointwise) product; this follows immediately from the fact that the convolution $\mu = \mu_1 * \mu_2$ of two measures with support in $[0,T]$ is a measure with support in $[0,2T]$ and from well known properties of the Laplace transform (2.1). Moreover,

$$\hat{\mu}(A) = (\hat{\mu}_1 \hat{\mu}_2)(A) = \hat{\mu}_1(A)\hat{\mu}_2(A). \tag{2.5}$$

The proof of (2.5) follows from the definition of convolution of measures and from the semigroup properties of $S(\cdot)$.

LEMMA 2.1. *Let* $g \in \mathcal{E}_R$ $(g \neq 0)$ *for some* $R > 0$. *Assume that*

$$g(A)E \subseteq K_T(m) \tag{2.6}$$

for some $T > 0$. *Then there exists* $T' > 0$ *such that: for every* $u \in E$ *we can find a* $f = f_u \in \mathcal{E}_{T'}$, $f \neq 0$ *with*

$$f(A)u = 0.$$

PROOF. Let $u_1, \ldots, u_n, u_{n+1}$ be arbitrary elements of E. Then, for every j, $1 \le j \le n+1$ there exist functions $f_{j1}, \ldots, f_{j,n-1}, f_{j,n} \in \mathcal{E}_T$ such that

$$g(A)u_j = \sum_{k=1}^{n} f_{jk}(A)b_k. \tag{2.7}$$

Assume for the moment that not all of the $g(A)u_1, \ldots$ $\ldots, g(A)u_{n+1}$ are zero. Then the matrix

$$\mathcal{F} = \{f_{jk}(\lambda);\ 1 \le j \le n+1,\ 1 \le k \le n\}$$

does not vanish identically. Since the determinant of any of the minors of $\mathcal{F}$ vanishes identically or does not vanish outside of a (at most) countable set we can always assume (if necessary re-ordering the indices) that there exists m, $1 \le m \le n$ such that

i) The matrix $\Delta = \{f_{jk}(\lambda);\ 1 \le j, k \le m\}$ is non-singular (except perhaps for λ in a countable set Λ).

ii) The matrix $\mathcal{F}$ has rank m if $\lambda \notin \Lambda$.

Given j, $1 \le j \le m$ denote by Δ_j the matrix $\{f_{jk}(\lambda);\ 1 \le j, k \le m\}$ with the j-th column replaced by the column vector $\{-f_{j,m+1}(\lambda);\ 1 \le j \le m\}$. Define, finally,

$$f_j(\lambda) = \det \Delta_j(\lambda) \qquad (1 \le j \le m),$$

$$f_{m+1}(\lambda) = \det \Delta,$$

$$f_j(\lambda) = 0 \qquad (m+1 < j \le n+1).$$

Then $f_j \in \mathcal{E}_{mT}$ $(1 \le j \le n+1)$, not all f_j are zero,

and

$$\sum_{j=1}^{n+1} f_j(\lambda) f_{jk}(\lambda) = 0$$

for $1 \le k \le m$ and then, in view of ii), for $1 \le k \le n$. Taking this and (2.7) into account, we see that

$$\sum_{j=1}^{n+1} h_j(A) u_j = 0 \qquad (2.8)$$

where $h_j = g f_j \in \mathcal{E}_{mT+R}$ and not all the h_j are zero. (Note that if $g(A)u_1 = \ldots = g(A)u_{n+1} = 0$, then (2.8) holds trivially taking $h_j = g$, $1 \le j \le n+1$.)

Let next $u_1, \ldots, u_n$ be n arbitrary elements of E, $T_1 > mT + R$, $h_1, \ldots, h_n, h_{n+1}$ the functions in (2.8) for the (n+1)-ple

$$(u_1, \ldots, u_n, -\hat{\delta}(A) u_n)$$

where δ is the Dirac measure centered at T_1 (which means $\hat{\delta}(\lambda) = e^{\lambda T_1}$, $\hat{\delta}(A) = S(T_1)$). Then

$$\sum_{k=1}^{n} p_k(A) u_k = 0 \qquad (2.9)$$

where $p_1 = h_1$, $p_2 = h_2, \ldots, p_n(\lambda) = h_n(\lambda) -$ $- e^{\lambda T_1} h_{n+1}(\lambda)$. Assume that $p_1 = p_2 = \ldots = p_n = 0$. Then, in particular,

$$h_n(\lambda) = e^{\lambda T_1} h_{n+1}(\lambda) .$$

Since $h_n \in \mathcal{E}_{mT+R}$, $|h_n(\lambda)| \le C e^{(mT+R)\mathrm{Re}\lambda}$. But then

$$e^{T_1 \mathrm{Re}\lambda} |h_{n+1}(\lambda)| = |h_n(\lambda)| \le C e^{(mT+R)\mathrm{Re}\lambda}$$

hence h_{n+1}, by Liouville's theorem, vanishes identically. Then $h_1 = h_2 = \ldots = h_{n+1} = 0$, which contradicts the way in which they were chosen. We have thus shown that if (2.8) holds for an arbitrary (n+1)-ple with $h_1, \ldots, h_{n+1}$ not all zero, then the same relation holds for arbitrary n-ples. Repeating this inductive step $n - 1$ more times we obtain the desired conclusion.

COROLLARY 2.2. *Assume the hypotheses in Lemma* 2.1 *hold. Then there exists* $f \in E$, $f \neq 0$ *such that*

$$f(A) = 0. \tag{2.10}$$

PROOF. By Lemma 2.1 there exist $f_{b_1}, f_{b_2}, \ldots, f_{b_n} \in E_{T'}$, none of them zero and such that

$$f_{b_1}(A)b_1 = \ldots = f_{b_n}(A)b_n = 0.$$

We obtain the desired result by observing that, for any $u \in E$ we have

$$g(A)u = \sum_{k=1}^{n} f_k(A)b_k$$

for some $f_1, \ldots, f_n \in E$. Then, if we set $f = f_{b_1} f_{b_2} \ldots f_{b_n} g$ it is clear that $f(A)u = 0$ for all $u \in E$.

3. THE CONTROLLABILITY RESULTS

LEMMA 3.1. *Let* $g \in E$. *Assume*

$$g(A)E \subseteq K_\infty(m). \tag{3.1}$$

Then

$$g(A)E \subseteq K_T(m) \tag{3.2}$$

for some $T > 0$.

The proof can be deduced from Theorem 1.1 in [6] as that of Theorem 3.1 therein. We sketch below, however, a direct proof. Define, for $M,N = 1,2,\ldots$ $K_{M,N}(m)$ as the set of all $u \in E$ such that

$$g(A)u = \sum_{k=1}^{n} \int_0^M S(M-s)b_k\mu_k(ds) \tag{3.3}$$

with

$$\|u_k\| = \int_0^M |\mu_k|(ds) \leq N \quad (1 \leq k \leq n). \tag{3.4}$$

Clearly,

$$\bigcup_{M,N} K_{M,N}(m) = K(m).$$

Let $\{u_m\}$ be a sequence in some $K_{M,N}$, and let $\{\mu_{km};\ 1 \leq k \leq n;\ m = 1,2,\ldots\}$ be the measures associated with each m by (3.3). Assume $u_m \longrightarrow u \in E$. By the Riesz and Alaoglu theorems we may assume, if necessary passing to a subsequence, that there exist measures $\mu_1,\ldots,\mu_n$ such that

$$\int_0^M \xi(s)\mu_{km}(ds) \longrightarrow \int_0^M \xi(s)\mu_k(ds)$$

for any (scalar-valued) continuous function ξ, where the measures μ_k satisfy (3.4). Then, if $u^* \in E^*$,

$$\langle u^*, g(A)u_m\rangle = \sum_{k=1}^{n} \int_0^M \langle u^*, S(M-s)b_k\rangle\mu_{km}(ds).$$

Letting $m \to \infty$ we see that $u \in K_{M,N}$, so that each

$K_{M,N}$ is closed. It follows then from Baire's theorem that some $K_{M,N}$ must have an interior point. This is easily seen to imply that K_M must coincide with the whole space, thus proving Lemma 3.1.

LEMMA 3.2. *Assume* (3.1) *holds for some* $g \in E$, $g \neq 0$. *Then, either* $\sigma(A) = \emptyset$ *or* $\sigma(A)$ *consists of a (finite or countable) sequence* $\Lambda = \{\lambda_j\}$ *of points where* $R(\lambda;A) = (\lambda I - A)^{-1}$ *has poles of order* $m_j < \infty$. *Moreover,*

$$\sum_{j=1}^{\infty} \frac{|\mathrm{Re}(\lambda_j)|m_j}{1 + |\lambda_j|^2} < \infty. \tag{3.5}$$

PROOF. In view of Lemma 3.1, $g(A)E \subseteq K_T(m)$ for some $T > 0$. Then, Corollary 2.2 shows that there exists $f \in E$ such that $f \neq 0$,

$$f(A) = 0. \tag{3.6}$$

The proof then follows from the next Lemma, whose proof is substantially similar to that of Proposition (2.1) in [3] and is therefore omitted.

LEMMA 3.3. *Let* A *be an infinitesimal generator. Assume that* (3.6) *holds for some* $f \in E$, $f \neq 0$. *Then,* $\sigma(A)$ *is contained in* Z_f, *the set of zeros of* f. *If* $\lambda_0 \in \sigma(A)$, *then* $R(\lambda;A) = (\lambda I - A)^{-1}$ *has a pole at* λ_0 *of order* $m_0 \leq n_0$ = *multiplicity of the zero of* f *at* λ_0.

Inequality (3.5) is a consequence of this result and of the fact that (3.5) holds for any function $f \in E$ if $\{\lambda_j\}$ are its zeros, m_j their multiplicities

([8], Chapter 8).

We recall some properties of A that arise from the conclusion of Theorem 3.2. If $\lambda_n \in \Lambda$, the fact that $R(\lambda;A)$ has a pole of order $m_n < \infty$ at λ_n implies that λ_n is an <u>eigenvalue</u> of A. If P_n is the projection corresponding to λ_n, that is, if

$$P_n = \frac{1}{2\pi i} \int_C R(\lambda;A)\,d\lambda$$

(C a sufficiently small circle around λ_n) and $E_n = P_n E$, then E consists of all generalized eigenvectors of A corresponding to λ_n; more precisely, $E_n \subseteq D(A^\infty)$,

$$(\lambda_n I - A)^{m_n} u = 0 \qquad (u \in E_n),$$

and any $u \in D(A^m)$ for which $(\lambda_n I - A)^m u = 0$ for some $m > 0$ belongs to E_n. This is essentially all we can deduce from (3.6), besides (3.5).

REMARK 3.3. We can deduce from Theorem 3.2 the two particular cases mentioned in Section 1. In fact, if $D(A^\infty) \subseteq K_\infty(\mathcal{m})$, we may take

$$g(\lambda) = \int e^{\lambda t}\,\psi(t)\,dt$$

where ψ is a non-null C^∞ function with compact support in $(0,\infty)$. It is well known that

$$g(A)u = \int S(t)u\psi(t)\,dt \in D(A^\infty), \qquad u \in E.$$

The case $T(\xi)E \subseteq K_\infty(\mathcal{m})$ is even simpler, for we may take $g(\lambda) = e^{\lambda\xi}$ $(g(A) = T(\xi))$.

REMARK 3.4. In actual practice, it is not usually enough to know, say, that $u \in K_T(\mathcal{F})$ for the control system

$$u'(t) = Au(t) + Bf(t) \tag{3.7}$$

since there may be limitations to the "size" of the controls f to be used. Assume, for the sake of simplicity, that these limitations take the form

$$\|f\|_T = \operatorname*{ess\,sup}_{0 \le t \le T} |f(t)| \le 1. \tag{3.8}$$

Assume that

$$K_T(F) = E$$

for some $T > 0$. Then, for any $u \in E$ we define

$$|||u||| = \inf \|f\|_T ,$$

the infimum taken over all f with

$$u = \int_0^T S(T - s)Bf(s)ds .$$

It is not difficult to see ([3]) that, under mild restrictions on the control space, $|||u|||$ is a norm that makes E complete; since $\|u\| \le C|||u|||$ the closed graph theorem implies that $|||\cdot|||$ and $\|\cdot\|$ are equivalent. <u>We can then make sure that</u> (3.8) <u>will hold if we take</u> $\|u\|$ <u>small enough</u>. In the case where the control space is finite dimensional, however, we have seen in Section 1 that we can never have $K_\infty = E$ so that the happy situation outlined above can never arise!

We can try to remedy the problem by looking for a normed space $(E_1 , \|\cdot\|_1)$ such that

a) $E_1 \subseteq K_T(F)$

b) $|||u||| \le C\|u\|_1 \quad (u \in E_1)$

for then (3.8) will hold if we take $\|u\|_1$ small enough. Obviously, it would be desirable to choose an $(E_1 , \|\cdot\|_1)$

which is related to (3.7) in some "natural" way; for instance,

$$E_1 = D(A^m), \ \|u\|_1 = \|u\| + \|A^m u\|$$

or, in the case of an analytic semigroup,

$$E_1 = S(\xi)E, \ \|S(\xi)u\|_1 = \|u\| .$$

However, Theorem 3.2 establishes that these two choices - as well as many other possible ones - will not do if A does not satisfy the conclusion of Theorem 3.2.

We point out as an example that the Laplacian Δ (with Dirichlet or more general boundary conditions) in a smooth, bounded domain of n-dimensional Euclidean space satisfies the conclusion of Theorem 2.3 if $n = 1$ but not if $n > 1$ (at least if the multiplicity of the eigenvalues is bounded). Probably the same is true of any second-order uniformly elliptic operator.

REFERENCES

1. BUTKOVSKII, A. G. and POLTAVSKII, L. N., *The optimal control of a distributed oscillating system* (Russian), Automatics and Remote Control (1965).

2. DUNFORD, N. and SCHWARTZ, J. T., *Linear Operators*, Part I, Interscience Publishers, New York, 1957.

3. FATTORINI, H. O., *Control in finite time of differential equations in Banach space*, Communications on Pure and Applied Mathematics XIX (1966), 17-34.

4. FATTORINI, H. O., *On Jordan operators and rigidity of linear control systems*, Revista de la Unión Matemática Argentina XXIII (1966), 67-75.

5. FATTORINI, H. O. and RUSSELL, D. L., *Exact controllability theorems for linear parabolic equations in one space dimension*, Archive for Rational Mechanics and Analysis 43 (1971), 272-292.

6. FUHRMANN, P. A., *On weak and strong reachability and controllability of infinite-dimensional systems*, J. Optimization Theory and Applications 2 (1972), 77-89.

7. HILLE, E. and PHILLIPS, R. S., *Functional Analysis and Semi-groups*, American Math. Soc. Colloquium Publications, 31, Providence, R. I. 1957.

8. HOFFMAN, K. *Banach Spaces of Analytic Functions*, Prentice-Hall, Englewood Cliffs, N. J., 1962.

9. KUPERMAN, L. M. and REPIN, Ju. M., *On controllability in infinite-dimensional spaces*, Doklady Akademii Nauk SSSR 200 (1971), 767-769 (Russian). English translation.

10. RUSSELL, D. L., *Nonharmonic Fourier series in the theory of distributed parameter systems*, J. Math. Analysis and Applications 18 (1967), 542-560.

ON THE STATISTICAL STUDY OF THE NAVIER-STOKES EQUATIONS

by

CIPRIAN FOIAS

Faculty of Mathematics
University of Bucharest
Bucharest, Romania

and

Courant Institute of Mathematical Sciences
New York University
New York, New York 10012

The aim of this note is to give some statistical estimates connected with the time existence of the regular solutions for the Navier-Stokes equations.

1. Let Ω be a bounded domain in R^3 with a C^2 boundary $\partial\Omega$. We shall consider the initial value problem (on $[0,\infty) \times \Omega$) for the Navier-Stokes equations

$$(1.1) \qquad \begin{cases} \partial_t u - \nu \, \Delta u + (u \cdot \text{grad})u = -\text{grad } p + F \\ \\ \text{div } u = 0, \quad u|\partial\Omega = 0 \end{cases}$$

where $\nu > 0$ and $F \in (L^2(\Omega))^3$. An adequate framework for this problem can be obtained in the following way (see [6], Ch.I, §6). Define the spaces

$$(1.2)\quad \begin{cases} H^1 = \{u \in (H_0^1(\Omega))^3 : \operatorname{div} u = 0\} \\ H = \text{closure in } (L^2(\Omega))^3 \text{ of } H^1, \end{cases}$$

and the operators

$$(1.3)\quad \begin{cases} Au = -P\,\Delta u \quad \text{for } u \in \mathcal{D}_A = H^1 \cap (H^2(\Omega))^3, \\ B(u,v) = P[(u\cdot \operatorname{grad})v] \quad \text{for } u,v \in \mathcal{D}_A \quad , \end{cases}$$

where P denotes the orthogonal projection of $(L^2(\Omega))^3$ $(= L^2(\Omega) \oplus L^2(\Omega) \oplus L^2(\Omega))$ onto H. The scalar product and norm in H, respectively H^1, will be

$$(u,v) = \int (\sum_1^3 u_i v_i)dx, \; |u| = (u,u)^{1/2} \quad (u,v \in H), \text{ resp.}$$

$$((u,v)) = \int (\sum_1^3 \partial_i u_j \cdot \partial_i v_j)dx, \; \|u\| = ((u,u))^{1/2} \quad (u,v \in H^1).$$

By definition, a <u>solution</u> of (1.1) with initial data $u_0 \in H$ is an H-valued function $u(t)$, which is weakly continuous at any $t_0 \in [0,\infty)$ and strongly continuous from the right at any $t_0 \in [0,\infty) \setminus \Theta$ (where $\Theta \subset (0,\infty)$ depends on the solution and $\text{meas}(\Theta) = 0$), and which satisfies

$$(1.4)\begin{cases} u(t) \in L^2(0,t;H^1) \quad (t \in (0,\infty)) \\ \frac{1}{2}|u(t)|^2 + \nu\int_{t_0}^{t}\|u(\tau)\|^2\, d\tau \le \frac{1}{2}|u(t_0)|^2 \\ \qquad + \int_{t_0}^{t}(f,u(\tau))d\tau \quad (t \in (0,\infty), t_0 \in [0,t]\setminus\Theta), \\ u(t) = u_0 + \int_0^t u'(\tau)d\tau,\ u'(\tau) \in L^1(0,t;H^{-1})\,(t \in (0,\infty)) \\ u'(t) + \nu\, Au(t) + B(u(t),u(t)) = f \text{ a.e. on } (0,\infty) \end{cases}$$

where $f = PF$ and the last two relations are considered in the dual H^{-1} of H^1; this makes sense since H is embedded in H^{-1} by the identification $\langle h,u\rangle = (h,u)$ $(h \in H,\ u \in H^1)$ and $AH^1 \subset H^{-1}$, $B(H^1,H^1) \subset H^{-1}$. For any $u \in H$ there exists at least one solution $u(t)$ with initial data $u \in H$ (see [5], §3). Moreover, if $u \in H^1$, there exists a largest interval $[0,t(u)) \subset [0,\infty)$ such that on $[0, t(u))$ the solution $u(t)$ is uniquely determined and continuous as an H^1-valued function (i.e. $u(t)$ is <u>regular</u> on $[0,t(u))$); actually

$$(1.5) \qquad t(u) \ge \kappa_1\nu^3[\max\{\nu^{-1}\lambda_1^{-1/2}|f|,\ \|u\|\}]^{-4}$$

where the constants κ_1 and

$$(1.6) \qquad \lambda_1 = \min\{\|u\|^2 : |u| = 1\} > 0$$

depend only on Ω (for a proof of these statements see [8], §1 or [2], §4; Sec. 4).

2. By definition a <u>stationary statistical solution of the Navier-Stokes equations</u> (SSSNSE) is a Borel

probability measure μ on H satisfying the following conditions:

$$(2.1)\quad \begin{cases} \int \|u\|^2 \, d\mu(u) < \infty \\ \int_{\{E_1 \le |u|^2 \le E_2\}} (\nu\|u\|^2 - (f,u))\, d\mu(u) \le 0 \quad (0 \le E_1 \le E_2 \le \infty), \end{cases}$$

$$(2.2)\quad \int \langle \nu\, Au + B(u,u) - f, \Phi'(u) \rangle \, d\mu(u) = 0$$

for all real functionals Φ on H with a Fréchet derivative $\Phi'(u)$ at any $u \in H^1$, such that $\Phi'(\cdot)$ is continuous and bounded from H^1 into H^1. It follows at once that

$$(2.3)\quad \begin{cases} \text{supp}\, \mu \subset \{u \in H : |u| \le \nu^{-1}\lambda_1^{-1}|f|\} \\ \int \|u\|^2 \, d\mu(u) \le \nu^{-2}\lambda_1^{-1}|f|^2 . \end{cases}$$

A Borel probability measure μ on H will be called <u>accretive</u> if

$$(2.4)\quad \mu(\omega(t)) \ge \mu(\omega) \quad \text{for all } t \ge 0 \text{ and Borel subsets } \omega \subset H,$$

where $\omega(t)$ denotes the set of the values at time t of all solutions starting (at time = 0) from ω; it can be shown that if ω is Borel in H, then $\omega(t)$ is measurable with respect to any Borel measure on H (see [5], §4) so that the relation (2.4) is meaningful.

The following fact results readily from [4], §3:

Let $u(t)$ be a solution of (1.1) and $\{t_j\}_{j=1}^{\infty}$ a sequence in $(0,\infty)$ converging to ∞. Let $\mathcal{C}$ be the set of all real functionals on H weakly continuous on the bounded subsets of H. Then there exists an accretive SSSNSE μ such that

$$\int \Phi \, d\mu(u) = \lim \frac{1}{t_j}\int_0^{t_j} \Phi(u(\tau))d\tau \tag{2.5}$$

for any $\Phi \in \mathcal{C}$ for which the limit exists.

Thus accretive SSSNSE are intimately connected with the asymptotic behavior of the individual solutions of the Navier-Stokes equations. Our main concern will be to connect the function $t(u)$ (defined in Section 1 for $u \in H^1$ and $= 0$ for $u \in H^1 \setminus H$ (this is a Borel $[0,\infty]$-valued function on H since $\{u \in H : t_0 < t(u)\}$ is open in H^1 for any $t_0 \in [0,\infty)$!)) to the accretive SSSNSE.

3. Let μ be a fixed SSSNSE (accretive or not). Then for $0 \le E_1 \le E_2 \le \nu^{-2}\lambda_1^{-2}|f|^2$ (see (2.3)), we deduce readily from (2.1) the relation

$$\int_{\{E_1 \le |u|^2 \le E_2\}} \|u\|^2 d\mu(u) \le \nu^{-1}|f|E_2^{1/2}\mu(\{E_1 \le |u|^2 \le E_2\}). \tag{3.1}$$

If we set

$$t_0 = \kappa_1 \nu^7 \lambda_1^2 |f|^{-4} \ , \ t = \delta t_0 \ , \ 0 \le \delta \le 1 \ , \tag{3.2}$$

then, by (1.5) and (3.1), we will have

$$\mu(\{u \in H : t(u) \le t,\ E_1 \le |u|^2 \le E_2\})$$
$$\le \mu(\{u \in H^1 : \kappa_1 \nu^3 [\max\{\nu^{-1}\lambda_1^{-1/2}|f|, \|u\|\}]^{-4} \le t,$$
$$E_1 \le |u|^2 \le E_2\})$$
$$= \mu(\{u \in H^1 : \kappa_1 \nu^3 \|u\|^{-4} \le t,\ E_1 \le |u|^2 \le E_2\})$$
$$\le \kappa_1^{-1/2}\nu^{-3/2}t^{-1/2} \int_{\{\kappa_1\nu^3 \le \|u\|^4 t, E_1 \le |u|^2 \le E_2\}} \|u\|^2 d\mu(u)$$
$$\le \kappa_1^{-1/2}\nu^{-3/2}t^{1/2} \int_{\{E_1 \le |u|^2 \le E_2\}} \|u\|^2 \, d\mu(u)$$
$$\le |f| E_2^{1/2}\kappa_1^{-1/2}\nu^{-5/2}t^{1/2}\ \mu(\{E_1 \le |u|^2 \le E_2\}),$$

that is

$$(3.3)\quad \mu(\{t(u) \le t,\ E_1 \le |u|^2 \le E_2\}) \le |f| E_2^{1/2}\kappa_1^{-1/2}\nu^{-5/2}t^{1/2} \cdot \mu(\{E_1 \le |u|^2 \le E_2\}).$$

Let us introduce now, for any Borel set $\omega \subset H$, the expectation $\mu(\omega \mid |u|^2 = E)$ by

$$(3.4)\quad \mu(\omega \cap \{|u|^2 \le E_0\}) = \int_0^{E_0} \mu(\omega \mid |u|^2 = E) d_E \mu(\{|u|^2 \le E\})$$
$$\text{(for all } 0 \le E_0 \le \infty).$$

Then (3.3) implies that (actually that we can assume that)

$$(3.5)\quad \mu(t(u) \le t \mid |u|^2 = E) \le |f| E^{1/2}\kappa_1^{-1/2}\nu^{-5/2}t^{1/2}.$$

Taking into account (3.2), we obtain the following

PROPOSITION 1. *For any SSSNSE* μ *and any* $E \in [0,\nu^{-2}\lambda_1^{-2} f^{2}]$, $\delta \in [0,1]$ *we have*

$$(3.6) \quad \mu(t(u) \leq \delta\kappa_1\nu^7\lambda_1^2|f|^{-4} \mid |u|^2 = E) \leq \delta^{1/2}E^{1/2}/\nu^{-1}\lambda_1^{-1}|f|.$$

Among the consequences of this proposition we point out the following

PROPOSITION 2. *Let* μ *be an accretive SSSNSE. Then the random variables* $t(u)$ *and* $|u|^2$ *are independent if and only if* $t(u) = \infty$ *almost surely.*

Plainly, we have to prove the case "only if". Let thus $t(u)$ and $|u|^2$ be independent (with respect to μ). Then for any $t \geq 0$, we can assume that the function $\mu(t(u) \leq t \mid |u|^2 = E)$ is a constant $c(t)$; in particular, for $t = t_0$ (see (3.2)), in virtue of (3.6), $c(t_0)$ must be 0. Therefore, by (3.4) we can now infer that

$$\mu(\{t(u) \leq t_0\}) = 0 ,$$

and consequently, that

$$(3.7) \qquad \int \frac{1}{t(u)}\, d\mu(u) \leq \frac{1}{t_0} < \infty.$$

But in virtue of [4], §3; Sec. 3.7, the relation (3.7) implies $t(u) = \infty$, μ a.e.

4. We shall give now another condition on accretive SSSNSE (actually on any accretive Borel probability measure μ satisfying the first relation (2.1)) in order that $t(u) = \infty$ almost surely.

To this aim let μ be a fixed accretive SSSNSE and let, for $t \geq 0$

$$\Theta(t) = \mu(\{t(u) \leq t\}) \ , \tag{4.1}$$

i.e. Θ is the distribution function of $t(u)$ with respect to μ. For $0 \leq t_1 \leq t_2 \leq \infty$ and $t \geq 0$ we have

$$\{u \in H : t_1 + t < t(u) \leq t_2 + t\}(t) \subset \{u \in H : t_1 < t(u) \leq t_2\},$$

whence, since μ is accretive

$$\begin{aligned} \Theta(t_2+t) - \Theta(t_1+t) &= \mu(\{t_1+t < t(u) \leq t_2+t\ ) \\ &\leq \mu(\{t_1+t < t(u) \leq t_2+t\}(t)) \\ &\leq \mu(\{t_1 < t(u) \leq t_2\}) = \Theta(t_2) - \Theta(t_1). \end{aligned} \tag{4.2}$$

Since $\Theta(t)$ is nondecreasing on $[0,\infty)$, $\Theta'(t)$ exists for all $t \in (0,\infty)\setminus\sigma$, where $\text{meas}(\sigma) = 0$. From (4.2) we infer readily that $\Theta'(t)$ is a nonincreasing function on $(0,\infty) \setminus \sigma$. Moreover we can also conclude that if

$$\theta_1(t) = \lim_{s \notin \sigma, s \to t+0} \Theta'(s)$$

then

$$\Theta(t) = \int_0^t \theta_1(\tau) d\tau \qquad (0 \leq t < \infty); \tag{4.3}$$

it is plain that θ_1 is the unique nonincreasing function on $(0,\infty)$, continuous at the right, satisfying (4.3). Using (1.5), (4.1), (4.3) and the notation (3.2) we have

$$2\sqrt{t}\,\theta_1(t) = \int_0^t \frac{\theta_1(t)}{\sqrt{\tau}}\,d\tau \le \int_0^t \frac{\theta_1(\tau)}{\sqrt{\tau}}\,d\tau$$

$$= \int_{\{t(u)\le t\}} (t(u))^{-1/2}\,d\mu(u)$$

$$\le \kappa_1^{-1/2}\nu^{-3/2} \int_{\{\kappa_1^{1/2}\nu^{3/2}t^{-1/2}\le \|u\|^2\}} \|u\|^2 d\mu(u),$$

whence, if $\gamma = (2\kappa_1\nu^3)^{-1}$,

$$\lim_{t\to+0} \theta_1(t) \le \liminf_{\varepsilon\to\infty} \varepsilon\,\gamma \int_{\{\|u\|^2\ge\varepsilon\}} \|u\|^2\,d\mu(u)\ . \tag{4.4}$$

If the second term in (4.4) is $=0$ then $\theta_1(t) \equiv 0$, because $\theta_1(t)$ is nonincreasing; consequently,

$$\mu(\{t(u) < \infty\}) = \int_0^\infty \theta_1(t)dt = 0\ .$$

Thus we can conclude with the following

PROPOSITION 3. *Let* μ *be an accretive SSSNSE such that*

$$\liminf_{\varepsilon\to\infty} \varepsilon \int_{\{\|u\|^2\ge\varepsilon\}} \|u\|^2\,d\mu(u) = 0\ . \tag{4.5}$$

Then $t(u) = \infty$ *almost surely. (Plainly,* (4.5) *is satisfied if* $\int\|u\|^4\,d\mu(u) < \infty$.)

REMARK. An analogous result to the preceding one, in a different but related frame, was obtained a long time ago by G. Prodi (see [7], Part IV).

5. We will conclude this note by proving the following

PROPOSITION 4. *There exists an accretive SSSNSE* μ *satisfying* (4.5) *if and only if there exists an individual solution* $u(t)$ *such that*

$$(5.1)\quad \liminf_{\varepsilon \to \infty} \varepsilon [\limsup_{t \to \infty} \frac{1}{t} \int_{\{\tau \le t, \|u(\tau)\|^2 \ge \varepsilon\}} \|u(\tau)\|^2 \, d\tau] = 0.$$

Let us first suppose that μ is an accretive SSSNSE satisfying (4.5). Since $t(u) = \infty$, μ a.e. the map $RS(t_0) : u_0 \to u(t_0)$ is well-defined, where $u_0 \in \{u \in H^1 : t(u) = \infty\}$ and $u(t)$ is the unique solution with initial data u_0, is defined μ a.e. By [4], §3, Sec. 3.7, μ is invariant with respect to $\{RS(t_0)\}_{t_0 \ge 0}$, which forms a semigroup of measurable transformations. Let

$$f_\varepsilon(u) = \|u\|^2 \chi_{\{\|u\|^2 \ge \varepsilon\}}(u) \qquad (u \in H^1).$$

Then $f_\varepsilon \in L^1(\mu)$, thus by Birkhoff's ergodic theorem (see [1], Ch. VIII) we have

$$(5.2)\begin{cases} \lim\limits_{t\to\infty} \dfrac{1}{t} \displaystyle\int_{\{\tau \le t, \|u(\tau)\|^2 \ge \varepsilon\}} \|u(\tau)\|^2 d\tau = \lim\limits_{t\to\infty} \dfrac{1}{t} \displaystyle\int_0^t f_\varepsilon(RS(\tau)u) d\tau \\ \qquad = f_\varepsilon^*(u) \quad \text{exists } \mu \text{ a.e.,} \\ \displaystyle\int_{\{\|u\|^2 \ge \varepsilon\}} \|u\|^2 \, d\mu(u) = \int f_\varepsilon(u) d\mu(u) = \int f_\varepsilon^*(u) d\mu(u). \end{cases}$$

Let $\varepsilon_j \downarrow 0$ be such that

$$\sum_{j=1}^{\infty} \varepsilon_j \int_{\{\|u\|^2 \geq \varepsilon_j\}} \|u\|^2 \, d\mu(u) < \infty .$$

From (5.2) we infer first,

$$\sum_{j=1}^{\infty} \varepsilon_j \, f^*_{\varepsilon_j}(u) \, d\mu(u) = \sum_{j=1}^{\infty} \varepsilon_j \int f^*_{\varepsilon_j}(u) \, d\mu(u) < \infty ,$$

and second,

$$(5.3) \qquad \lim_{j\to\infty} \varepsilon_j \, f^*_{\varepsilon_j}(u) = 0$$

μ a.e.; therefore we can pick a u_0 such that (5.3) holds for $u = u_0$. It is now clear that (5.1) will be satisfied by $u(t) \equiv RS(t)u_0$ $(t \geq 0)$.

Let us now suppose conversely that (5.1) is satisfied by a certain individual solution $u(t)$, and let us denote

$$(5.4) \quad c(\varepsilon) = \limsup_{t \to \infty} \frac{1}{t} \int_{\{\tau \leq t, \|u(\tau)\|^2 \geq \varepsilon\}} \|u(\tau)\|^2 \, d\tau .$$

Let μ be any time average of $u(t)$, that is an accretive SSSNSE μ such that for an adequate sequence $\{t_j\}_{j=1}^{\infty}$ converging to ∞, the relation (2.5) holds for all $\Phi \in \mathcal{C}$. (For the existence of such a μ see [2], §6 or [4], §3, Sec. 3.2 and 3.4.) We shall show that μ satisfies (4.5). To this aim let us recall some more properties of the operator A (see for instance [3], §2), namely that A is selfadjoint ≥ 0, A^{-1} is compact, $H^1 = D_{A^{1/2}}$, and $((u,v)) = (A^{1/2}u, A^{1/2}v)$, $u,v \in H^1$.

Therefore there exists an orthonormal basis $\{w_m\}_{m=1}$ in H such that $Aw_m = \lambda_m w_m$ $(m = 1,2,\ldots)$ and $0 < \lambda_1 \le \lambda_2 \le \ldots$. (Notice that the notation is consistent with (1.6).) Let P_m $(m = 1,2,\ldots)$ denote the orthogonal projection of H onto the space spanned by $w_1, w_2, \ldots, w_m$ and for $\eta \in (0,1)$ let $\phi_\eta(\zeta)$ be the continuous function on $[0,\infty)$, linear on $[\varepsilon,(1+\eta)\varepsilon]$ and such that

$$\phi_\eta(\zeta) = 0 \quad \text{for} \quad 0 \le \zeta \le \varepsilon, \; \phi_\eta(\zeta) = 1 \quad \text{for} \quad (1+\eta)\varepsilon \le \zeta.$$

Then the functional Φ defined by

$$\Phi(u) = \phi_\eta(\|P_m u\|^2)\|P_m u\|^2 \qquad (u \in H)$$

belongs to C and therefore (by (2.5))

$$\int \phi_\eta(\|P_m u\|^2)d\mu(u) = \lim \frac{1}{t_j}\int_0^{t_j} \phi_\eta(\|P_m u(\tau)\|^2)\|P_m u\|^2 d\tau$$

$$\le \limsup \frac{1}{t_j}\int_{\{\tau\le t_j, \|P_m u(\tau)\|^2\ge\varepsilon\}} \|P_m u(\tau)\|^2\, d\tau$$

$$\le \limsup \frac{1}{t_j}\int_{\{\tau\le t_j, \|u(\tau)\|^2\ge\varepsilon\}} \|u(\tau)\|^2\, d\tau\ ,$$

because $\|P_m u\| \le \|u\|$ $(u \in H^1;\ m = 1,2,\ldots)$. Consequently, we can conclude that

$$\text{(5.5)} \qquad \int \phi_\eta(\|P_m u\|^2)\|P_m u\|^2\, d\mu(u) \le c(\varepsilon).$$

Since $\|P_m u\|^2 \uparrow \|u\|^2$ for $m \to \infty$, we can let $m \to \infty$ in (5.5) and apply B. Levi's convergence theorem, obtaining

$$\int \phi_\eta(\|u\|^2)\|u\|^2 \, d\mu(u) \le c(\varepsilon) \ ;$$

letting now $\eta \to +0$ we infer, as before, that

$$\int_{\{\|u\|^2 \ge \varepsilon\}} \|u\|^2 \, d\mu(u) \le c(\varepsilon) \ . \tag{5.6}$$

Finally, it is now obvious that the relations (5.1), (5.4) and (5.6) yield (4.5).

ACKNOWLEDGEMENT. The author gratefully acknowledges that this paper was written while he enjoyed a Fulbright lecturer award at the Courant Institute of Mathematical Sciences (New York University).

REFERENCES

1. DUNFORD, N. and SCHWARTZ, J. T., *Linear Operators, Part I. General Theory,* Interscience-Wiley, New York, 1958.

2. FOIAS, C., *Statistical Study of Navier-Stokes Equations, Part I.* Rend. Sem. Mat. Univ. Padova 48 (1973), 219-348; *Part II.* idem. 49 (1973) 9-123.

3. FOIAS, C., *A functional framework for the theory of turbulence* (in Russian), Uspehi Mat. Nauk (to appear).

4. FOIAS, C., *On the stationary statistical solutions of the Navier-Stokes equations* (to appear).

5. FOIAS, C. and PRODI, G., *Sur les solutions statistiques des équations de Navier-Stokes* (to appear).

6. LIONS, J. L., *Quelques Methods de Résolution des Problèmes aux Limites Non Linéaires*, Dunod, Gauthiers-Villars, Paris, 1969.

7. PRODI, G., *Résultats rêcents dans la théorie des équations de Navier-Stokes. Les équations aux dérivées partièlles*, Colloques Intern. CNRS, Paris, (1962), 181-196.

8. PRODI, G., *Teoremi di tipo locale per il sistema di Navier-Stokes e stabilità delle soluzioni stazionarie*, Rend. Sem. Mat. Univ. Padova 32 (1962), 374-397.

ASYMPTOTIC BEHAVIOR OF SOLUTIONS TO THE QUASILINEAR WAVE EQUATION

by

J. M. GREENBERG*

Courant Institute of Mathematical Sciences
New York University
251 Mercer Street
New York, New York 10012

and

Department of Mathematics
State University of New York at Buffalo
Amherst, New York 14226

1. INTRODUCTION

My goal in this paper is asymptotic estimates for the the quasilinear wave equation which generalize, in a natural way, the estimate that for a single conservation law

(1) $$\frac{\partial u}{\partial t} + \frac{\partial}{\partial x} f(u) = 0, \quad \text{with}$$

(2) $$\frac{d^2 f}{du^2} \geq k_o > 0, \quad -\infty < u < \infty,$$

N waves decay as $1/t^{1/2}$ as time t tends to plus in-

* This research was partially supported by the National Science Foundation, Grant No. NSF-PO 36486-001.

finity.

I shall write the wave equation as

$$\text{(WE)} \qquad \frac{1}{c(q)}\frac{\partial q}{\partial t} - \frac{\partial v}{\partial x} = 0 \quad \text{and} \quad \frac{\partial v}{\partial t} - c(q)\,\frac{\partial q}{\partial x} = 0.$$

The function $c(q)$ represents the local sound speed of the medium and is assumed to satisfy:

$$(3) \qquad 0 < c(q) \quad \text{and} \quad \frac{dc(q)}{dq} < 0, \quad -\infty < q < \infty.$$

This form of the wave equation is chosen because of the particularly simple way in which the Riemann Invariants α and β are related to v and q. The result is

$$(4) \qquad \alpha = \frac{v-q}{2} \quad \text{and} \quad \beta = \frac{v+q}{2}.$$

In regions of $t > 0$ where the solution of (WE) is smooth, the invariants satisfy

$$\text{(WE)}_{(\alpha,\beta)} \quad \frac{\partial \alpha}{\partial t} + c(\beta-\alpha)\,\frac{\partial \alpha}{\partial x} = 0 \quad \text{and} \quad \frac{\partial \beta}{\partial t} - c(\beta-\alpha)\,\frac{\partial \beta}{\partial x} = 0.$$

The potentials

$$(5) \quad \begin{cases} U(q) \overset{\text{def}}{=} \displaystyle\int_0^q \frac{ds}{c(s)}, \quad \Sigma(q) \overset{\text{def}}{=} \int_0^q c(s)\,ds, \\ e(q) \overset{\text{def}}{=} \displaystyle\int_0^q \frac{\Sigma(s)}{c(s)}\,ds, \quad T(q,v) \overset{\text{def}}{=} \frac{v^2}{2} + e(q), \quad \text{and} \\ P(q,v) \overset{\text{def}}{=} \Sigma(q)\,v \end{cases}$$

will be of particular importance in what follows. They, together with v, have the following physical interpretations when evaluated at a mass point x at time t:

U is the specific volume or strain,

v is the particle velocity,

Σ is the stress,

e is the internal energy,

T is the total energy (kinetic + internal), and

P is the power.

The original equation (WE) may be rewritten in terms of U, Σ, and v as

$$(\mathrm{WE})_{(U,\Sigma)} \qquad \frac{\partial U(q)}{\partial t} - \frac{\partial v}{\partial x} = 0 \quad \text{and} \quad \frac{\partial v}{\partial t} - \frac{\partial \Sigma(q)}{\partial x} = 0.$$

The former equation is the Lagrangian form of the continuity equation and the latter is the momentum balance for the medium. In regions of $t > 0$ where q and v are smooth we also have the energy equation:

$$(\mathrm{En}) \qquad \frac{\partial T(q,v)}{\partial t} - \frac{\partial P(q,v)}{\partial x} = 0.$$

I shall consider the situation where the medium of interest occupies the interval $x > 0$. Thus we seek a solution of (WE) in the quarter plane

$$(6) \qquad Q = \{(x,t) \mid x > 0 \text{ and } t > 0\}.$$

At time $t = 0$, I shall assume that the medium is at rest in homogeneous equilibrium. This assumption translates into the initial condition

$$(\mathrm{IC}) \qquad (q,v)(x,0) = (0,0), \quad x > 0.$$

At time $t = 0$ an impulsive, compressive load is applied to the end $x = 0$ and this load is maintained over the interval $0 < t < T_1$. This loading condition and the one to one character of the function $\Sigma(\cdot)$ implies that the following boundary condition must prevail over the interval $(0,T_1)$:

$(BC)_{(0,T_1)}$ $\qquad q(0,t) = q_1 < 0, \quad 0 < t < T_1.$

At the time T_1 it is assumed that either the end load is removed or the end $x = 0$ is brought abruptly to rest and held there. The former condition implies that $q(0,t)$ must satisfy

$(BC)^1_{(T_1,\infty)}$ $\qquad q(0,t) = 0, \quad t > T_1,$

while the latter yields

$(BC)^2_{(T_1,\infty)}$ $\qquad v(0,t) = 0, \quad t > T_1.$

In the sequel P_i, $i = 1$ or 2, will refer to the problem of solving (WE) in the quarter plane Q subject to the conditions (IC), $(BC)_{(0,T_1)}$, and $(BC)^i_{(T_1,\infty)}$. Slightly more general initial and boundary conditions could be considered but these tend to complicate the analysis and the associated solutions do not exhibit phenomena not already observable in the solutions of P_1 and P_2.

Our results for the problem P_1 are as follows. The forward facing invariant, α, decays as $1/t^{1/2}$ and the back facing invariant, β, decays as $1/t^{3/2}$. Moreover, the forward facing invariant α behaves qualitatively like the solution of the single conservation law:

(E) $$\frac{\partial A}{\partial t} + \frac{\partial}{\partial x}\left(c(0)A + \frac{|c'(0)|A^2}{2}\right) = 0,$$

(IC) $\qquad A(x,0) = 0, \quad x > 0, \quad$ and

$(BC)_1$ $\qquad A(0,t) = \mathcal{A}(t), \quad t > 0,$

where

$$(7)\quad \begin{cases} A(t) = A_o > 0 \quad \text{for} \quad 0 < t < T_1; \\ 0 < A(t) < A_o \quad \text{and} \quad \dfrac{dA(t)}{dt} \leq 0 \quad \text{for} \quad t > T_1; \text{ and} \\ A(t) = O(1/t^{3/2}) \quad \text{as} \quad t \to \infty. \end{cases}$$

A similar result obtains for the problem P_2. The forward facing invariant, α, decays as $1/t^{1/2}$ and the back facing invariant, β, goes as $1/t^{3/2}$. But, associated with the solution of P_2 there is a wake region where α decays as $1/t^{3/4}$. In this wake region the medium is in tension, that is $q > 0$. Here, the forward facing invariant, α, behaves qualitatively like the solution of the conservation law:

$$(E)\quad \frac{\partial A}{\partial t} + \frac{\partial}{\partial x}\left(c(0)A + \frac{|c'(0)|A^2}{2}\right) = 0, \quad x > 0$$

$$(IC)\quad A(x,0) = 0, \quad x > 0, \quad \text{and}$$

$$(BC)_2\quad A(0,t) = A(t), \quad t > 0$$

where

$$(8)\quad \begin{cases} A(t) = A_o > 0, \quad 0 < t < T_1; \\ A(t) < 0 \quad \text{for} \quad t > T_1; \quad \text{and} \\ |A(t)| = O(1/t^{3/2}) \quad \text{as} \quad t \to \infty.^{\dagger} \end{cases}$$

† Results of a similar nature have been obtained recently by DiPerna [5].

2. PRELIMINARIES

The nature of the initial and boundary conditions for P_1 and P_2 forces one to look at a weak formulation of the problem. The solutions will satisfy the integral identities

$$(\mathrm{WE})^{\#} \qquad \oint_C (U dx + v d\tau) = 0 \quad \text{and} \quad \oint_C (v dx + \Sigma d\tau) = 0$$

for every piecewise smooth, simple, closed curve C in the quarter plane Q. These conditions imply that the original equations (WE) or $(\mathrm{WE})_{(\alpha,\beta)}$ or $(\mathrm{WE})_{(U,\Sigma)}$ hold where q and v are smooth and that the shocks or curves where q and v experience jump discontinuities satisfy the Rankine-Hugoniot condition:

$$(\mathrm{RH}) \qquad \frac{ds}{dt}(U(q_+) - U(q_-)) + (v_+ - v_-) = 0$$

and

$$\frac{ds}{dt}(v_+ - v_-) + (\Sigma(q_+) - \Sigma(q_-)) = 0.$$

Here, $x = s(t)$ is a typical shock and we are employing the notation that $+$ denotes a limit from the right and $-$ a limit from the left (in x at fixed t).

We shall also insist that our solutions satisfy the dissipation inequality

$$(\mathrm{En})^{\#} \qquad \oint_C (T(q,v)dx + P(q,v)d\tau) \geq 0$$

for every piecewise smooth, simple, closed curve C in Q. The energy identity (En) of Section 1 implies that $(\mathrm{En})^{\#}$ is really a constraint on shock waves. This constraint and the conditions (RH) imply that on a forward shock

$$(1)\qquad \frac{ds}{dt} = c(q_+, q_-) \overset{\text{def}}{=} \sqrt{\frac{\Sigma(q_+) - \Sigma(q_-)}{U(q_+) - U(q_-)}}\ ,$$

$$(2)\qquad v_- - v_+ = J(q_+, q_-) \overset{\text{def}}{=} \sqrt{(\Sigma(q_+)-\Sigma(q_-))(U(q_+)-U(q_-))}$$

and

$$(3)\qquad q_+ > q_-\ ,$$

while on a back shock

$$(4)\qquad \frac{ds}{dt} = -c(q_-, q_+)\ ,$$

$$(5)\qquad v_- - v_+ = J(q_-, q_+)\ ,$$

and

$$(6)\qquad q_- > q_+\ .$$

Finally, the following strengthened version of $(En)^{\#}$ is valid for the solutions of P_1 and P_2 on the interval $t > T_1$:

$$(En)^{\#\#} \int_0^\infty T(q,v)(x,t)\,dx + \int_{T_1}^t \sum_{x \in S(\tau)} D(q_+, q_-)(x,\tau)\,d\tau$$

$$= \int_0^\infty T(q,v)(x,T_1)\,dx, \quad \text{where}$$

(7) $S(\tau) = \{x > 0 \mid x$ is the location of a shock wave at time $\tau\}$,

$$(8)\quad \begin{cases} D(q_+ , q_-)(x,\tau) \\ \displaystyle = \frac{ds(\tau)}{d\tau} \int_{q_-(s(\tau),\tau)}^{q_+(s(\tau),\tau)} \left(\Sigma(\eta) - \frac{\Sigma(q_+)+\Sigma(q_-)}{2} \right) \frac{d\eta}{c(\eta)} \end{cases},$$

and

$$(9)\qquad x = s(\tau) \in S(\tau).$$

The fact that shocks satisfying (2.1)-(2.6) satisfy

$$(10)\qquad D(q_+ , q_-) = O\left(\frac{|c'(q_\pm)|\,|q_+-q_-|^3}{6} \right)$$

and the fact that across a forward (backward) shock the change in $\beta(\alpha)$ may be written as

$$(11)\qquad \beta_- - \beta_+(\alpha_- - \alpha_+) = O\left(\frac{(c')^2(q_\pm)\,|q_+-q_-|^3}{48c^2(q_\pm)} \right)$$

imply that the solutions of P_1 and P_2 satisfy

$$(12)\qquad \int_{T_1}^{\infty} \left[\sum_{x\in F(\tau)} (\beta_- - \beta_+)(x,\tau) + \sum_{x\in B(\tau)} (\alpha_- - \alpha_+)(x,\tau) \right] d\tau$$

$$\le O\left(\frac{|c'(0)| \int_0^{\infty} T(q,v)(x,T_1)\,dx}{8c^2(0)} \right).$$

Here,

(13) $F(\tau) = \{x > 0 \,|\, x$ is the location of a forward shock at time $\tau\}$,

and

(14) $\mathcal{B}(\tau) = \{x > 0 \mid x$ is the location of a back shock at time $\tau\}$.

3. ESTIMATES FOR THE PROBLEM P_1

Throughout this section the reader is advised to consult Figure 1.

The early stages of the solution to P_1 and P_2 are identical. Starting from the origin we have a forward shock

(1) $$x = R(t) \overset{\text{def}}{=} c(0,q_1)t, \quad 0 \le t \le T_1 ,$$

and the solution is given by

(2) $$(q,v)(x,t) = \begin{cases} (q_1, J(0,q_1)), & 0 < x < c(0,q_1)t \\ & \text{and } 0 \le t \le T_1 , \\ (0,0), & c(0,q_1)\, t < x \\ & \text{and } 0 \le t \le T_1 . \end{cases}$$

Again $c(\cdot,\cdot)$ and $J(\cdot,\cdot)$ are the functions defined in (2.1) and (2.2).

If we let

(3) $$T_2 \overset{\text{def}}{=} \frac{c(q_1)T_1}{c(q_1)-c(0,q_1)} ,$$

then the solution to P_1 on the interval $T_1 \le t \le T_2$ is given by

$$(4)\quad (q,v)(x,t) = \begin{cases} (0,\ J(0,q_1)+q_1), & 0 \le x \le c(0)(t-T_1), \\ \left(\bar{q}\left(\frac{x}{t-T_1}\right),\ J(0,q_1)+q_1-\bar{q}\left(\frac{x}{t-T_1}\right)\right), & \\ \qquad c(0)(t-T_1) < x \le c(q_1)(t-T_1), & \\ (q_1\ ,J(0,q_1)), & c(q_1)(t-T_1) < x < c(0,q_1)t, \\ (0,0), & c(0,q_1)t < x, \end{cases}$$

where $\bar{q}\left(\frac{x}{t-T_1}\right)$ is the unique solution of

$$(5)\qquad c(\bar{q}) = \frac{x}{t-T_1}\ ,\quad c(0)(t-T_1) < x \le c(q_1)(t-T_1)\ .$$

To continue the solution to P_1 past $t = T_2$ we are forced to look for a solution of $(WE)^{\#}$ in $\Omega \cap \{t > T_2\}$ satisfying

$$(IC)^1_{T_2}\ (q,v)(x,T_2) = \begin{cases} (0,J(0,q_1)+q_1), & 0 \le x \le c(0)(T_2-T_1), \\ \left(\bar{q}\left(\frac{x}{T_2-T_1}\right), J(0,q_1)+q_1-\bar{q}\left(\frac{x}{T_2-T_1}\right)\right), & \\ \qquad c(0)(T_2-T_1)<x<c(q_1)(T_2-T_1), & \\ (0,0), & c(0,q_1)T_2 < x, \end{cases}$$

and the homogeneous boundary condition

$$(BC)^1_{(T_2\ ,\ \infty)}\qquad q(0,t) = 0,\quad t > T_2\ .$$

That this problem has a solution follows from the arguments employed by the author in [1]; in fact the procedure employed there to obtain piecewise smooth approximate solutions carries over without modification.

Certain relevant facts about the solution are summarized below. The solution has exactly one forward shock, $x = R(t)$, namely the one eminating from $c(0,q_1)T_2$ at time T_2. In addition, for any time $\bar{t} > T_2$, q and β are decreasing in x on $[0, R(\bar{t}))$ and for any $0 < x_1 < x_2 < R(\bar{t})$:

(a) The Riemann Problem for $(WE)^{\#}$ with initial data $(q,v)(x_2,\bar{t})$ for $x < 0$ and $(q,v) = (0,0)$ for $x > 0$ is solvable with a forward and a backward shock.

(b) The Riemann Problem for $(WE)^{\#}$ with initial data $(q,v)(x_1,\bar{t})$ for $x < 0$ and $(q,v)(x_2,\bar{t})$ for $x > 0$ is solvable with a back shock and a forward rarefaction wave.

(c) The Riemann Problem for $(WE)^{\#}$ with initial data $(q,v)(x_1,\bar{t})$ for $x > 0$ and the homogeneous boundary condition $(q,v)(0,t) = (0,0)$ is solvable with a forward rarefaction wave.

The preceding properties (a)-(c) and (3.2)-(3.4) in turn yield the pointwise bounds

$$(6)\quad \begin{cases} q_1 \le q \le 0,\ 0 \le v \le J(0,q_1),\ 0 \le \beta \le \dfrac{J(0,q_1)+q_1}{2}, \\ \text{and}\quad 0 \le \alpha \le \dfrac{J(0,q_1)-q_1}{2} \end{cases}$$

for $T_1 \le t$ and $x \ge 0$. We also have the following qualitative estimates which are valid on $t \ge T_1$:

$$(7)\quad \begin{cases} \dfrac{d}{dt} q_-(R(t),t) \ge 0,\ \dfrac{\partial \alpha}{\partial t}(0,t) = \dfrac{\partial \beta}{\partial t}(0,t) \le 0, \\ \text{and}\quad \dfrac{d\alpha}{dt}(f(t),t) \le 0 \quad \text{if}\quad \dfrac{df}{dt} = c(q). \end{cases}$$

The task now is to obtain the decay estimates described in Section 1. The following parameters will play an important role in the sequel:

$$(8)\qquad V_1 = \int_0^{c(0,q_1)T_1} v(x,T_1)\,dx = c(0,q_1)J(0,q_1)T_1 ,$$

and

$$(9)\qquad \begin{aligned} \mathcal{T}_1 &= \int_0^{c(0,q_1)T_1} T(q,v)(x,T_1)\,dx \\ &= c(0,q_1)\left(\frac{J^2(0,q_1)}{2} + e(q_1)\right)T_1 . \end{aligned}$$

V_1 is the total momentum of the system at time T_1 and is conserved; i.e. for all t's exceeding T_1

$$(10)\qquad V_1 = \int_0^{R(t)} v(x,t)\,dx .$$

$\mathcal{T}_1$ is the total energy of the system at time T_1.

Through points $(R(t),t)$, $t > T_2$, we construct the characteristics $x = f_t(\eta)$, $\eta \leq t$, as solutions of

$$(11)\qquad \frac{df}{d\eta} = c(q(f,\eta)), \quad f(t) = R(t).$$

Clearly, for t's close to T_2, $f_t(T_1) = 0$. We define $T_3 > T_2$ as the largest time t such that $f_{T_3}(T_1) = 0$.

To obtain the estimates for times $t \in [T_2, T_3]$ we apply the conservation law

$$\frac{\partial v}{\partial t} - \frac{\partial \Sigma}{\partial x} = 0$$

to the region bounded on the left by $f_t(\cdot)$, on the

right by the shock $R(\cdot)$, and on the bottom by the horizontal line $t = T_1$. The result is the identity

$$\text{(12)}\qquad \int_{T_1}^{t} (c(q)v + \Sigma(q))(f_t(\eta),\eta)d\eta = V_1 .$$

If we now make use of the identities

$$c(q)v + \Sigma(q) = 2c(q)v - \int_0^q c'(s)sds ,$$

$$-\int_0^q c'(s)sds = O\left(\frac{|c'(0)|q^2}{2}\right),$$

and

$$q = \beta-\alpha$$

we obtain

$$\text{(13)}\qquad \begin{aligned} c(q)v + \Sigma(q) = \beta\left(2c(q) + O\left(|c'(0)|\left(\frac{\beta-2\alpha}{2}\right)\right)\right) \\ + O\left(\frac{|c'(0)|\alpha^2}{2}\right). \end{aligned}$$

The last identity, the inequalities

$$\text{(14)}\qquad 0 \le \beta,\ c(0) \le c(q) \quad \text{for} \quad q \le 0, \quad \text{and} \quad \alpha \le \frac{J(0,q_1)-q_1}{2} ,$$

and (3.12) imply that if $q_1 < 0$ satisfies

$$\text{(15)}\qquad 2c(0) + O\left(|c'(0)|\left(\frac{J(0,q_1)-q_1}{2}\right)\right) > 0 ,$$

then $\alpha(f_t(\cdot),')$ satisfies

$$\text{(16)}\qquad \frac{|c'(0)|}{2}\int_{T_1}^{t} \alpha^2(f_t(\eta),\eta)d\eta \le O(V_1).$$

The fact that $\alpha^2(f_t(\eta),\eta)$ is nonincreasing on $[T_1 , t]$ and (3.16) in turn yield

LEMMA 1. *On the shock* $x = R(t)$ *the following inequality prevails:*

$$(17) \quad 0 \le \alpha_-(R(t),t) \le O\left(\frac{2V_1}{|c'(0)|(t-T_1)}\right)^{1/2}, \quad T_2 \le t \le T_3 .$$

We now turn our attention to what happens on the interval $t \quad T_3$. We shall first require an estimate for $\beta(0,t)$ for time $t > T_1$.

LEMMA 2. *The following inequalities prevail for time* $t > T_1$:

$$(18) \quad \int_0^t \beta(0,\tau)\,d\tau \le O\left((J(0,q_1)+q_1)T_1 + \frac{|c'(0)|T_1}{4c^2(0)}\right),$$

and

$$(19) \quad \alpha(0,t) = \beta(0,t) \le O\left(\frac{1}{t}\left((J(0,q_1)+q_1)T_1 + \frac{|c'(0)|T_1}{4c^2(0)}\right)\right).$$

PROOF. Since $\beta(0,\cdot)$ is nonincreasing and satisfies $0 \le \beta(0,t) \le \frac{J(0,q_1)+q_1}{2}$, $t \ge 0$, we know that $\beta(0,\cdot)$ has at most a countable set of points of discontinuity $0 < t_1 < \dots < t$ on any interval $[0,t]$ and that

$$\int_0^t \beta(0,\tau)\,d\tau = \sum_i \int_{t_i}^{t_{i+1}} \beta(0,\tau)\,d\tau .$$

Moreover, since the solution to P_1 has only one forward shock, $x = R(\cdot)$, we have

$$\beta(b(\eta,\tau),\eta) = \beta(0,\tau)$$

on the back characteristic $x = b(\eta,\tau)$, $\eta < \tau$, defined by

$$\frac{db}{d\eta} = -c(\beta(0,\tau) - \alpha(b,\eta)), \quad b(\tau,\tau) = 0, \tag{20}$$

provided τ is a point of continuity of $\beta(0,\cdot)$. We define $\hat{\eta}(\tau)$ as that time such that

$$b(\hat{\eta}(\tau),\tau) = R(\hat{\eta}(\tau)) . \tag{21}$$

It is easily checked that $\hat{\eta}(\cdot)$ is increasing, continuous on the intervals (t_i , t_{i+1}), and satisfies $0 < \hat{\eta}(\tau) < \tau$ for $\tau > 0$. Moreover,

$$\int_0^t \beta(0,\tau)d\tau = \sum_i \int_{\eta_i=\hat{\eta}(t_i)}^{\eta_{i+1}=\hat{\eta}(t_{i+1})} \beta_-(R(\eta),\eta)d\hat{\tau}_i(\eta).$$

Here $\hat{\tau}_i(\cdot)$ is the inverse of $\hat{\eta}(\cdot)$ on (t_i , t_{i+1}) and $\beta_-(R(\eta),\eta) = \lim_{\substack{x\to R(\eta)\\ x<R(\eta)}} \beta(x,\eta)$.

We shall now show that for $\eta \in (\hat{\eta}(t_i), \hat{\eta}(t_{i+1}))$

$$\frac{d\hat{\tau}_i}{d\eta} \leq 2\left(\frac{c(q_1)}{c(0)}\right)^{1/2} .$$

This will allow us to replace the integral of β on $x = 0$ by an integral of β on the shock.

Equations (2.1), (3.20), and (3.21) imply that

$$\left\{\begin{aligned} &\frac{d\hat{\eta}}{d\tau} = \mathcal{B}(\hat{\eta}(\tau),\tau)/(c(0,q_-) + c(q_-)) \quad \text{and} \\ &\mathcal{B}(\eta,\tau) = \frac{\partial b}{\partial \tau}(\eta,\tau) \end{aligned}\right. \tag{22}$$

where $q_- = q_-(R(\hat{\eta}(\tau)), \hat{\eta}(\tau))$, while (3.20) and the inequalities $\frac{dc(q)}{dq} < 0$ and $\frac{\partial \beta(0,\tau)}{\partial \tau} \leq 0$ imply that

$$\text{(23)} \quad \begin{cases} \frac{d\mathcal{B}}{d\eta} \leq c'(q(b(\eta,\tau),\eta)) \frac{\partial \alpha}{\partial b}(b(\eta,\tau),\eta)\mathcal{B} \quad \text{and} \\ \mathcal{B}(\tau,\tau) = c(0). \end{cases}$$

The equations $(WE)_{(\alpha,\beta)}$ of Section 1 and the fact that

$$\beta(b(\eta,\tau),\eta) = \beta(0,\tau), \quad \hat{\eta}(\tau) < \eta < \tau,$$

imply that

$$\frac{\partial \alpha}{\partial b}(b(\eta,\tau)\eta) = \frac{1}{2c(q(b(\eta,\tau),\eta))} \frac{dq}{d\eta}(b(\eta,\tau),\eta)$$

and this combines with (3.23) to give

$$\frac{d}{d\eta}\left(\frac{B(\eta,\tau)}{c^{1/2}(q(b(\eta,\tau),\eta))}\right) \leq 0$$

or equivalently

$$B(\hat{\eta}(\tau),\tau) \geq c^{1/2}(q_-(R(\hat{\eta}(\tau)),\hat{\eta}(\tau)))c^{1/2}(0).$$

The last identity, (3.22), and the inequalities

$$c(0) < c(0,q_-) < c(q_-) \leq c(q_1)$$

imply that

$$\frac{d\hat{\tau}_i}{d\eta} \leq 2\left(\frac{c(q_1)}{c(0)}\right)^{1/2}.$$

This inequality in turn yields

$$\int_0^t \beta(0,\tau)d\tau \leq 2\left(\frac{c(q_1)}{c(0)}\right)^{1/2} \int_0^\infty \beta_-(R(\eta),\eta)d\eta.$$

The estimate (3.18) now follows from the fact that

$$\beta_-(R(t),t) = \frac{J(0,q_1)+q_1}{2}, \quad 0 \leq t \leq T_1,$$

and from (2.11) which implies that

$$\int_{T_1}^{\infty} \beta_-(R(\eta),\eta)\,d\eta \le O\left(\frac{|c'(0)|T_1}{8c^2(0)}\right).$$

The estimate (3.19) is a consequence of (3.18), the fact that $\alpha(0,t) = \beta(0,t)$ for $t > T_1$, and $\frac{\partial\beta}{\partial t}(0,t) \le 0$. □

To obtain the desired estimates for times $t > T_3$ we integrate the conservation law

$$\frac{\partial v}{\partial t} - \frac{\partial \Sigma(q)}{\partial x} = 0$$

around the circuit bounded on the left by $f_t(\cdot)$ (see (3.11)), on the right by the forward shock $R(\cdot)$, and on the bottom by the horizontal line $\tau = \tau(t)$. Here, $\tau(t) > T_1$ is defined by $f_t(\tau(t)) = 0$. The result is

$$(24) \qquad \int_{\tau(t)}^{t} (c(q)v + \Sigma(q))(f_t(\eta),\eta)\,d\eta = V_1 .$$

We also find that if (3.15) holds, then (3.24) implies that

$$(25) \qquad \frac{|c'(0)|\alpha_-^2(R(t),t)}{2}(t-\tau(t)) \le O(V_1).$$

But, (3.25) may be rewritten as

$$(26) \qquad \frac{|c'(0)|}{2}\alpha_-^2(R(t),t)(t-T_1) \le O(V_1) + \frac{|c'(0)|\alpha_-^2(R(t),t)(\tau(t)-T_1)}{2}$$

and this inequality together with

$$(27)\quad \begin{cases} 0 \le \alpha_-(R(t),t) \le \alpha(0,\tau(t)) \\ \le \min\left(\dfrac{J(0,q_1)+q_1}{2}, O\left(\dfrac{1}{\tau}\left((J(0,q_1)+q_1)T_1 + \dfrac{|c'(0)|T_1}{4c^2(0)}\right)\right)\right) \end{cases}$$

implies:

LEMMA 3. *For times* $t > T_3$, *the following estimate holds on* $x = R(\cdot)$:

$$(28)\qquad \alpha_-(R(t),t) \le O\left(\frac{2(V_1+K_1)}{|c'(0)|(t-T_1)}\right)^{1/2},$$

where

$$(29)\quad K_1 = O\left(\frac{|c'(0)|(J(0,q_1)+q_1)\left((J(0,q_1)+q_1)T_1 + \dfrac{|c'(0)|T_1}{4c^2(0)}\right)}{4}\right).$$

The estimates (3.17) and (3.28) and the facts that $\beta \ge 0$ and q is decreasing on $(0,R(t))$ combine to give

$$(30)\quad 0 \ge q(x,t) \ge q_-(R(t),t) \ge -O\left(\frac{2(V_1+K_1)}{|c'(0)|(t-T_1)}\right)^{1/2}$$

for $T_2 \le t$. This inequality, together with the shock relation

$$(31)\quad \begin{aligned} \frac{dR}{dt} &= c(0,q_-(R(t),t)) \\ &= c(0) + O\left(\frac{c'(0)q_-(R(t),t)}{2}\right) \end{aligned}$$

and (3.3) imply that

$$(32)\quad R(t) \le \begin{cases} c(0,q_1)t, & 0 < t < T_2 \\ c(0,q_1)T_2 + O\left((2|c'(0)|(V_1+K_1))^{\frac{1}{2}}\left((t-T_1)^{\frac{1}{2}} - (T_2-T_1)^{\frac{1}{2}}\right)\right), & T_2 \le t. \end{cases}$$

The inequalities (3.17) and (3.30) and the identity (2.10) in turn yield

$$(33)\qquad 0 \le \beta_-(R(t),t) \le O\left(\left(\frac{|c'(0)|}{48c^2(0)}\right)^{\frac{1}{2}}\left(\frac{2(V_1+K_1)}{t-T_1}\right)^{\frac{3}{2}}\right)$$

for $t > T_2$. If we then observe that a back characteristic from $x = 0$ at time t intersects the shock $x = R(\cdot)$ at $\tau \ge t/3$, we see that for large t

$$(34)\qquad 0 \le \alpha(0,t) = \beta(0,t) \le O\left(\frac{3^{\frac{1}{2}}|c'(0)|^{\frac{1}{2}}}{2^{5/2}c^2(0)}\left(\frac{V_1+K_1}{t-3T_1}\right)^{\frac{3}{2}}\right).$$

This represents an improvement over the estimate (3.19). Since $\beta \ge 0$ is decreasing on $(0,R(t))$, the estimate (3.34) provides a global bound for β. The estimate for α follows from (3.30) and (3.34) and the identity $\alpha = \beta - q$.

We now take up the question of regularity of the solution to P_1 in the region

$$(35)\qquad \mathcal{R} = \{(x,t) \mid 0 < x < R(t),\ \ t > T_1\}.$$

What we shall obtain are estimates for

$$(36)\quad A(x,t) \overset{\text{def}}{=} c^{1/2}(\beta-\alpha)\frac{\partial\alpha}{\partial x} \quad\text{and}\quad B(x,t) \overset{\text{def}}{=} c^{1/2}(\beta-\alpha)\frac{\partial\beta}{\partial x}\,.$$

We shall show that if the parameter q_1 appearing in the boundary condition $(BC)_{(0,T_1)}$ is sufficiently small, then A and B are bounded in $\mathcal{R}$ and decay to zero as t tends to plus infinity.

It is easily checked that A and B evolve as

$$(37)\qquad \frac{\partial A}{\partial t} + c(\beta-\alpha)\frac{\partial A}{\partial x} + p(\beta-\alpha)A^2 = 0,$$

and

$$(38)\qquad \frac{\partial B}{\partial t} - c(\beta-\alpha)\frac{\partial B}{\partial x} + p(\beta-\alpha)B^2 = 0,$$

where

$$(39)\qquad p(q) = |c'(q)|/c^{1/2}(q).$$

These formulas imply that for any pair of times $\tau < t$

$$(40)\qquad A(x,t) = \frac{A(f(\tau),\tau)}{1 + A(f(\tau),\tau)\int_\tau^t p(q)(f(s),s)ds}$$

and

$$(41)\qquad B(x,t) = \frac{B(b(\tau),\tau)}{1 + B(b(\tau),\tau)\int_\tau^t p(q)(b(s),s)ds}.$$

Here $\xi = f(\eta)$, $\tau < \eta < t$, is the forward characteristic

$$(42)\qquad \frac{df}{d\eta} = c(q), \quad f(t) = x,$$

and $\xi = b(\eta)$, $\tau < \eta < t$, is the back characteristic

$$(43)\qquad \frac{db}{d\eta} = -c(q), \quad b(t) = x.$$

The formulas (3.40)-(3.43) are valid provided the denominators of (3.40) and (3.41) are not zero.

On the interval $T_1 \le t \le T_2 = \dfrac{c(q_1)T_1}{c(q_1)-c(0,q_1)}$

$$(44) \qquad B(x,t) = 0,$$

and $A(x,t)$ is given by

$$(45) \qquad A(x,t) = \begin{cases} 0, & 0 < \dfrac{x}{t-T_1} < \dfrac{c(0)}{t-T_1} \\ \dfrac{1}{p\left(\bar{q}\left(\dfrac{x}{t-T_1}\right)\right)(t-T_1)}, & \dfrac{c(0)}{t-T_1} < \dfrac{x}{t-T_1} < \dfrac{c(q_1)}{t-T_1}, \\ 0, & c(q_1)(t-T_1) < x < c(0,q_1)t \end{cases}$$

where $\bar{q}\left(\dfrac{x}{t-T_1}\right)$ is defined in (3.5). The identity (3.45) implies

$$(46) \qquad \begin{cases} 0 \le A(x,t) < \dfrac{1}{\underline{p}(t-T_1)} & \text{for } 0<x<c(0,q_1)t \text{ and} \\ & T_1 < t < T_2 \end{cases}$$

where

$$(47) \qquad \underline{p} = \min_{q_1 \le q \le 0} p(q) \quad \text{and} \quad \bar{p} = \max_{q_1 \le q \le 0} p(q).$$

The identity (3.40) implies that the estimate (3.46) persists on the strip

$$(48) \qquad R_{T_3} = \{(x,t) \mid f_{0,T_2}(\eta) < x < R(\eta), \quad T_2 \le \eta \le \bar{T}\}.$$

Here $\xi = f_{0,T_2}(\eta)$, $T_2 < \eta < \bar{T}$, is the forward characteristic

$$(49) \qquad \frac{df}{d\eta} = c(q)(f(\eta),\eta), \quad f(T_2) = 0$$

and $\overline{T}$ is the minimum of the following two numbers:

(i) the first time such that B tends to $-\infty$ or

(ii) the solution of the equation $f_{0,T_2}(\overline{T}) = R(\overline{T})$.

We shall now show that if $q_1 < 0$ is sufficiently small then the number $\overline{T}$ of (3.48) is the solution of $f_{0,T_2}(\overline{T}) = R(\overline{T})$. The identity

$$(50)\quad 0 \geq B_-(R(t),t) \geq \mathcal{O}\left(\frac{|c'(0)|^3 q_-^3(R(t),t)A_-(R(t),t)}{64c^3(0)}\right)^*$$

and (3.30), (3.44), and (3.46) imply that $B_-(R(t),t)$ satisfies

$$(51)\quad 0 \geq B_-(R(t),t) \geq \begin{cases} 0, & 0 \leq t \leq T_2, \\ -\mathcal{O}\left(\dfrac{|c'(0)|(V_1+K_1)}{8c^2(0)}\right)^{\frac{3}{2}} \dfrac{1}{\underline{p}(t-T_1)^{5/2}}, & T_2 \leq t \leq \overline{T}. \end{cases}$$

If we now choose $q_1 < 0$ so that

$$(52)\quad 1 - \mathcal{O}\left(\frac{3\overline{p}}{\underline{p}}\left(\frac{|c'(0)|(V_1+K_1)}{8c^2(0)}\right)^{3/2} \frac{T_2^{5/2}}{T_1^{3/2}(T_2-T_1)^{5/2}}\right) \geq 1/2,$$

then it follows that

$$(53)\quad D(\eta,t) = 1 - \mathcal{O}\left(\frac{\overline{p}}{\underline{p}}\left(\frac{|c'(0)|(V_1+K_1)}{8c^2(0)}\right)^{3/2} \frac{(\eta-t)}{(t-T_1)^{5/2}}\right) \geq 1/2$$

---

* For details see [2, p. 216].

for times $t \geq T_2$ and $t \leq \eta \leq 3t$. (3.41), (3.51), and (3.53) together with the observation that the smallest time that any back characteristic issuing from (x,t) could hit the line $x = R(\cdot)$ is bounded from below by $t/3$ imply that for $(x,t) \in \{(x,t) \mid 0 < x < R(t),\ T_2 \leq t \leq \bar{T}\}$, $B(x,t)$ satisfies

$$(54)\quad 0 \geq B(x,t) \geq -O\left(2\left(\frac{|c'(0)|(V_1+K_1)}{8c^2(0)}\right)^{3/2} \frac{3^{5/2}}{\underline{p}(t-3T_1)^{5/2}}\right).$$

The inequality (3.54) gives the desired result for the constant $\bar{T}$. Moreover, (3.54) and the identity $A(0,t) = -B(0,t)$ imply the following estimate for $A(0,t)$ for $T_2 \leq t \leq \bar{T}$:

$$(55)\quad 0 \leq A(0,t) \leq O\left(\frac{3^{5/2}}{2^{7/2}} \frac{|c'(0)|^{3/2}(V_1+K_1)^{3/2}}{c^3(0)} \frac{1}{\underline{p}(t-3T_1)^{5/2}}\right).$$

The formula (3.55) in turn yields

$$(56)\qquad A(0,t) \leq \frac{1}{\underline{p}(t-T_1)}, \quad T_2 \leq t \leq \bar{T}$$

provided q_1 satisfies

$$(57)\qquad O\left(\frac{3^{5/2}}{2^{7/2}} \frac{|c'(0)|^{3/2}(V_1+K_1)^{3/2}}{c^3(0)(T_2-3T_1)^{3/2}}\right) < 1.$$

The boundary estimate (3.57) and a boot strapping argument allow us to extend the estimates (3.46), (3.51), (3.54)-(3.56) to the whole strip $0 < x < R(t)$, $t > T_2$.

That the forward facing invariant behaves qualitatively like the solution to the scalar problem (E), (IC), and $(BC)_1$ of Section 1 follows from the estimates (3.7) and (3.34), the fact that on any forward facing characteristic issuing from the line $x = 0$ the back facing invariant is integrable, and the fact that if $q_1 < 0$ is small then no backwaves break in the strip $0 < x < R(t)$, $t > T_2$.

4. ESTIMATES FOR THE PROBLEM P_2

Throughout this section the reader is advised to consult Figures 2 and 3.

We shall assume throughout that $q_1 < 0$ is small enough so that P_2 has a solution defined in Q. That this assumption is not vacuous follows from Glimm's results (for details see [3]). We shall also assume that q_1 satisfies (3.15) and (3.52). Certain additional constraints will have to be placed on the parameter as we go along.

The following notation will be used throughout. The curve, $x = R(t)$, $t \geq 0$, is the forward shock through $(0,0)$; T_2 is the time the rarefaction wave from $(0,T_1)$ hits the shock and

$$(1) \qquad T_2 = \frac{c(q_1)T_1}{c(q_1)-c(0,q_1)} = O\left(\frac{2c(0)T_1}{|c'(0)||q_1|}\right);$$

T_3 is the time that the back characteristic through $(R(T_2),T_2)$ hits $x = 0$ and

$$(2) \qquad T_3 = O(2T_2) = O\left(\frac{4c(0)T_1}{|c'(0)||q_1|}\right);$$

and $x = L(t)$, $t \geq T_3$, is the forward characteristic

(continued as a shock) through $(0,T_3)$.

Our program is as follows. We first seek estimates in $\{L, 0, T_2, R\}$, the set bounded on the right by $x = R(\cdot)$, on the bottom by $t = T_2$, and on the left by $x = 0$ for $T_2 \le t \le T_3$ and then by $x = L(t)$, $t \ge T_3$. These estimates are similar to those obtained in Section 3. The key result here is that if (3.52) is valid, then no backwaves break in $\{L, 0, T_2, R\}$.

We next turn to the region $L_{T_3} = \{0 < x < L(t),$ $t \ge T_3\}$. Our first result is:

LEMMA 4. *The first time,* T^*, *that a wave can break in* L_{T_3} *satisfies*

$$T^* \ge T_3 + k_1/|q_1|^4, \quad k_1 > 0. \dagger \tag{3}$$

Moreover, the breaking occurs on a forward wave.

From the last result and estimates for the solution in $\{L, 0, T_2, R\}$ we obtain:

LEMMA 5. *For times* $t \ge T^*$ *the curve* $x = L(\cdot)$ *satisfies*

$$c(0)(t-T_3) - k_3|q_1|(t-T_3)^{\frac{1}{2}} \le L(t) \tag{4}$$

$$\le c(0)(t-T_3) - k_2|q_1|(t-T_3)^{\frac{1}{2}}.$$

The big result of this section is:

---

† Throughout the remainder of this section the k_i's will be positive constants which are independent of $|q_1|$.

LEMMA 6. *Suppose that no back waves break in* L_{T_3} *and that for points* $(x,t) \in L_{T_3}$

$$(5)\quad 0 \leq -\alpha(0,t) \quad \text{and} \quad \beta(x,t) \leq \min\left(k_4|q_1|^3, \frac{k_5|q_1|^{3/2}}{t^{3/2}}\right).$$

Then, for points $(x,t) \in L_{T_3}$

$$(6)\qquad -\min\left(k_4|q_1|^3, \frac{k_6}{t^{3/4}}\right) \leq \alpha(x,t) \leq 0.$$

This is the wake estimate alluded to in the Introduction.

We conclude this section by showing that if $|q_1|$ is suitably restricted, then the solution of P_2 satisfies the hypotheses of Lemma 6.

<u>Estimates in</u> $\{L, 0, T_2, R\}$

It is easily checked that on $\{L, 0, T_2, R\}$ $\cap \{t = \bar{t} > T_2\}$ the functions q and β are decreasing. We also have the pointwise estimates

$$(7)\quad q_1 \leq q(x,t), \quad 0 \leq \beta(x,t) \leq \frac{q_1 + J(0,q_1)}{2}, \text{ and}$$

$$\alpha(x,t) \leq \frac{J(0,q_1) - q_1}{2}.$$

As long as no back waves have broken in $\{L, 0, T_2, R\}$ we have the additional bounds

$$(8)\qquad \begin{aligned} q(x,t) &\leq q_1 + J(0,q_1) \quad \text{and} \\ -\frac{q_1 + J(0,q_1)}{2} &\leq \alpha(x,t) . \end{aligned}$$

Moreover, on the shock $x = R(t)$, $t \geq T_2$, the

following bounds prevail:

$$(9)\qquad 0 \le \alpha_-(R(t),t) \le O\left(\frac{2V_1}{|c'(0)|(t-T_1)}\right)^{1/2},$$

$$(10)\qquad 0 \ge q_1(R(t),t) \ge -O\left(\frac{2V_1}{|c'(0)|(t-T_1)}\right)^{1/2},$$

$$(11)\qquad 0 \le \beta_-(R(t),t) \le O\left(\frac{|c'(0)|^{1/2}}{48c^2(0)}\left(\frac{2V_1}{t-T_1}\right)^{3/2}\right),$$

and

$$(12)\quad R(t) \le \begin{cases} c(0,q_1)t, & 0 \le t \le T_2, \text{ and} \\ c(0,q_1)T_2 + O\left((2|c'(0)|V_1)^{\frac{1}{2}}((t-T_1)^{\frac{1}{2}} - (T_2-T_1)^{\frac{1}{2}})\right), & T_2 \le t, \end{cases}$$

where V_1 is the total momentum at time T_1 and is given by (3.8). These estimates are independent of whether or not back waves have broken in $\{L, 0, T_2, R\}$. They follow from the hypothesis that q_1 satisfies (3.15) and from arguments similar to those employed in Section 3. The estimates are also independent of the boundary values of α. This fact follows from the observation that for all times $t > T_2$ the forward characteristics, $x = f_t(\eta)$, $\eta < t$, defined by $\frac{df}{d\eta} = c(q)(f(\eta),\eta)$, $\eta < t$, and $f(t) = R(t)$ terminate in the corner $(0,T_1)$.

The arguments employed in Section 3 and the hypothesis (3.52) imply that if no back waves have broken in $\{L, 0, T_2, R\} \cap \{T_2 \le t < \overline{T}\}$, then $A = c^{1/2}(\beta-\alpha)\alpha_x$ and $B = c^{1/2}(\beta-\alpha)\beta_x$ satisfy

$$(13)\qquad 0 \le A(x,t) \le \frac{1}{\underline{p}(t-T_1)}\,,\quad \text{and}$$

$$(14)\begin{cases} 0 \ge B_-(R(t),t) \ge -O\left(\left(\dfrac{|c'(0)|V_1}{8c^2(0)}\right)^{3/2} \dfrac{1}{\underline{p}(t-T_1)^{5/2}}\right) \quad \text{and} \\ 0 \ge B(x,t) \ge -O\left(\dfrac{3^{5/2}|c'(0)|^{3/2}V_1^{3/2}}{2^{7/2}c^3(0)\underline{p}(t-3T_1)^{5/2}}\right) \end{cases}$$

for $(x,t) \in \{0, L, T_2, R\} \cap \{T_2 \le t < \overline{T}\}$. Again,

$$(15)\begin{cases} p(q) = |c'(q)|/c^{\frac{1}{2}}(q), \quad \underline{p} = \min\limits_{q_1 \le q \le 0} p(q), \quad \text{and} \\ \overline{p} = \max\limits_{q_1 \le q \le 0} p(q). \end{cases}$$

The inequality $(4.14)_2$ implies that if (3.52) is valid, then $\overline{T} = +\infty$.

The following result describes the behavior of q on the curve $x = L(t)$, $t \ge T_3$.

LEMMA 7. *The function* $q_+(L(t),t)$ *satisfies*

$$(16)\qquad 0 \le q_+(L(t),t) \le \frac{k_7|q_1|}{(t-T_3)^{1/2}}\,.$$

PROOF. For any $t > T_3$, we let $x = \ell_t(\eta)$, $\eta < t$, be the forward characteristic defined by

$$(17)\qquad \frac{d\ell}{d\eta} = c(q)(\ell(\eta),\eta), \quad \eta < t, \quad \text{and} \quad \ell(t) = L(t).$$

The number $\overline{\eta}(t) < T_3$ is defined by

(18) $$\ell_t(\overline{\eta}(t)) = 0.$$

Since $x = \ell_t(\cdot)$ is in $\{L, 0, T_2, R\}$ and since no back waves break in $\{L, 0, T_2, R\}$ (by virtue of (3.52)) we have

(19) $$q(\ell_t(\eta),\eta) > 0 \quad \text{and} \quad \frac{d}{d\eta} q(\ell_t(\eta),\eta) = 2c(q)\beta_x \leq 0.$$

To obtain the desired estimate for $q_+(L(t),t)$ we integrate the continuity equation

$$\frac{\partial U(q)}{\partial t} - \frac{\partial v}{\partial x} = 0$$

around the circuit bounded on the right by the curve $x = \ell_t(\eta)$, $\overline{\eta}(t) < \eta < t$, on the left by the line $x = 0$, and on the top by the horizontal line with time ordinate t. The result is

(20) $$\int_0^{L(t)} U(q)(x,t)\,dx = \int_{\overline{\eta}(t)}^{t} [v+c(q)U(q)](\ell_t(\eta),\eta)\,d\eta.$$

The inequality $\beta \geq 0$ for all (x,t) in Q, and the boundary conditon $v(0,t) = \beta(0,t) + \alpha(0,t) = 0$ imply that $\alpha \leq 0$ and $q = \beta - \alpha \geq 0$ in L_{T_3}. These in turn imply that the left hand side of (4.20) is non-negative. If we now make use of the identities

(21) $$\begin{cases} v + c(q)U(q) = 2\beta + \displaystyle\int_0^q \frac{c(q)-c(s)}{c(s)}\,ds, \quad \text{and} \\ \displaystyle\int_0^q \frac{c(q)-c(s)}{c(s)}\,ds = -O\left(\frac{|c'(0)|q^2}{2c(0)}\right), \end{cases}$$

and (4.19) and (4.20) we obtain

$$(22)\quad \left\{ \begin{aligned} &\frac{|c'(0)|q_+^2(L(t),t)(t-\bar{\eta}(t))}{2c^2(0)} + \int_0^{L(t)} U(q)(x,t)\,dx \\ &\qquad \le O\left(2\int_{\bar{\eta}(t)}^{t} \beta(\ell_t(\eta),\eta)\,d\eta\right). \end{aligned} \right.$$

But, the estimates $(4.7)_2$ and (4.11) and the fact that any back characteristic from $(\ell_t(\eta),\eta)$ crosses $x = R(\cdot)$ at $\tau' > \frac{\eta}{3}$ imply that $\beta(\ell_t(\eta),\eta)$ satisfies:

$$(23)\quad \left\{ \begin{aligned} &0 < \beta(\ell_t(\eta),\eta) \\ &\quad \le \min\left(\frac{q_1+J(0,q_1)}{2},\, O\left(\frac{3^{\frac{1}{2}}|c'(0)|^{1/2}}{2^{5/2}c^2(0)}\left(\frac{V_1}{\eta-3T_1}\right)^{3/2}\right)\right), \end{aligned} \right.$$

and

$$(24)\quad \left\{ \begin{aligned} &0 < \int_{\bar{\eta}(t)}^{t} \beta(\ell_t(\eta),\eta)\,d\eta \\ &\quad \le \left(\frac{q_1+J(0,q_1)}{2}\right)(T_2-T_1) \\ &\quad + O\left(\frac{3^{1/2}|c'(0)|^{1/2}V_1^{\ 3/2}}{2^{3/2}c^2(0)(T_2-3T_1)^{1/2}}\right) \stackrel{\text{def}}{=} B_1^2 . \end{aligned} \right.$$

But,

$$(25)\quad \left\{ \begin{aligned} &\frac{q_1+J(0,q_1)}{2} = O\left(\frac{|c'(0)|^2|q_1|^3}{48c^2(0)}\right) \quad \text{and} \\ &V_1 = c(0,q_1)J(0,q_1)T_1 \end{aligned} \right.$$

and (4.1) imply that $B_1 \le k_8|q_1|^2$. Finally, (4.22) and (4.24) imply

$$(26) \qquad 0 \le q_+(L(t)t) \le \frac{2c(0)B_1}{|c'(0)|^{1/2}(t-\bar{\eta}(t))^{1/2}}$$

where $T_1 \le \bar{\eta}(t) \le T_3 \le t$. The lemma now follows from (4.26).

<u>Estimates in</u> L_{T_3}

We start by proving Lemma 4. The hypotheses (3.15) and (3.52) and the arguments of Section 3 imply the following estimates for β and $B = c^{1/2}(\beta-\alpha)\beta_x$ in $L_{T_3} \cap \{T_3 \le t < T^*\}$:

$$(27) \qquad 0 \le \beta(x,t) \le O\left(\frac{3^{1/2}|c'(0)|^{1/2}V_1^{3/2}}{2^{5/2}c^2(0)(t-3T_1)^{3/2}}\right),$$

and

$$(28) \quad 0 \ge B(x,t) \ge -O\left(\frac{3^{5/2}|c'(0)|^{3/2}V_1^{3/2}}{2^{7/2}c^3(0)} \cdot \frac{1}{\underline{p}(t-3T_1)^{5/2}}\right),$$

where again T^* is the first time any wave breaks in L_{T_3}. The inequality (4.28) in turn implies that back waves are not the first to break in $L_{T_3} \cap \{T_3 \le t < T^*\}$.

The boundary condition $v(0,t) = \beta(0,t) + \alpha(0,t) = 0$, $t > T_1$, implies

$$\alpha(0,t) = -\beta(0,t) \quad \text{and} \quad A(0,t) = B(0,t),$$

and this identity together with (4.27) and (4.28) imply

$$(29) \qquad 0 \geq \alpha(0,t) \geq -O\left(\frac{3^{1/2}|c'(0)|^{1/2}V_1^{3/2}}{2^{5/2}c^2(0)(t-3T_1)^{3/2}}\right),$$

and

$$(30) \qquad 0 \geq A(0,t) \geq -O\left(\frac{3^{5/2}|c'(0)|^{3/2}V_1^{3/2}}{2^{7/2}c^3(0)} \cdot \frac{1}{\underline{p}(t-3T_1)^{5/2}}\right).$$

The evolution equation for A (see (3.40)) then tells us that on forward characteristics $x = f_t(\eta)$, $\eta > t$, defined by

$$\frac{df}{d\eta} = c(q)(f(\eta),\eta), \quad \eta > t \quad \text{and} \quad f(t) = 0$$

A is given by

$$(31) \quad A(f_t(\eta),\eta) = \frac{A(0,t)}{1+A(0,t)\int_t^\eta p(q(f_t(s),s))ds}, \quad t < \eta$$

provided

$$(32) \quad 1 + A(0,t)\int_t^{\eta'} p(q(f_t(s),s))ds > 0, \quad t \leq \eta' < \eta,$$

and $x = f_t(\cdot)$ has not hit a forward shock. The formulas (4.31) and (4.32) and the estimate (4.30) yield the bound

$$(33) \qquad T^* \geq T_3 + O\left(\frac{2^{7/2}c^3(0)\underline{p}(T_3-3T_1)^{5/2}}{3^{5/2}|c'(0)|^{3/2}V_1^{3/2}}\right).$$

This inequality, and

$$T_3 = O\left(\frac{4c(0)T_1}{|c'(0)||q_1|}\right) \quad \text{and} \quad V_1 = c(0,q_1)J(0,q_1)T_1$$

$$= O(c(0)|q_1|T_1)$$

yield the desired result. □

PROOF OF LEMMA 5. The definition of the curve $x = L(t)$, $t \geq T_3$, implies

$$\text{(34)} \quad \frac{dL}{dt} = c(q_+, q_-)(L(t),t), \quad t \geq T_3 \quad \text{and} \quad L(T_3) = 0.$$

We also have

$$\text{(35)} \quad q_-(L(t),t) = q_+(L(t),t), \quad T_3 \leq t \leq T^*,$$

and for time $t \geq T^*$

$$\text{(36)} \quad \begin{cases} 0 \leq q_-(L(t),t) \leq q_+(L(t),t) \quad \text{and} \\ c(q_+) \leq c(q_+, q_-) \leq c(q_+, 0). \end{cases}$$

To obtain the desired lower bound we use (4.8), (4.16), (4.34), and (4.36) and

$$c(q) \geq c(0) - |c'|_{max}\, q ,$$

where

$$\text{(37)} \quad |c'|_{max} = \max_{0 \leq q \leq q_1 + J(0,q_1)} |c'(q)| \quad \text{and} \quad |c'|_{min} = \min_{0 \leq q \leq q_1 + J(0,q_1)} |c'(q)|.$$

The result is

$$L(t) \geq c(0)(t-T_3)$$

$$- |c'|_{max} \min\left((q_1 + J(0,q_1))(t-T_3),\ 2k_5|q_1|(t-T_3)^{1/2}\right).$$

To obtain the upper bound we make use of (4.34), (4.36), and

$$c(q,0) \le c(0) - \frac{|c'(0)|q}{2} + \frac{k_9 q^2}{2}, \quad 0 \le q \le q_1 + J(0,q_1),$$

where

$$k_9 \overset{\text{def}}{=} \max_{0\le q\le q_1+J(0,q_1)} \left| \frac{\partial^2 c(q,0)}{\partial q^2} \right|,$$

to obtain

$$\text{(38)} \qquad \frac{dL}{dt} \le c(0) - \frac{|c'(0)|q_+}{2} + \frac{k_9 q_+^2}{2}, \quad t \ge T^*.$$

The characteristic equation (4.17) and (4.20)-(4.25) imply that

$$\text{(39)} \qquad |c'(0)|\alpha_+(L(t),t) \le \frac{L(t)}{(t-T_3)} - c(0) + \frac{k_{10}|q_1|^2}{(t-T_3)}$$

while (4.16), (4.38), (4.39) and $q_+ = \beta_+ - \alpha_+$ and $\beta_+ \ge 0$ imply

$$\frac{dL}{dt} \le c(0) + \frac{1}{2}\left(\frac{L(t)}{t-T_3} - c(0)\right) + \frac{k_{11}|q_1|^2}{(t-T_3)}, \quad t > T^*.$$

Integrating this inequality we obtain

$$\text{(40)} \qquad L(t) \le c(0)(t-T_3) + \frac{(t-T_3)^{1/2}}{(T^*-T_3)^{1/2}}(W(T^*) + 2k_{11}|q_1|^2)$$

where

$$\text{(41)} \qquad W(T^*) = L(T^*) - c(0)(T^*-T_3).$$

It is easily checked that $W(T^*)$ satisfies an estimate of the form

$$W(T^*) \le -k_{12}|q_1|^3(T^* - T_3)$$

and this, together with (4.3) and (4.40) imply

$$\frac{W(T^*)}{(T^*-T_3)^{1/2}} + \frac{2k_{11}|q_1|^2}{(T^*-T_3)^{1/2}}$$

$$\le -k_{12}|q_1|^3(T^*-T_3)^{1/2} + \frac{2k_{11}|q_1|^2}{(T^*-T_3)^{1/2}}$$

$$\le -k_1^{1/2}k_{12}|q_1| + \frac{2k_{11}|q_1|^4}{k_1^{1/2}}$$

$$\le -k_1^{1/2}k_{12}|q_1|\big/2$$

provided $|q_1|$ is sufficiently small. This is the desired result.

REMARK. The estimates for the curve $x = L(\cdot)$ are independent of the function $q_-(L(\cdot),\cdot)$.

PROOF OF LEMMA 6. We shall only prove the estimate (4.6) for points which lie on the curve $x = L(t)$ when $t \ge T^*$. It is clear that the same arguments yield the estimate for points (x,t) with $x < L(t)$.

Through points $(L(t),t)$ we construct the characteristic $x = f_t(\eta)$, $\eta < t$, as the solution of

$$\frac{df}{d\eta} = c(q)(f(\eta),\eta), \quad \eta < t \quad \text{and} \quad f(t) = L(t).$$

And we define $\bar{\eta}(t) > T_3$ as the solution of

$$f_t(\bar{\eta}(t)) = 0. \tag{42}$$

The assumption that no back waves break in L_{T_3} guarantees that

$$\alpha(0,\bar{\eta}(t)) = \alpha_{-}(L(t),t)$$

and that

$$(43) \quad L(t) = c(0)(t-\bar{\eta}(t)) + \alpha(0,\bar{\eta}(t)) \int_{\bar{\eta}(t)}^{t} \gamma(t,s)\,ds - \int_{\bar{\eta}(t)}^{t} \gamma(t,s)\beta(f_t(s),s)\,ds.$$

Here, $\gamma(t,s)$ is defined by

$$\gamma(t,s) = |c'(q(t,s))| > 0$$

and $q(t,s)$ satisfies

$$c(\beta(f_t(s),s) - \alpha(0,\bar{\eta}(t))) - c(0) = c'(q(t,s))(\beta(f_t(s),s) - \alpha(0,\bar{\eta}(t))).$$

The hypothesis that $\beta(x,t)$ satisfies (4.5) throughout L_{T_3}, the fact that $\bar{\eta}(t) \geq T_3$, and the identity

$T_3 = O\left(\frac{4c(0)T_1}{|c'(0)||q_1|}\right)$ imply that $\int_{\bar{\eta}(t)}^{t} \gamma(t,s)\beta(f_t(s),s)\,ds$ may be written as $P_1(t,q_1)|q_1|^2$ where P_1 is positive and of order $|q_1|^0$. Similarly, $\int_{\bar{\eta}(t)}^{t} \gamma(t,s)\,ds$ may be written as $\bar{\gamma}(t,q_1)(t-\bar{\eta}(t))$ where $\bar{\gamma}(t,q_1)$ satisfies

$$|c'|_{min} \leq \bar{\gamma}(t,q_1) \leq |c'|_{max}$$

and $|c'|_{max}$ and $|c'|_{min}$ are defined (4.37).

Thus (4.43) takes the form

$$(44)\quad \begin{cases} L(t) = c(0)(t-\bar{\eta}(t)) + \bar{\gamma}(t,q_1)(t-\bar{\eta}(t))\alpha(0,\bar{\eta}(t)) \\ \qquad\qquad - P_1(t,q_1)|q_1|^2. \end{cases}$$

But the results of Lemma 5 imply that for $t \geq T^*$

$$(45)\qquad L(t) = c(0)(t-T_3) - k(t,q_1)|q_1|(t-T_3)^{1/2}$$

where $k_2 \leq k(t,q_1) \leq k_3$ and k_1 and k_3 are the constants of (4.4). Combining equations (4.44) and (4.45) we find that

$$\bar{\gamma}(t,q_1)|\alpha(0,\bar{\eta}(t))|(t-T_3) - k(t,q_1)|q_1|(t-T_3)^{1/2}$$

$$+ ((c(0) - \bar{\gamma}(t,q_1)|\alpha(0,\bar{\eta}(t))|)(\bar{\eta}(t) - T_3)$$

$$+ P_1(t,q_1)|q_1|^2) = 0.$$

But this implies

$$(t-T_3)^{1/2} = \frac{k(t,q_1)|q_1|\left[1 - \sqrt{1 - \dfrac{4\bar{\gamma}|\alpha|D}{k^2|q_1|^2}}\right]}{2\gamma(t,q_1)|\alpha(0,\bar{\eta}(t))|}$$

where

$$0 < D = ((c(0) - \gamma(t,q_1)|\alpha(0,\bar{\eta}(t))|)(\bar{\eta}(t) - T_3)$$

$$+ P_1(t,q_1)|q_1|^2.$$

The last two relations imply that

$$(46)\qquad \frac{D}{k(t,q_1)|q_1|} \leq (t-T_3)^{1/2} \leq \frac{2D}{k(t,q_1)|q_1|}$$

and this together with (4.5) yields the desired result.□

Our task now is to show that the solution to P_2 satisfies the hypotheses of Lemma 6 for times $t \geq T_3$. The reader is advised at this point to consult Figure 3.

The proof of Lemma 4 guarantees that the estimates (4.27)-(4.30) are valid in $L_{T_3} \cap \{T_3 \leq t \leq T^*\}$. Again T^* is the first time that a forward wave breaks in L_{T_3} and it satisfies (4.3).

In the sequel $x = F(t)$, $t \geq T^*$, will denote the curve defined by

$$\frac{dF}{dt} = c(q_+, q_-)(F(t),t) \quad \text{and} \quad F(T^*) = 0; \tag{47}$$

and $\{F, T^*, L\}$ will denote the region bounded on the left by $x = F(t)$, $t \geq T^*$, on the bottom by the interval $0 < x < L(T^*)$ and $t = T^*$, and on the right $x = L(t)$, $t \geq T^*$. We shall let T_{LF} denote the solution of

$$L(T_{LF}) = F(T_{LF}), \tag{48}$$

and T^{**} will be T_{LF} if $B = c^{1/2}(\beta-\alpha)\beta_x > -\infty$ in $\{F, T^*, L\}$ and will be the first time that B tends to $-\infty$ in $\{F, T^*, L\}$ otherwise.

LEMMA 8. $T^{**} = T_{LF}$. *Moreover, on the curve* $x = F(t)$, $T^* \leq t \leq T_{LF}$

$$0 \leq \beta_+(F(t),t) \leq O\left(\frac{3^{1/2}|c'(0)|^{1/2}V_1^{3/2}}{2^{3/2}c^2(0)(t-3T_1)^{3/2}}\right) \tag{49}$$

and

$$(50)\quad B_+(F(t),t) \geq -O\left(\frac{3^{5/2}|c'(0)|^{3/2}V_1^{3/2}}{2^{5/2}c^3(0)} \frac{1}{\underline{p}(t-3T_1)^{5/2}}\right).^{\dagger}$$

PROOF. The identity (2.11) and the fact that β is locally constant on back characteristics imply that for any point $(x,t) \in \{F, T^*, L\}_{int} \cap \{T^* \leq t < T^{**}\}$

$$(51)\qquad 0 \leq \beta(x,t) \leq \beta_-(R(t_R),t_R) + M(x,t).$$

Here, $\xi = b(\eta)$, $\eta < t$, is the back characteristic defined by

$$\frac{db}{d\eta} = -c(q)(b(\eta),\eta), \quad \eta < t, \quad \text{and} \quad b(t) = x;$$

the numbers $\frac{t}{3} < t_R < t_L$ are respectively the solutions of

$$(52)\qquad b(t_L) = L(t_L) \quad \text{and} \quad b(t_R) = R(t_R);$$

and $M(x,t)$ satisfies

$$(53)\qquad M(x,t) \leq k_{13}(\Sigma(\alpha_- - \alpha_+))^3$$

where the sum in (4.53) extends over all forward shocks other than $R(\cdot)$ cut by the back characteristic $\xi = b(\eta)$, $\eta < t$.

To obtain the desired estimate we must examine $\Sigma(\alpha_- - \alpha_+)$ more closely. The fact that $A(0,t) \leq 0$, $T_3 \leq t \leq T^*$, and the evolution equation

$$A_t + c(\beta-\alpha)A_x + p(\beta-\alpha)A^2 = 0 \quad \text{and} \quad p(q) = |c'(q)|/c^{1/2}(q)$$

---

† The bounds (4.49) and (4.50) are to be interpreted as twice the bounds (4.27) and (4.28).

together imply that

(54) $A(x,t) \leq 0, \quad (x,t) \in \{F,T^*,L\}_{int} \cap \{T^* \leq t < T^{**}\}.$

This tells us that α decreases along $\xi = b(\eta)$, $\bar{t} \leq \eta \leq t$, where

(55) $$\bar{t} = \max\,(t_L\,,\ T^*)$$

which in turn yields

(56) $$0 < \sum_{\{F,T^*,L\}_{int} \cap \{\bar{t} \leq \eta \leq t\}} (\alpha_- - \alpha_+) \leq |\alpha(b(\bar{t}),\bar{t}|.$$

The fact that $\alpha(0,t)$ satisfies (4.29) on $[T_3, T^*]$ and the hypotheses that no back waves break in $\{F, T^*, L\}_{int} \cap \{T^* \leq t < T^{**}\}$ imply that the conclusion of Lemma 6 is valid there and hence that

(57) $$|\alpha(b(\bar{t}),\bar{t})| \leq \frac{k_6}{\bar{t}^{3/4}}.$$

The inequalities $\bar{t} \geq \max(T^*, t/3)$ and $T^* \geq T_3 + k_1/|q_1|^4$ imply that $\frac{1}{\bar{t}^{3/4}} \leq \frac{3^{1/2}}{k_1^{1/4}} \frac{|q_1|}{t^{1/2}}$ and this and (4.56) and (4.57) imply

(58) $$0 < \sum_{\{F,T^*,L\}_{int} \cap \{\bar{t} \leq \eta \leq t\}} (\alpha_- - \alpha_+) \leq \frac{k_{14}|q_1|}{t^{1/2}}.$$

The only other term in the sum is the contribution from the curve $x = L(\cdot)$. A bound for this term is obtained from (4.16) and the result is an expression of the same form as (4.58). These results, when combined with (4.11), (4.53) and $t_R > t/3$ imply that for points

$(x,t) \in \{F,\ T^*,\ L\}_{int} \cap \{T^* \le t < T^{**}\}$

$$0 \le \beta(x,t) \le O\left(\frac{3^{1/2}|c'(0)|^{1/2}V_1^{\ 3/2}}{2^{5/2}c^2(0)(t-3T_1)^{3/2}}\right) + \frac{k_{15}|q_1|^3}{t^{3/2}} .$$

Since $V_1 = O(c(0)|q_1|T_1)$ this is the desired estimate if $k_{15}|q_1|^{3/2}$ is suitably restricted.

We now seek an estimate for B. It was demonstrated in [4, pp. 994, 995] that across any forward shock the values of A and B were related by the identity

(59) $B_- = e_1(\alpha_-,\alpha_+)B_+ - e_2(\alpha_-,\alpha_+)A_+ - e_3(\alpha_-,\alpha_+)A_-$

where the e_i's are positive and satisfy

(60) $|e_1 - 1|,\ e_2,$ and $e_3 \le k_{16}(\alpha_- - \alpha_+)^3$.

The inequality (4.60) and (4.54) imply that across any forward shock in $\{F,\ T^*,\ L\}_{int} \cap \{T^* \le t < T^{**}\}$ the following inequality prevails:

$(59)_{int}$ $B_- \ge e_1(\alpha_-,\alpha_+)B_+$.

If we combine $(4.59)_{int}$ with the evolution equation

$$B_t - c(\beta-\alpha)B_x + p(\beta-\alpha)B^2 = 0 \quad \text{and} \quad p(q) = |c'(q)|/c^{1/2}(q)$$

we obtain the lower bound

(61) $$B(x,t) \ge \frac{(1+M^*(x,t))B_-(L(t_L),t_L)}{1+(1+M^*(x,t))B_-(L(t_L),t_L)\displaystyle\int_{t_L}^{t} p(q(b(s),s))\,ds}$$

provided

$$1 + (1+M^*(x,t))B_-(L(t_L),t_L)\int_{t_L}^{t} p(q(b(s),s))ds > 0.$$

Again, $\xi = b(\eta)$, $\eta < t$, is back characteristic through the point (x,t), t_L is defined in (4.52), and $M^*(x,t)$ satisfies

$$(62) \qquad M^*(x,t) \le k_{17}[\Sigma(\alpha_- - \alpha_+)]^3 \le \frac{k_{17}|q_1|^3}{t^{3/2}},$$

where the index set for the summation is $\{F, T^*, L\}_{int} \cap \{\bar{t} \le \eta \le t\}$.

The assertion that $T^{**} = T_{LF}$ and (4.50) now follow from (4.59), the observation that (4.13), (4.16) and (4.60) imply

$$(63) \quad e_2(\alpha_-, \alpha_+)A_+(L(t_L),t_L) \le \frac{k_{18}|q_1|^3}{(t-T_3)^{3/2}} \cdot \frac{1}{\underline{p}(t-T_1)},$$

and (4.14) and $t > t_L > t_R > t/3$.

We shall also require an upper bound for the number T_{LF}. The result is:

LEMMA 9. $T_{LF} \le T_3 + \dfrac{k_{19}(T^*-T_3)^2}{|q_1|^2}$.

PROOF. The definition of the curve, $x = F(t)$, $t \ge T^*$, and the inequality $0 \le q_-(F(t),t) \le q_+(F(t),t)$ imply that F satisfies

$$\frac{dF}{dt} \ge c(q_+)(F(t),t) \ge c(0) - |c'|_{max}q_+(F(t),t)$$

where $|c'|_{max}$ is defined in (4.37). But the estimate (4.49) and the fact that α satisfies the conclusion of Lemma 6 in $\{F, T^*, L\}$ imply that for $|q_1|$ small

$$0 \leq q_+(F(t),t) \leq \frac{2k_6}{t^{3/4}}$$

where k_6 is the constant of Lemma 6. Thus we have

$$(64) \quad F(t) \geq c(0)(t-T^*) - 8k_6|c'|_{max}(t^{1/4}-(T^*)^{1/4}).$$

The fact that $L(\cdot)$ satisfies (4.4) and (4.64) now yield the desired result. □

We shall now conclude this section. As was mentioned earlier, the proof of Lemma 4 guarantees that the solution of P_2 satisfies (4.27)-(4.30) on $T_3 \leq t \leq T^*$. A closer examination of this proof yields the result that (4.27)-(4.30) are valid in the triangle

$$(65) \quad \Delta = \left\{ 0 \leq x \leq \frac{c(0)}{2}(T^*-T_3) - c(0)(t-T^*), \quad T^* \leq t \leq T^* + \frac{(T^*-T_3)}{2} \right\}.$$

Since q is nonnegative in $\mathcal{L}_{T_3}$, all wave speeds s satisfy

$$(66) \quad |s| \leq c(0).$$

This fact and the observation that (4.27)-(4.30) are valid in Δ imply that the earliest $A(0,t) = B(0,t)$ could be positive on $x = 0$ is bounded from below by $T^* + \frac{(T^*-T_3)}{2}$. Moreover, (4.64) and (4.66) imply that the earliest any forward wave from $(0,t')$ with $t' \geq T^* + \frac{(T^*-T_3)}{2}$ could intersect $x = F(\cdot)$ is bounded from below by

(67) $$T_I = \left(\frac{c(0)}{16k_6}\right)^4 (T^*-T_3)^4 ,$$

and T_I is greater that T_{LF} for $|q_1|$ sufficiently small.

It is now a simple matter to show that subject to the constraint that $T_I > T_{LF}$:

(68) $$A_-(F(t),t) \leq 0, \quad T^* \leq t \leq T_{LF}$$

and for points $(x,t) \in L_{T_3} \cap \{T^* \leq t \leq T_{LF}\}$

(69) $$0 \leq \beta(x,t) \quad \text{and} \quad -\alpha(0,t) \leq O\left(\frac{3^{1/2}|c'(0)|^{1/2}V_1^{3/2}}{2^{3/2}c^2(0)(t-3T_1)^{3/2}}\right),$$

and

(70) $$\begin{cases} B(x,t) \text{ and } A(0,t) \leq \\ -O\left(\frac{3^{5/2}|c'(0)|^{3/2}V_1^{3/2}}{2^{5/2}c^3(0)} \cdot \frac{1}{\underline{p}(t-3T_1)^{5/2}}\right) . \end{cases}$$

The formula (4.31) and Lemma 9 also imply that at $t = T_{LF}$, A is bounded from below on $0 \leq x \leq L(T_{LF})$ and that $x = L(T_{LF})$ is the only forward shock in this interval.

One now boot straps to obtain the desired result that no back waves break in L_{T_3} and that (4.69) is valid for $t \geq T_{LF}$. It is clear that if we can establish that B satisfies (4.70) for times $t \geq T_{LF}$, then $A(0,t)$ will do the same. In addition, (4.4), (4.31), and (4.46) guarantee that if $A(0,t)$ satisfies (4.70), then $A(x,t)$ will be bounded from below in $L_{T_3} \cap \{t \geq T_{LF}\}$ and hence $x = L(t)$, $t \geq T_{LF}$, will be the only shock in this region. The estimates (4.69) for α and β will then follow trivially from the characteristic relations for β, and (2.11) and (4.16).

To show that (4.70) persists, it suffices to show that if no back waves have broken in $T_{LF} \leq t < T'$ then $B_-(L(t),t)$ and $B_+(L(t),t)$ differ by a term of the form $k|q_1|^3/t^{5/2}$. This will guarantee that $T' = +\infty$. That this is so when $A_-(L(t),t) \leq 0$ follows from (4.59) and (4.63). To see that the same result is valid when $A_-(L(t),t) \geq 0$ we observe that (4.46) implies that $A_-(L(t),t)$ satisfies

$$(71) \quad A_-(L(t),t) \leq \frac{1}{\underline{p}\left(t - \frac{2k_3|q_1|}{c(0)}(T_3+(t-T_3))^{1/2}\right)}$$

where k_3 is the constant of (4.4). This is the upper bound that would obtain if there were a forward center at $(0,\overline{\eta}(t))$ where $\overline{\eta}(t)$ is defined (4.42). The formulas (4.16), (4.59) and (4.71) now yield the desired result.

REFERENCES

1. GREENBERG, J. M., *On the interaction of shocks and simple waves of the same family*, Arch. Rational Mech. Anal. 37 (1970), 136-160.

2. GREENBERG, J. M., *On the interaction of shocks and simple waves of the same family, II*, ibid. 51 (1973), 209-217.

3. GLIMM, J., *Solutions in the large for nonlinear hyperbolic systems of equations*, Comm. Pure Appl. Math., 18 (1975), 697-715.

4. GREENBERG, J. M., *Estimates for fully developed shock solutions to the equation*

$$\frac{\partial u}{\partial t} - \frac{\partial v}{\partial x} = 0 \quad \textit{and} \quad \frac{\partial v}{\partial t} - \frac{\partial \sigma(u)}{\partial x} = 0,$$

Indiana Univ. Math. Jour. 22 (1973), 989-1003.

5. DIPERNA, R. J., *Decay and asymptotic behavior of solutions to nonlinear hyperbolic systems of conservations laws*, to appear.

Figures 1 - 3 are on the following three pages.

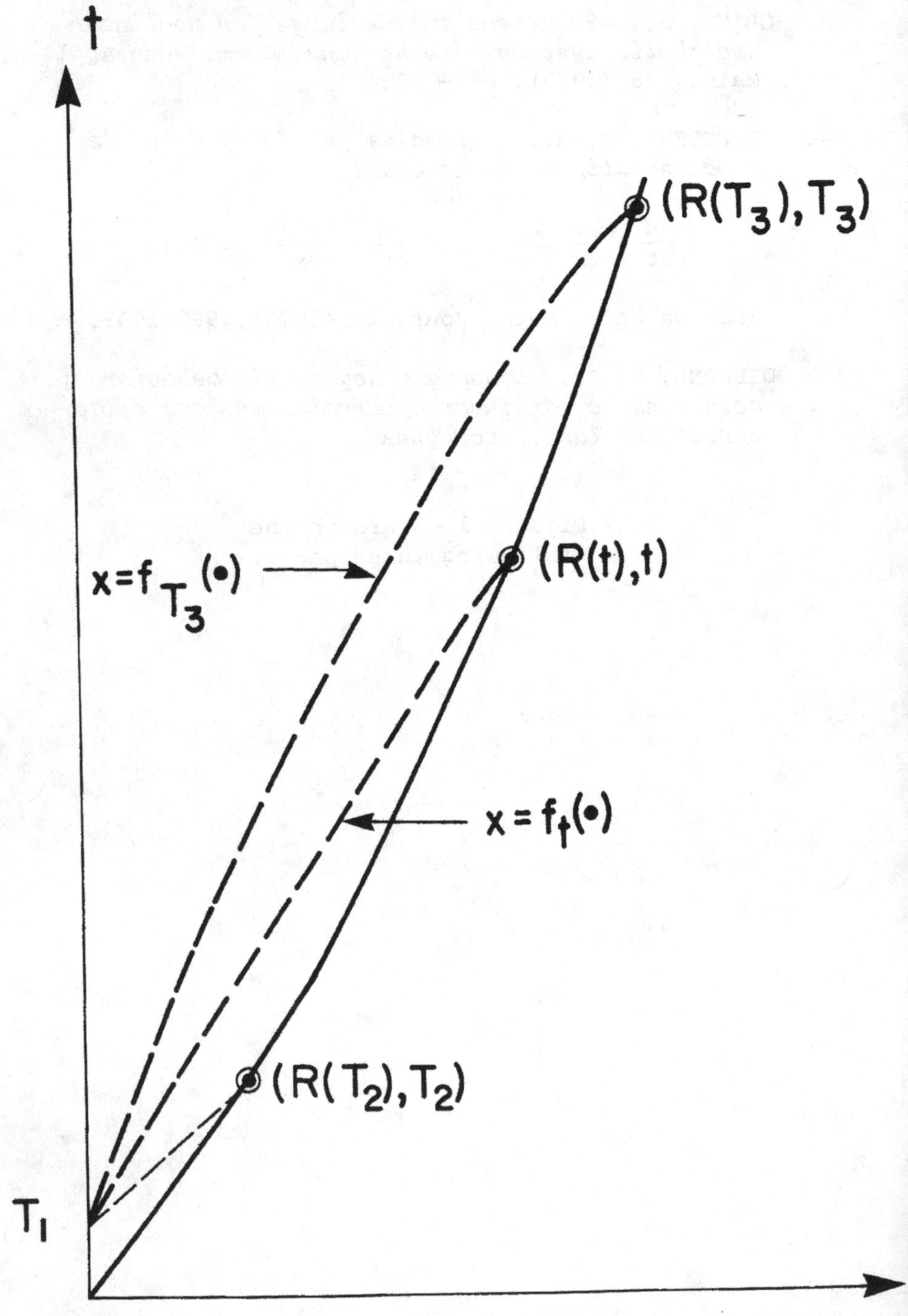

fig. 1

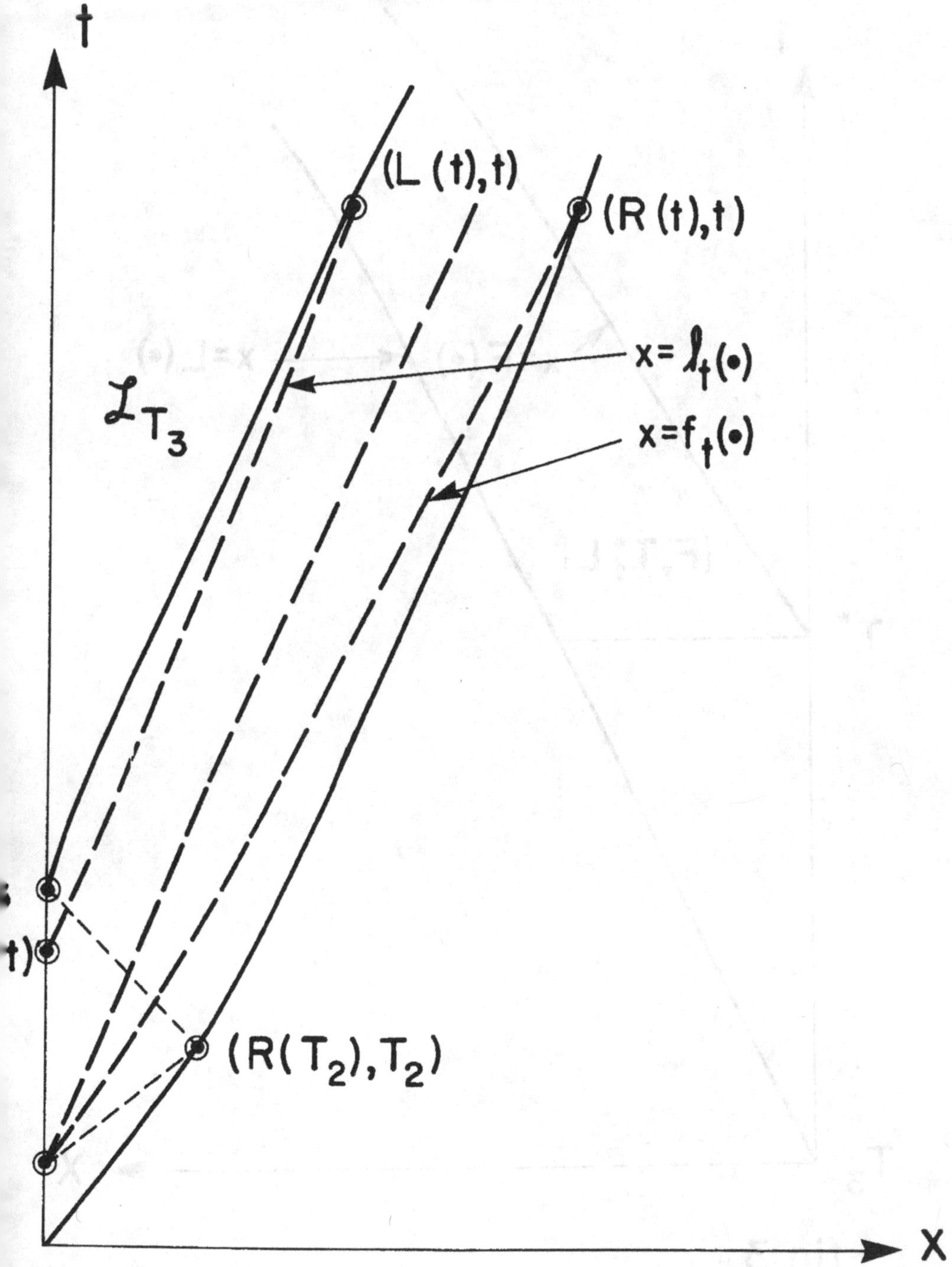

fig. 2

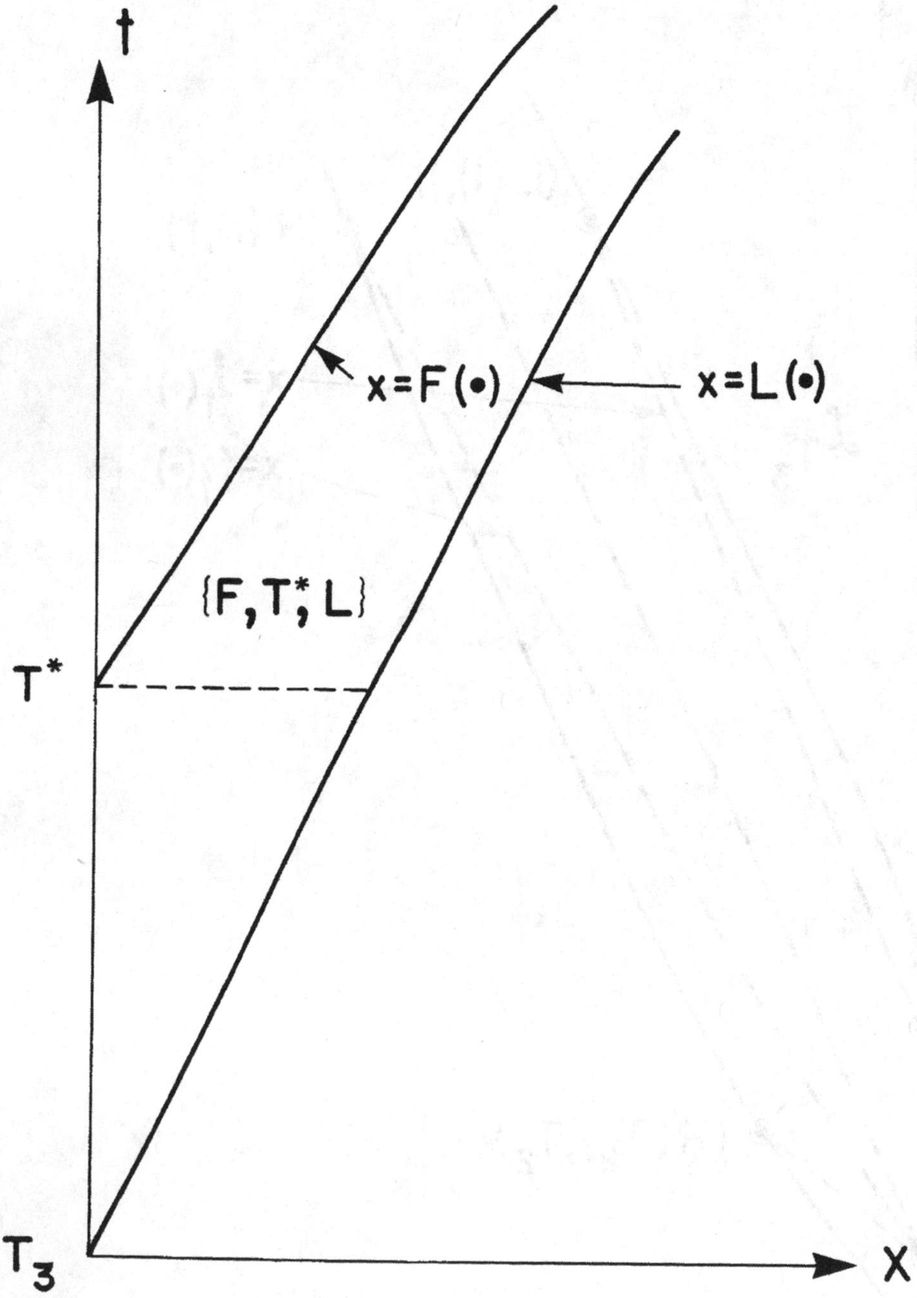

fig. 3

INVERSE PROBLEMS FOR NONLINEAR RANDOM SYSTEMS

by

F. ALBERTO GRÜNBAUM

Department of Applied Mathematics
California Institute of Technology
Pasadena, California 91109

(Present address:

Department of Mathematics
University of California
Berkeley, California 94720)

The purpose of this paper is to discuss some problems and methods that arise in the analysis of nonlinear random systems.

We begin by mentioning a problem of obvious practical importance.

Consider $x(t)$ to be the response of a nonlinear system when excited with white noise. For concreteness, take $x(t)$ to be the solution of the second order differential equation

$$\ddot{x}(t) + F(x(t), x(t)) = \text{white noise.}$$

The problem at hand is that of getting as much information as possible regarding the unknown function F from the distribution of the observable process $x(t)$. The problem posed in this generality is beyond the reach of our methods.

The process $x(t)$, though, can be expressed as a non-instantaneous nonlinear transformation of white noise. In this paper we will be concerned mainly with instantaneous transformations of random processes and some of the inverse problems associated with them. The techniques discussed here should be looked at as tools for dealing with the more complex and realistic situations mentioned above.

The paper is divided into four sections. The first one deals with the study of the distribution of cubic forms in Gaussian variables, and is independent of the rest. The second one contains some of the methods that one uses over and over again in dealing with nonlinear transformations of Gaussian processes. The third and fourth sections contain different applications of some of those techniques as well as some related results.

1. CUBIC FORMS IN GAUSSIAN VARIABLES

While it is a simple exercise to compute the distribution function for either a linear or a quadratic function of n Gaussian variables, this computation becomes impossible for forms of higher degree. The results in the linear and quadratic cases can be best summed up as follows: If $P(X_1,\dots,X_n)$ and $Q(X_1,\dots,X_n)$ have the same distribution, then there exists an orthogonal transformation O in $\mathbb{R}^n$ such that

$$P = Q \circ O. \tag{1}$$

Here we give some partial results for the case of cubic homogeneous polynomials.

First we look at the case $n = 2$ which is well understood, see [10]. It is very easy to see that the polynomials

$$P_1 = x^3-3xy^2 \;, \; P_2 = x^3+xy^2$$

have the same distribution - if x and y are independent $N(0,1)$ variables - while they are not equivalent under an orthogonal transformation as in (1).

To get a better picture one should notice that P_1 and P_2 belong to different irreducible representations of $O(2)$, the group of orthogonal transformations in $\mathbb{R}^2$. An arbitrary homogeneous cubic polynomial in x and y is a linear combination of P_1 and P_2. The main result is

THEOREM: *Let* P *and* Q *be two homogeneous cubic polynomials. If their distribution functions coincide, then either they are equivalent under an appropriate rotation* $P = Q \circ O$, *or else one is equivalent to (a scalar multiple of)* P_1 *and the other one to (the same scalar multiple of)* P_2.

Thus (1) tells the whole story except when P and Q lie in different irreducible subspaces for the action of $O(2)$. It would be nice and simple if this were the picture for any number of variables. The next family of examples gives us adequate warning.

Consider the polynomials

$$Z = x_1(x_3^2 - c^2x_4^2) + 2cx_2x_3x_4$$

$$W = x_1(x_3^2 + c^2x_4^2) .$$

Bring in the decomposition

$$W = (x_1(x_3^2+c^2x_4^2) - \frac{1+c^2}{6} x_1r^2) + \frac{1+c}{6} x_1r^2$$

$$Z = (x_1(x_3^2-c^2x_4^2) + 2cx_2x_3x_4 - \frac{1-c}{6} x_1r^2) + \frac{1-c}{6} x_1r^2$$

into an harmonic cubic polynomial plus r^2 times a linear polynomial. Since we have linear parts of different length it is clear that (for $c \neq 0$) W and Z are not equivalent under a rotation. On the other hand their distribution is the same, if $x_1, \ldots, x_4$ are independent $N(0,1)$, for each c.

There is good evidence to indicate that the case $n \geq 4$ is particularly tricky to handle.

2. HERMITE POLYNOMIALS AND GRAPHS

In dealing with Gaussian variables it is advantageous to use Hermite polynomials instead of ordinary powers. Here we prove some results which serve to illustrate this point.

Define the Hermite polynomials

$$H_n(x,t) = (-t)^n e^{x^2/2t} D^n e^{-x^2/2t} .$$

When the "variance" t vanishes we recover the ordinary

powers, i.e. $H_n(x,0) = x^n$. These polynomials satisfy the relation

$$\sum_0^\infty \frac{\mu^n}{n!} H_n(x,t) = e^{\mu x - \frac{t\mu^2}{2}} .$$

Take X to be a Gaussian variable with mean m and variance R, i.e.

$$EX = m \ , \ E(X-m)^2 = R.$$

Then we have

$$E\left(e^{\mu X - \frac{t\mu^2}{2}}\right) = e^{-\frac{t\mu^2}{2}} \int_{-\infty}^{\infty} e^{\mu x} \frac{e^{-\frac{(x-m)^2}{2R}}}{\sqrt{2\pi R}} dx =$$

$$= e^{m\mu} e^{-\frac{1}{2}(t-R)\mu^2} = \sum \frac{\mu^n}{n!} H_n(m,t-R),$$

and on the other hand

$$E\left(e^{\mu X - \frac{t\mu^2}{2}}\right) = \sum \frac{\mu^n}{n!} E(H_n(X,t)).$$

Thus, we conclude

$$E(H_n(X,t)) = H_n(m,t-R). \qquad (2)$$

This formula can be extended to the multivariate case in a convenient fashion by defining the Hermite tensors

$$\mathcal{H}^{(n)}(x,T) = (-T)^n e^{\frac{1}{2}\langle T^{-1}x,x\rangle} \nabla^n e^{-\frac{1}{2}\langle T^{-1}x,x\rangle} .$$

Here T is an arbitrary symmetric operator, x a

vector in $\mathbb{R}^m$ and n a multi-index of length m.

If X denotes a vector Gaussian random variable, with mean vector m and correlation matrix R, the formula corresponding to (2) reads

$$EH^{(n)}(X,T) = H^{(n)}(m,T-R). \tag{3}$$

If one introduces a basis of eigenvectors of T, then $H^{(n)}(x,T)$ is a product of ordinary one dimensional Hermite polynomials in the components of x, with variance parameters given by the eigenvalues of T. In such a case (3) gives a formula for the computation of the expression

$$E(H_{n_1}(x_1,t_1)\, H_{n_2}(x_2,t_2) \ldots H_{n_p}(x_p,t_p)). \tag{4}$$

Such a formula is most useful when interpreted as a sum of contributions corresponding to a variety of <u>allowed graphs</u>. This is to be explained below.

RULES FOR THE COMPUTATION OF (4)

Corresponding to each variable X_i in (4) $(i=1,\ldots,p)$ draw a vertex. For each i draw n_i legs stemming out of vertex i. These legs are to be used to form "allowed graphs" which we now describe.

Legs can be connected in pairs either to legs from a different vertex to form an <u>edge</u>, or to a leg from the same vertex to form a <u>loop</u>. They can finally be left alone; this last kind of legs we call <u>external legs</u>. To complete the rules of the game we have to specify the contribution from each "allowed graph" described above.

For each external leg from vertex i write m_i, from each edge joining vertices i and j write R_{ij}, for each loop on vertex i write $R_{ii} - t_i$. Now multiply all of these factors together to get the contribution from the graph in question. The value of (4) is obtained by adding all the contributions corresponding to all possible allowed graphs.

For instance we have

$$E(H_2(x_1,t_1)H_2(x_2,t_2)) = m_1^2 m_2^2 + 4m_1 m_2 R_{12} + (R_{11}-t_1)m_2^2$$

$$+ (R_{22}-t_2)m_1^2 + (R_{11}-t_1)(R_{22}-t_2) + 2R_{12}^2$$

in perfect correspondence with the graphs

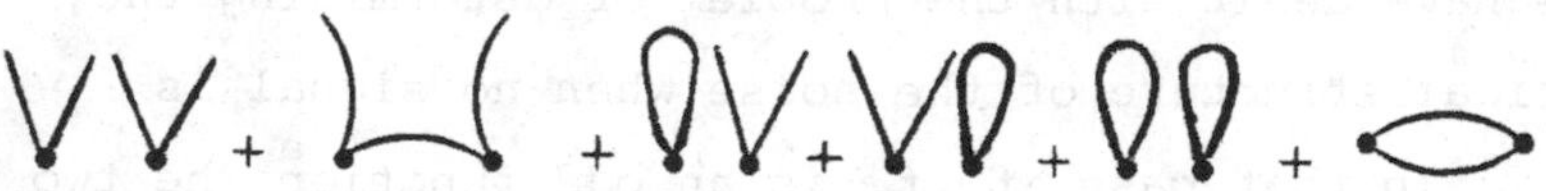

One should notice that some combinatorial factors appear due to the many ways in which some graphs can be formed.

3. THE IMPORTANCE OF BEING NOISY

A signal $s(t)$, corrupted by additive noise $n(t)$, is encoded by means of an instantaneous transformation f in such a way that the receiver gets the distorted message

$$y(t) = f(s(t) + n(t)). \tag{5}$$

The problem at hand is that of reconstructing the original signal $s(t)$ from sample averages of $y(t)$, if both the statistical structure of the noise and the function f are known.

The noise will be taken to be a nonstationary mean square continuous Gaussian process with mean zero and correlation function

$$R(t,s) = E(n(t)n(s)).$$

For convenience we assume that $R(t,t) \leq 1$ for all t, and then it is natural to take $f \in L^2(e^{-x^2/2})$.

Before stating any results, we give a brief review of what is known on a related problem. In [3], [8] and [9], we have dealt with the problem of determining the statistical structure of the noise when no signal is present. In that case if f is an odd function the two dimensional distribution of $f(n(t))$ allows one to determine R. The situation gets harder the more even the function f is, as the next example illustrates rather well.

THEOREM. *The correlation of the (centered) Gaussian process* $n(t)$ *can be read off from the* m-*dimensional distribution of the process*

$$x(t) = n^2(t) + an(t)$$

with

$$m = 2 \quad \textit{if} \quad a > a_0$$

$$m = 3 \quad \textit{if} \quad 0 < a \le a_0$$

$$m = \infty \quad \textit{if} \quad a = 0.$$

We have shown previously that the class of <u>non-constant even functions</u> f allowing for the reconstruction of R from the distribution of $f(n(t))$ includes for instance

a) any positive definite function f,

b) any function whose Hermite expansion has non-negative coefficients,

c) the characteristic function of an interval,

d) Dirac's delta,

and <u>finally</u>, any function f such that there exists another function h so that $g \equiv h \circ f$ belongs to one of the classes given above.

Going back to the problem of determining the signal $s(t)$ in (5) we will limit ourselves to the case $f(\xi) = \xi^2$, i.e. the problem is to recover $s(t)$ from

$$y(t) = (s(t) + n(t))^2. \tag{6}$$

Start with the observation that

$$Ey(t) = s^2(t) + R(t,t)$$

and since R is assumed to be known, $s^2(t)$ is readily available and the only problem is getting its sign straight. One should make here the obvious remark that if the noise were totally absent in (6) there would clearly be no chance of settling this problem.

Set

$$x(t) = s(t) + n(t)$$

to get a Gaussian process with mean $s(t)$ and variance R. To use the formulas in the previous section it is convenient to look at

$$\tilde{y}(t) = y(t) - R(t,t) = H_2(x(t), R(t,t)).$$

One has

$$E(\tilde{y}(t_1)\tilde{y}(t_2)) = s^2(t_1)s^2(t_2) + 2R^2(t_1,t_2) + 4s(t_1)s(t_2)R(t_1,t_2)$$

showing that one can read off the product $s(t_1)s(t_2)$ as long as $R(t_1,t_2) \neq 0$.

For the rest of the discussion we make the important <u>assumption</u>

$$R(t,t) \neq 0 \quad \text{for all} \quad t. \tag{7}$$

From the computation made above and the continuity of R it follows that

$$Q(t_1,t_2) \equiv s(t_1)s(t_2)$$

is known if t_1 and t_2 are close enough. Our aim is to show that $Q(t_1,t_2)$ is <u>determined everywhere</u> by the quantities

$$E(\tilde{y}(t_1) \ldots \tilde{y}(t_n)). \tag{8}$$

This would clearly show that <u>the signal</u> $s(t)$ <u>is determined from</u> (8) <u>up to a global sign</u>.

DETERMINING $Q(t,s)$

Take $t_1 < t_2 < \dots < t_n$ and assume that $Q(t_i,t_j)$ is known except for the pair (t_1,t_n). Two kinds of graphs enter in (8), connected and disconnected ones. The first class involves only R and thus can be ignored. Among those in the second class we can concentrate on those containing the product $Q(t_1,t_n)$. This class can be further subdivided and we can conclude that the new information contained in (8) - as we go from $n-1$ to n t_i's - comes from those graphs with one path going through all the vertices in an arbitrary order beginning at 1 and ending at n. These paths would become closed cycles if t_1 and t_n were connected to each other. For instance, for $n = 3$, we have

$$E(\tilde{y}(t_1)\tilde{y}(t_2)\tilde{y}(t_3))=s^2(t_1)s^2(t_2)s^2(t_3)+s^2(t_1)E(\tilde{y}(t_2)\tilde{y}(t_3))$$
$$+s^2(t_2)E(\tilde{y}(t_1)\tilde{y}(t_3))+s^2(t_3)E(\tilde{y}(t_1)\tilde{y}(t_2)) +$$
$$+R(t_1,t_2)R(t_2,t_3)s(t_1)s(t_3) + R(t_1,t_3)R(t_3,t_2)s(t_1)s(t_2)$$
$$+R(t_2,t_1)R(t_1,t_3)s(t_2)s(t_3)$$

and the new piece of information is

$$R(t_1,t_2)R(t_2,t_3)s(t_1)s(t_3).$$

We could find $s(t_1)s(t_3)$ from the 3-dimensional distribution of $\tilde{y}(t)$ if we knew that

$$R(t_1,t_2)R(t_2,t_3) \neq 0$$

in the same way that from the 2-dimensional distribution we had $s(t_1)s(t_3)$ if $R(t_1,t_3) \neq 0$.

Proceeding by induction one can show that if $t_1 < t_2 < \ldots < t_n$ and $Q(t_i,t_j)$ is already known except for $Q(t_1,t_n)$, then the n-dimensional distribution of $\tilde{y}(t)$ will give $Q(t_1,t_n)$ unless

$$R(t_1,t_2)R(t_2,t_3) \ldots R(t_{n-1},t_n) = 0. \qquad (9)$$

Now we can clinch the proof by observing that if (a,b) is fixed, and we take

$$a = t_1 < t_2 < \ldots < t_n = b$$

we cannot have (9) for every n, since R was continuous and we assumed a noisy channel, (7). Thus $Q(a,b) = s(a)s(b)$ can eventually be found from the quantities (8). Although we have used here information of arbitrarily large order, it is clear from the details of the proof, which can be adapted from [3], that if the noise is any realistic process we need only use third order information.

MISCELLANEOUS RESULTS

In this last section we mention briefly some problems which are either similar to those given above or whose solution depends on similar techniques.

4.a. TRANSLATION INVARIANTS

In [1] the following result is proved.

THEOREM. *If* $f \in L^1(\mathbb{R})$, *then its* n*-order correlation functions*

$$M^n(f)(t_1,\ldots,t_n) = \int_{-\infty}^{\infty} f(x+t_1)\ldots f(x+t_n)f(x)dx, \quad n=1,2,\ldots$$

suffice to determine f *up to a rigid translation* $f(x+c)$.

In terms of its Fourier transform $\hat{f}$, it is clear that the information given above is

$$\prod_1^n \hat{f}(\lambda_i) \quad , \quad \sum_1^n \lambda_i = 0.$$

It is worthwhile to remark that in [3] and in the previous section - if $n(t)$ were stationary - we were given

$$R(\xi_1-\xi_2)R(\xi_2-\xi_3) \ \ldots \ R(\xi_n-\xi_1)$$

<u>only as a symmetrized function</u> of its arguments. R is real valued and this prevents the symmetrization to cause any loss of information.

In [4] we have shown that in general no bound can be put on the order of the information required to get R. For the question of translation invariants a similar result is given in [2].

4.b. ORTHOGONAL INVARIANTS FOR VECTORS IN HILBERT SPACE

Consider a family x_i, $i \in I$, of vectors in a real Hilbert space H, and assume that the volume of the parallelogram spanned by any finite subset of them is known. In [6] we have shown that this information determines the family except for the obvious freedom of a sign for each vector and a common partial isometry of H.

If we were dealing with a continuous curve in H - which misses the origin - these volumes are enough to determine the curve up to a partial isometry in H, see [5].

Both of these results can be obtained by using arguments similar to those in Section 3.

4.c. THE SQUARE OF SHOT NOISE

Shot noise is usually taken as a model for the fluctuating part of the plate current in a vacuum tube due to the random emission of electrons from the cathode. The process is defined as

$$s_\lambda(t) = \int_{-\infty}^{\infty} f(t-s)dN_\lambda(s).$$

Here $N_\lambda(s)$ stands for the Poisson process with fixed rate λ and f is an arbitrary function which decays fast enough at infinity and represents the "current pulse" due to a single electron.

In [7] we have shown that if the only available

information about $s_\lambda(t)$ consists of measurements of its modulus, one can still recover f up to a global sign, provided one has the means to speed up the Poisson process. The tools involved in proving this assertion have some similarities with those in Sections 2 and 3, and are developed from scratch in [7].

4.d. IDENTIFICATION IN QUEUING SYSTEMS

In [11] the problem of identifying the interarrival and the service time distributions is cleverly reduced to that of identifying a function $\Psi(t)$ from the knowledge of the product

$$\Psi(t_1)\Psi(t_2)\Psi(t_3) \quad \text{if} \quad t_1 + t_2 + t_3 = 0.$$

While in general this is not enough to give $\Psi(t)$, it is plenty in this set-up. This is strongly related to the comment made at the end of Section 3.

4.e. THE PHASE PROBLEM IN CHRYSTALLOGRAPHY

A central problem in chrystallography is that of determining a 3-dimensional structure from the data of the modulus of the Fourier transform of its density. This is clearly equivalent to having

$$F(\lambda)F(-\lambda)$$

with $F = \hat{\rho}$. In the case of a discrete configuration this is equivalent to giving the set of mutual difference vectors. If one had not only this, but pairs of such differences to a common point, one would have

$$F(\lambda_1)F(\lambda_2)F(\lambda_3) \quad \text{if} \quad \lambda_1 + \lambda_2 + \lambda_3 = 0$$

as in Section 3. The reader could look at [13] and [14] for some interesting points, as well as to a forthcoming publication by J. Franklin. For a different line of research leading to the same mathematical problems look at [12] and [15].

REFERENCES

[1] ADLER, R. L. and KONHEIM, A. G., Proc. Amer. Math. Soc. 13 (1962), 425.

[2] CHAZAN, D. and WEISS, B., Information and Control 16 (1970), 378.

[3] GRÜNBAUM, F. A., Z. Wahrscheinlichkeitstheorie und Verw. Geb. 23 (1972), 121.

[4] GRÜNBAUM, F. A., Bull. Amer. Math. Soc. 78 (1972), 615.

[5] GRÜNBAUM, F. A., Proc. Amer. Math. Soc. 42 (1974), 268.

[6] GRÜNBAUM, F. A., Proc. Amer. Math. Soc. 43 (1974), 331.

[7] GRÜNBAUM, F. A., to appear in Z. Wahrscheinlichkeitstheorie und Verw. Geb.

[8] GRÜNBAUM, F. A., to appear in Advances in Mathematics.

[9] GRÜNBAUM, F. A., to appear in International Journal of Multivariate Analysis.

[10] GRÜNBAUM, F. A., submitted for publication.

[11] KENDALL, D. G. and LEWIS, T., Z. Wahrscheinlichkeitstheorie und Verw. Geb. 4 (1965), 144.

[12] KOTLARSKI, I., Pacific Journal of Math. 20 (1967), 69.

[13] PAULING, L. and SHAPPEL, M., Z. Kristall 75 (1930), 128.

[14] PATTERSON, A. L., Physical Review 65 (1944), 195.

[15] RAO, C. R., Sankhya Ser. A. 33 (1971), 255.

THE METHOD OF TRANSMUTATIONS

by

REUBEN HERSH

Department of Mathematics
and Statistics
University of New Mexico
Albuquerque, New Mexico 87131

1. INTRODUCTION

A standard mathematical strategy, when faced with a new problem, is to reduce it to a previously solved problem, or at least to a simpler problem. For example, to reduce a problem with a singular coefficient to one with regular coefficients; to reduce a problem containing a small parameter to one independent of the parameter; to transform a second-order equation into a first-order equation, or vice versa; to transform a Goursat problem into a Cauchy problem, or vice versa.

Usually it is not hard to verify the properties of such a transformation, once it has been found. The problem is to find the right transformation into an old problem. In this note we will show that there is a systematic method to find such a transformation. This method often makes the task of constructing the transformation no harder than the task of verifying its properties. We will also discuss the connection of our method with probability theory. It often happens that the transformation we seek can be expressed as the expected value of a suitable random variable; indeed, it has sometimes been in a probabilistic context that they have first been found. We start out by listing examples. Our method includes as particular examples many formulas scattered through the literature: we mention the classical reduction of the Euler-Poisson-Darboux equation to the wave equation, the "transmutation operators" of Delsarte and Lions, the "related equations" of Bragg and Dettman, the "diffusion transform" of S. Rosencrans, and various formulas of A. Weinstein, J. Donaldson, W. Roth, M. Kac and S. Kaplan.

We will not discuss the rigorous verification of our formulas, or give precise conditions for their validity, for this has been done in the references we cite, and on this score we have nothing new to say. Our purpose here is to show that these seemingly scattered formulas all can be obtained by a single technique, which provides a uniform heuristic approach. This technique, as will be seen, is simply an operational version of the methods of classical transform theory including the Fourier transform, Hankel transform, and Laplace transform.

The paper falls into four sections. In Section 2, following this introduction, we collect five examples of transmutation formulas. In the third section, we use these examples to explain our general method for constructing transmutation formulas. In the last section we comment briefly on the probabilistic aspect of these formulas, and also on the application of transmutation formulas to problems on regular and singular perturbations.

2. EXAMPLES

EXAMPLE 1. *The method of spherical means, and Darboux equation*

In the second volume of Courant-Hilbert, Ch. VI, par. 13, it is shown, using spherical means and fractional differentiation, that if $u(x,t)$ and $v(x,t)$ are connected by the formula

$$v(x,t) = \int_{-1}^{1} u(x,t\mu)(1-\mu^2)^{(n-3)/2}\, d\mu,$$

where $x = (x_1, \ldots, x_n)$, and if u satisfies the n-dimensional wave equation,

$$u_{tt} = \Delta u, \quad u(x,0) = f, \quad u_t(x,0) = 0$$

then v satisfies the Darboux equation,

$$v_{tt} + \frac{n-1}{t} v_t = \Delta v, \quad v(x,0) = f, \quad v_t(x,0) = 0.$$

The Darboux equation is singular at $t = 0$, so in this example we solve a singular equation in terms of a regular equation.

EXAMPLE 2. *The transmutation operators of Delsarte and Lions* [2,3]

Let $D = d/dt$. Let $L = D^2 + r(t)D + q(t)$, where $r(t)$ and $q(t)$ are given functions. Let A be an operator independent of t (usually a partial differential operator with space-dependent coefficients). Then a solution of $D^2u + Au = 0$ is transformed into a solution of $Lv + Av = 0$ by setting $v = Hu$, where the operator H satisfies $HD^2 = LH$. In fact, if $D^2u + Au = 0$ and $v = Hu$, then, assuming A commutes with H, we have

$$Lv + Av = (L+A)Hu = H(D^2+A)u = 0.$$

To construct the "transmutation operator" H, Delsarte and Lions seek a kernel $h(s,t)$, so that the action of H is given by $Hf = \int h(s,t)f(t)dt$. Substituting this representation of H into the equation $HD^2 = LH$, and integrating by parts, one easily derives a necessary and sufficient condition for $h(s,t)$: $h_{tt} = h_{ss} + r(s)h_s + q(s)h$, with boundary conditions determined by the side conditions of the two operators D^2 and L. Since $h(s,t)$ satisfies a second-order hyperbolic equation in two independent variables, it can be expressed in terms of the Riemann function of that equation. In particular, if we specialize

$$A = -\Delta, \quad r(t) = \frac{n-1}{t}, \quad q(t) \equiv 0,$$

then the Riemann function is expressible explicitly in elementary terms, and we recover the same formula as in Example 1.

EXAMPLE 3. *From a second-order to a first-order equation*

If $u(t)$ is a vector-valued function, satisfying the abstract Cauchy problem $\{u_{tt} = Au,\ u(0) = f,\ u_t(0) = 0\}$, then

$$v(t) = 1/\sqrt{\pi t}\int_0^\infty u(s)e^{-s^2/4t}\,ds$$

satisfies the abstract Cauchy problem $\{v_t = Av, v(0) = f\}$. This formula has been rediscovered repeatedly; see [5-9].

If $A = d^2/dx^2$, then $u = \frac{1}{2}(f(x+t) + f(x-t))$, and our formula reduces to the classical Poisson solution of the heat equation,

$$v_t = v_{xx}, \quad v(0) = f,$$

$$v = \frac{1}{2\sqrt{\pi t}} \int_{-\infty}^{\infty} e^{-s^2/4t} f(s+x)\,ds.$$

EXAMPLE 4. *From a Cauchy problem to a Goursat problem, and back*

In the thesis of W. J. Roth [4] it is shown that if $u(t)$ is a vector-valued function and A is a closed linear operator, and $\{u_{tt} = Au,\ u(0) = f,\ u_t(0) = 0\}$, then

$$v(r,s) = \frac{2}{\pi} \int_0^{\pi/2} u(2\sqrt{rs}\sin\theta)\,d\theta$$

satisfies the Goursat problem $\{v_{rs} = Av$ if $r > 0$, $s > 0$, $v(0,s) = v(r,0) = f\}$ and moreover we recover u from v by the formula

$$u(t) = \frac{d}{dt} \int_0^{\pi/2} t \sin\theta\, v\left(\frac{t\sin\theta}{2}, \frac{t\sin\theta}{2}\right) d\theta.$$

(By a modification of Duhamel's formula, one can generalize to the case where $v(r,0)$ and $v(0,s)$ are unequal and arbitrary.)

EXAMPLE 5. *From a first-order equation to a family of higher-order equations*

Suppose $u(t)$ is a vector-valued function for $-\infty < t < \infty$ and satisfies $\{u_t = Au,\ u(0) = f\}$, where A is closed and $f \in \mathcal{D}(A^\infty)$. Suppose moreover that $P(d/dt, d/dx)$ is a hyperbolic differential operator, or is parabolic of positive genus. (Its coefficients may be t-dependent.) Let $\hat{g}_k(t,x)$ be a fundamental solution of $P(d/dt, -d/dx)\hat{g}_k = 0$, all of whose Cauchy data vanish except for the k'th; let $(d/dt)^k \hat{g}_k(0,x) = \delta(x)$. Then it is easily verified (see [10]) that

$$v(t) = \int_{-\infty}^{\infty} u(s)\hat{g}_k(t,s)\,ds$$

is a solution of

$$\{P(d/dt,A)v = 0,\ \left(\frac{d}{dt}\right)^k v(0) = f,\ \left(\frac{d}{dt}\right)^j v(0) = 0 \text{ if } j \neq k\}.$$

If $P(d/dt,A) = d/dt - A^2$, then one has $\hat{g}_0 = (1/2\sqrt{\pi t})e^{-s^2/4t}$, almost as in Example 3 above. If $P(d/dt,A) = d^2/dt^2 - A^2$, one has $\hat{g}_0(t,s) = \frac{1}{2}(\delta(s+t) + \delta(s-t))$,

$$\hat{g}_1(t,s) = \begin{cases} 1/2 & |s| \leq t \\ 0 & |s| > t \end{cases}$$

so that $\{v_{tt} = A^2 v,\ v(0) = v_0,\ v(0) = v_1\}$ is solved by

$$v(t) = \int_{-\infty}^{\infty} [u_0(s)g_0(t,s) + u_1(s)g_1(t,s)]ds$$

$$= \frac{1}{2}[u_0(t) + u_0(-t) + \int_{-t}^{t} u_1(s)\,ds]$$

where

$$\frac{du_0}{dt} = Au_0 \,, \qquad u_0(0) = v_0$$

$$\frac{du_1}{dt} = Au_1 \,, \qquad u_1(0) = v_1 \,.$$

3. A SYSTEMATIC APPROACH TO DISCOVERY OF TRANSMUTATION FORMULAS

In the references cited for the five examples given above (and in many other such works) there is provided rigorous proof that the function $v(t)$, expressed in terms of $u(t)$ by the given transmutation formula, does indeed satisfy the conditions on v. What is often lacking is a clue to explain how such formulas may be discovered. Or in some cases, as in Example 1, a derivation is given which is quite special, leaving no clue how to proceed to relate a different pair of problems for $u(t)$ and $v(t)$.

Our purpose here is to describe a uniform approach to <u>discovering</u> formulas such as those of Examples 1-5; the task of verifying the formula, once it is written down, is often straightforward. On this score, in any case, we have nothing to add to the cited references.

In general, a transmutation formula can be regarded in the following light:

We have two problems involving some operator A. The solution to the first problem, $u(t)$, we regard as a function of A, depending on t as a parameter: $u = u(t,A)$. Similarly, in the second problem, the solution v depends on A, as well as on a parameter

$s: v = v(s,A)$. Then, to transmute u to v -- to represent v in terms of u -- we must find a representation of the function $v(s,\cdot)$ in terms of the one-parameter family of functions $u(t,\cdot)$:

$$v(s,\lambda) = \int h(s,t)u(t,\lambda)dt$$

where the kernel $h(s,t)$ may be a genuine function or a distribution, and where λ is ultimately to be replaced by the given operator A. Let us work through a couple of our examples.

In Example 1, let us regard the operator Δ as a symbol on which we can operate as if it were a complex number--i.e., we assume we have a functional calculus for Δ. Then we can express $u(t)$ symbolically,

$$u(t) = \cos(t\sqrt{-\Delta})f.$$

To find a comparable expression for v, we can reduce the equation $v_{tt} + \frac{n-1}{t}v_t - \Delta v = 0$ to a formal "Bessel equation" of order $((n/2) - 1)$ by the "substitution"

$$w(z) = t^{((n/2)-1)} v(t), \quad z = t^{((n/2-1)} \sqrt{-\Delta}.$$

Therefore we have

$$v = ct^{1-(n/2)} J_{(n/2)-1} (t^{(n/2)-1} \sqrt{-\Delta})f$$

where $J_\nu(z)$ is as usual the Bessel function of order ν of the first kind. So far we have two formal expressions for u and v; the problem is to give a meaning to these expressions. The essence of the transmutation method is to use these two formal expressions to relate v to u; then v(t) is known if u(t) is known.

The problem of expressing v(t) in terms of u(t) is evidently the problem of expressing a Bessel function in terms of cosine; this, however, is well-known; we have the standard formula

$$J_\nu(z) = \frac{(z/2)^\nu}{\pi^{1/2}\Gamma(\nu+\frac{1}{2})}\int_0^1 (1-s^2)^{\nu-\frac{1}{2}} \cos(zs)\,ds$$

(see [11], formula 3.7).

Now, if we replace z by $t^{(n/2)-1}\sqrt{-\Delta}$ and $\cos(st^{(n/2)-1}\sqrt{-\Delta})$ by $u(st^{(n/2)-1})$ we get the formula cited in Example 1. This approach to the Euler-Poisson-Darboux equation is similar to that used by J. Donaldson [12].

Example 4 is closely related to the first example.

As Roth [4] points out, his formula for transforming v(r,s), the solution for an interior Goursat problem, into u(t), the solution of a Cauchy problem, can be obtained by formally "solving"

$$\{u_{tt} = Au, \ u(0) = f, \ u_t(0)\}$$

$$\text{by} \quad u = \cos(t\sqrt{A})f$$

and "solving"

$$\{v_{rs} = Av, \ v(0,s) = v(r,0) = f\}$$

$$\text{by} \quad v = J_0(2\sqrt{rs}\,A)f.$$

To express u and v in terms of each other, again we resort to classical identities from the theory of Bessel functions. That v can be expressed in terms of u is clear from the completeness of the cosine functions; essentially, one is seeking the cosine transform of J_0; and conversely, to express u in terms of v amounts to asking for a Hankel transform of $\cos \xi$. We obtain the representations given in Example 4 above by using the identities

$$J_0(\xi) = \frac{2}{\pi}\int_0^{\pi/2} \cos(\xi \sin \theta)\, d\theta,$$

$$\cos \xi = \frac{d}{d\xi}\int_0^{\pi/2} \xi \sin \theta \; J_0(\xi \sin \theta)\, d\theta.$$

If in these formulas we replace cos by u and J_0 by v, according to the formal "solutions" written above, we obtain the transmutations from u to v and back; the questionable symbolic expressions for functions of A drop out of our formulas.

These two examples may be enough to make clear the general pattern:

Given two problems involving an operator A, let $u(t,A)$ be the solution of Problem I, and let $v(t,A)$ be the solution of Problem II.

Let $u(s,\lambda)$ be the solution of Problem I in the special case where A is multiplication by a real or complex number λ. Similarly, define $v(t,\lambda)$. To construct a transmutation to represent $v(t,A)$ in terms of $u(s,A)$, it is enough to consider this special case where $A = \lambda$. For if we can solve I and II in this special case, and if the complex-valued functions $u(s,\lambda)$ and $v(t,\lambda)$, are related by a kernel,

$$v(t,\lambda) = \int_\Gamma g(s,t)u(s,\lambda)ds$$

where Γ is some curve, usually an interval on the real axis, then, if $u(s,A)$ exists for $s \in \Gamma$, the candidate for $v(t,A)$, the solution of Problem II, is evidently

$$\int_\Gamma g(s,t)u(s,A)ds.$$

The integral is a Bochner integral if the kernel $g(s,t)$ is a classical function. If $g(s,t)$ is a generalized function or "distribution", as in Example 5 above, the integral is "symbolic" and may be interpreted by a formal integration by parts.

It should be emphasized that the transmutation method is applicable even if neither Problem I nor Problem II is "well-posed".

The existence of a transmutation operator implies that the class of admissible data for Problem II includes the admissible data for Problem I. If this second class is a complete linear space--i.e., if Problem I is well-posed--then so is Problem II. On the other hand, the method retains its validity even if the data are highly restricted. For instance, by choosing $A = -\Delta$, we see from Example 3 that the admissible data for the backward heat equation include the admissible data for Cauchy's problem for the Laplace equation.

We have in this instance a transmutation of one improperly posed problem to a second improperly posed problem.

Let us go on to show how our method yields the formula of Example 3. Problem I is

$$\{u_{tt} = Au, \quad u(0) = f, \quad u_t(0) = 0\}.$$

Problem II is

$$\{v_t = Au, \quad v(0) = f\}.$$

Again we have $u = \cos t\sqrt{-A}\, f$, and evidently $v = e^{tA}f$, at least if $A = \lambda$ is a complex number.

Now, in the Fourier transform formula

$$e^{-\lambda^2 t} = \frac{1}{\sqrt{\pi t}} \int_0^\infty e^{\frac{-s^2}{4t}} \cos s\lambda \, ds$$

substitute $\lambda = \sqrt{-A}$, and we obtain the formula of Example 3.

A slightly different formula is given by Bragg and Dettman [6]; they choose the data differently in Problem I, setting $u(0) = 0$, $u_t(0) = f$.

As in the other examples, the verification is straightforward.

Example 5 is similar. We now have $u(t,s) = e^{tA}f$. For the sake of simplicity, assume $P(\tau,A)$ has simple roots $\tau_j(A)$; then

$$v(t,A) = \sum_j c_j e^{t\tau_j(A)} f$$

where

$$\sum_j \tau_j^k c_j = \delta_{j,k}.$$

Then

$$v(t,i\lambda) = \sum_j c_j e^{\tau_j(i\lambda)} f = \int e^{is\lambda} g(t,s)\, ds$$

where $g(t,s)$ is the Fourier transform of $\sum c_j e^{\tau_j(i\lambda)} f$. The existence of $g(t,s)$ as a Schwartz distribution follows from the assumed hyperbolicity of P, and substituting A for $i\lambda$, we get the desired formula.

Finally, we look at Example 2, the problem of Delsarte-Lions. Let $v(t,\lambda)$ be the solution of

$$\{Lv + \lambda^2 v = v_{tt} + r(t)v_t + q(t)v + \lambda^2 v = 0$$
$$v(0,\lambda) = 1, \quad v_t(0,\lambda) = 0\}.$$

Let $u(t,\lambda)$ be the solution of

$$\{D^2 u + \lambda^2 u = 0, \quad u(0,\lambda) = 1, \quad u_t(0,\lambda) = 0\}.$$

Then $u(t,\lambda) = \cos t\lambda$, and we have

$$v(t,\lambda) = \int h(t,s) u(s,\lambda) ds$$

where $h(t,s)$ is the Fourier cosine transform of $v(t,\lambda)$.

Since $Lv + \lambda^2 v = 0$, it follows that

$$Lh - h_{ss} = 0.$$

Define an operator H by

$$Hf = \int h(t,s) f(s) ds.$$

Then, from $Lh = h_{ss}$, it follows, on using two integrations by parts on the left, that

$$LH = HD^2.$$

This is the formula by which Lions and Delsarte define their transmutation operator H; thus we see that by our method we can recover their procedure as a special case.

4. PROBABILISTIC INTERPRETATIONS; EQUATIONS DEPENDING ON A SMALL PARAMETER

In some of the principal examples of transmutations, it is possible to rewrite the formula as $v(t) = E\{u(\tau)\}$ where $\tau(t)$ is, for each t, a random time, distributed according to some appropriate probability law, and E is the expected-value operator.

For instance, in Example 3, the appropriate τ is distributed as a Gaussian normal random variable, with mean 0 and variance t. Such a random variable has as its density function $\theta_t(s) = \frac{1}{2\sqrt{\pi t}} e^{-s^2/4t}$, and so, by the familiar elementary formula in probability for the expectation of a function of a random variable,

$$v(t) = E[u(\tau)] = \int_{-\infty}^{\infty} u(s)\theta_t(s)ds.$$

In fact, it was in this probabilistic representation that the formula in Example 3 arose in [7], in connection with a limit theorem on random evolutions.

An earlier example is due to M. Kac [13]. He found that if

$$\tau = \int_0^t (-1)^{N(s)} ds,$$

where $N(s)$ is a Poisson process with intensity a, and if u satisfies

$$\{u_{tt} = \Delta u, \quad u(0) = f, \quad u_t(0) = 0\}$$

and $v(t) = E\{u(\tau)\}$, then v satisfies

$$\{v_{tt} + 2a\, v_t = \Delta v,\ v(0) = f,\ v_t(0) = 0\}.$$

This formula was generalized by Kaplan [14] to the case where $a = a(t)$, a given function of t; Kaplan points out that by use of the density function as a kernel, $v(t)$ can be written in terms of $u(t)$ as an integral over the real axis, and in that form the Kac-Kaplan result is a special case of the Delsarte transmutation method.

In the work on random evolutions [7], these results were extended to more general operators and random variables. A systematic probabilistic approach which unifies the Kac-Kaplan and Hersh-Griego examples, and gives some new and more general results by systematic use of Ito's Lemma, was given by S. Rosencrans, in his work on the diffusion transform [15].

From the viewpoint of the present paper, the probabilistic representations comprise just those transmutations where the kernel $g(s,t)$ is, for each t, the derivative with respect to s of a finite measure. (g could be a delta function, if the probability measure has a singular part.) Any such kernel $g(s,t)$ can be regarded as the density of a suitably constructed random time $\tau(t)$.

Thus Rosencrans was able to give a probabilistic representation to the formulas of our Example 1; the solution of the Euler-Poisson-Darboux equation is obtained as the mean of solutions of the wave equation, evaluated at a certain random time.

A probabilistic representation makes it possible to use probabilistic limit theorems (laws of large numbers, central limit theorems) to prove asymptotic estimates for solutions of such equations as $\varepsilon u_{tt} + au_t = Au$; see [18].

However, for such purposes the method of transmutations itself provides a convenient and powerful approach, quite aside from its probabilistic connections. Given an equation in an operator A and a small parameter ε,

$$P_\varepsilon(d/dt,A)v = 0,$$

it may be possible to transmute v to some function $u(t,A)$, in such a way that the ε-dependence is entirely carried by the transmutation kernel $h(s,t)$:

$$v_\varepsilon(s,A) = \int h_\varepsilon(s,t)u(t,A)dt.$$

Thus the problem of sending ε to zero is reduced to the study of a real-valued function h_ε instead of a vector-valued function v_ε; the results are independent of the particular operator A which one happens to substitute into P. This program was carried out for a wide class of polynomials P in [16]; in particular, the singular perturbation problem

$$\varepsilon v_{tt} + v_t = Lv, \quad \varepsilon \to 0$$

is solved by transmuting $v_\varepsilon(t,L)$ to $u(t,L)$, the solution of

$$u_{tt} = Lu.$$

In [17], the singularly perturbed singular equations

$$u_{tt} + \frac{\varepsilon}{t} u_t = u_{xx}$$

and

$$\varepsilon u_{tt} + \frac{1}{t} u_t = u_{xx}$$

were studied for $t > 0$, $\varepsilon \to 0$.

In the second of these two equations, a transmutation method simultaneously overcomes two distinct complications: the loss of an initial condition as $\varepsilon \to 0$, and the singularity of the coefficient of u_t at the initial time $t = 0$.

This research was supported in part by NSF Grant GP-34188 A #1.

REFERENCES

1. COURANT, R. and HILBERT, D., *Methods of Mathematical Physics*, Vol. II, Wiley (Interscience), New York, 1962.

2. LIONS, J. L., *Operateurs de Delsarte et problemes mixtes*, Bull. Soc. Math. France, 81 (1956), 9-95.

3. LIONS, J. L., *On the generalized radiation problem of Weinstein*, J. of Math. and Mech. 8 (1959), 873-888.

4. ROTH, W. J., *Goursat problems for* $u_{rs} = Lu$, Indiana Univ. Math. J. 22 (1973), 779-788.

5. BALAKRISHNAN, A. V., *Abstract Cauchy problems of the elliptic type*, Bull. Amer. Math. Soc. 64 (1958), 290-291.

6. BRAGG, L. R. and DETTMAN, J. W., *Related problems in partial differential equations*, Bull. Amer. Soc. 74 (1968), 375-378.

7. GRIEGO, R. and HERSH, R., *Theory of random evolutions with applications to partial differential equations*, Trans. Amer. Math. Soc. 156 (1971), 405-418.

8. ROMANOV, N. P., *On one-parameter groups of linear transformation*, I, Ann. of Math. (2) 48 (1947), 216-233.

9. UNGAR, A., *On an integral transform related to the wave and to the heat equations*, A. M. S. Notices 18 (1971), 1100.

10. HERSH, R., *Explicit solution of a class of higher-order abstract Cauchy problems*, J. of Differential Equations 8 (1970), 570-579.

11. TRANTER, C. J., *Bessel Functions with some Physical Applications*, The English Universities Press Ltd., London, 1968.

12. DONALDSON, J. A., *An operational calculus for a class of abstract operator equations*, J. of Math. Anal. and Appl. 37 (1972), 167-184.

13. KAC, M., *Some stochastic problems in physics and mathematics*, Magnolia Petroleum Co. Colloquium Lectures, 2 (1956); reprinted in Rocky Mt. Math. J., Summer, 1974.

14. KAPLAN, S., *Differential equations in which the Poisson process plays a role*, Bull. Amer. Math. Soc. 70 (1964), 264-268.

15. ROSENCRANS, S. I., *Diffusion transforms*, J. of Differential Equations 13 (1973), 457-467.

16. BOBISUD, L. and HERSH, R., *Perturbation and approximation theory for higher-order abstract Cauchy problems*, Rocky Mt. J. of Math. 2 (1972), 57-73.

17. DONALDSON, J. A., *A singular Cauchy problem with a small parameter*, Howard University Preprint.

18. HERSH, R., *Stochastic solutions of hyperbolic equations*, this volume.

19. BRAGG, L. R. and DETTMAN, J. W., *An operator calculus for related partial differential equations*, J. of Math. Anal. and Appl. 22 (1968), 261-271.

20. BRAGG, L. R., *Hypergeometric operator series and related partial differential equations*, Trans. Amer. Math. Soc. 143 (1969), 319-336.

21. DETTMAN, J. W., *Initial-boundary value problems related through the Stieltjes transform*, J. of Math. Anal. and Appl. 25 (1969), 341-349.

STOCHASTIC SOLUTIONS OF HYPERBOLIC EQUATIONS

by

REUBEN HERSH*
Department of Mathematics
and Statistics
University of New Mexico
Albuquerque, New Mexico 87131

1. INTRODUCTION

It is well-known by now that workers in partial differential equations can sometimes draw powerful aid from the theory of probability. For problems in second-order elliptic or parabolic equations, we have at our disposal an elaborate theory, created by Wiener, Lévy, Feller, Doob, Kac, Dynkin, Ito, and their students and followers.

It is commonly thought that for hyperbolic equations there is no such probabilistic counterpart. In fact, a modest but significant beginning has been made, in recent years, in the stochastic solution of hyperbolic equations. In this paper we summarize the principal results that have been found so far. The theory is by no means so comprehensive or imposing as in the parabolic

---

*Research supported in part by NSF Grant GP-34188 A #1.

and elliptic cases, but it interesting and useful. Moreover, being of recent origin, it is presumably not yet full-grown. Most of the work is probably still to be done.

We begin, in the next section, by discussing the work of S. Goldstein, Kac, Kaplan, and Griego-Hersh on equations of second order. Then we discuss systems of first-order equations, including work of Pinsky and Heath. We conclude with a list of open problems and suggestions for further research.

A more comprehensive survey recently appeared in [10]. There the methods of proof are summarized, and applications to integral-differential equations are discussed. There is also a description of recent work of Ellis, Kertz and Keepler, which is relevant to partial differential equations.

2. SECOND-ORDER EQUATIONS

Following earlier work of S. Goldstein and G. I. Taylor, Kac in 1956 derived a probabilistic solution of the "telegrapher's equation",

$$u_{tt} + 2au_t = c^2 u_{xx}. \tag{1}$$

Let $N_a(t)$ be a Poisson process with constant intensity a. Define the random time $T(t)$ by

$$T(t) = \int_0^t (-1)^{N_a(\tau)} d\tau. \tag{2}$$

Let $v(x,t)$ be any function of two real variables such that

$$(3) \qquad \frac{\partial^2 v}{\partial t^2} = c^2 \frac{\partial^2 v}{\partial x^2} .$$

Then

$$(4) \qquad u(x,t) = E\{v(x,T(t))\}$$

satisfies (1), where $E\{\cdot\}$ denotes expected value with respect to the random variable $N_a(t)$. Moreover, $u(x,0) = v(x,0)$, and $u_t(x,0) = v_t(x,0)$. Thus the telegrapher's equation is solved by the expected value of a randomized solution of the wave equation.

Kac obtained this formula by a formal passage to the limit from a discrete random walk model, in which a particle moves to the right or the left at speed c, and reverses direction at the instants when the Poisson process performs a jump from N to $N + 1$.

The rigorous verification of (4) was by a direct computation; moreover, this computation turned out, surprisingly, to be valid even if $\partial^2/\partial x^2$ is replaced by $\sum \partial^2/(\partial x_j)^2$ in (1) and (3). For this n-dimensional "telegrapher-type" equation, there was no longer a stochastic model to motivate formula (4); so the formula had a certain aura of mystery about it.

The coefficient a in equation (1) is a frictional term, which represents dissipation of energy. In case $a = 0$, the Poisson process has intensity zero--that is, reversals of direction never occur--and the telegraph equation degenerates to the undamped wave equation. This shows that the possibility of a stochastic representation arises only because of the presence of a frictional or dissipative effect. We have a hyperbolic equation, but it has, so to speak, a "parabolic term" au_t, and the random time takes account of precisely this "parabolic" or dissipative part of the equation.

In 1963 Stanley Kaplan showed that one may permit the coefficient a in (1) to be time-dependent, $a = a(t)$. To do so, we redefine $N(t)$ to be a Poisson process with variable intensity $a(t)$, as follows: $N(0) = 0$, and for $m \geq 0$,

$$\text{(5)} \quad \begin{cases} \operatorname{Prob}\{N(t) - N(s) = m\} \\ = (m!)^{-1}\left[\int_s^t a(\tau)d\tau\right]^m \exp\left[-\int_s^t a(\tau)d\tau\right]. \end{cases}$$

With this choice of $N(t)$, (4) is again valid; the expectation E, of course, is now interpreted in terms of the variable-intensity process, and v is a solution of (1) with $a = a(t)$.

Kaplan pointed out also that one could replace $c^2\partial^2/\partial x^2$, in (1) and (3), by any linear operator L; and he mentioned that his formula was an example of Delsarte's transmutation operator. (See the article [11] on transmutation operators, in this volume, for further discussion of this point.)

The question arises, what use can be made of a representation such as (4)? By representing u stochastically, one makes available all the machinery of probability theory; in particular, the limit theorems known as laws of large numbers and central limit theorems.

It is not hard to prove, using a central limit theorem, that if in formula (2) a is replaced by a/ε^2, and the resulting random time is denoted by $T_\varepsilon(t)$, then

$1/\varepsilon\ T_\varepsilon$ converges in law, as $\varepsilon \to 0$, to a normal Gaussian random variable $Z(t)$, with mean zero and variance t/a. (See Griego-Hersh [8].)

Let $u^0(x,t)$ denote the expected value (with respect to this normal distribution) of the random function $v(x,Z(t))$. This function satisfies the heat equation

$$2au_t^0 = u_{xx}^0.$$

We thereby obtain a probabilistic proof that u^ε, the solution of

$$\{\varepsilon u_{tt}^\varepsilon + 2au_t^\varepsilon = u_{xx}^\varepsilon,\quad u^\varepsilon(x,0) = f,\quad u_t^\varepsilon(x,0) = g\}$$

converges to the solution of

$$\{2au_t^0 = u_{xx}^0,\ u^0(x,0) = f\}.$$

Notice that the influence of g, the second piece of initial data for $u^\varepsilon(t,x)$, disappears as $\varepsilon \to 0$.

A limit theorem of this kind was first proved by Hadamard, using a representation in terms of the Riemann function. But the probabilistic proof in [8] is valid in much greater generality; $c\ \partial/\partial x$ can be replaced, in (1) and (3), by an arbitrary group generator A, and one studies $\varepsilon u_{tt} + 2au_t = A^2 u$. In particular, A^2 can be any linear elliptic differential operator, with variable coefficients. In this generality, the singular perturbation result was new when it was obtained by this probabilistic argument; purely operator-theoretic methods were found later [24, 2, 19].

Presumably a similar perturbation theorem is true for $a = a(t)$, and could be proved by combining the methods of Kaplan with those of Griego-Hersh; but so far as I know, this is an untouched problem.

3. FIRST-ORDER SYSTEMS: REPRESENTATIONS

Let us now return to the model with which Goldstein and Kac worked, that is, a particle moving on the line, and changing velocity according to a random process. But instead of permitting only two velocities $\pm c$, we permit n of them, $v_1, \ldots, v_n$.

In terms of this model, we can solve the first order system

$$(6) \qquad U_t = VU_x + QU, \quad U_i(x,0) = f_i(x) \quad i = 1, \ldots, n$$

where $V = \text{diag} \langle v_1, \ldots, v_n \rangle$ and Q is a probability generating matrix (one whose row sums are all zero, and whose off-diagonal elements are all non-negative). The formula for $U(x,t)$ is

$$(7) \qquad U_i(x,t) = E_{x,i}\,(f_{v(t)}(X(t)))$$

where $X(t)$ is the random position, at time t, of a particle whose velocities $v(t)$ switch from one of the v_i to another according to the law $P_{ij}(t) = \exp(tQ)$, where

$$(8) \qquad P_{ij}(t) \stackrel{\text{def}}{\equiv} \text{Prob}\{v(t) = v_j \mid \text{given } v(0) = v_i\}$$

and where $E_{x,i}$ is the expected value operator, conditioned on the initial states $X(0) = x$, $v(0) = v_i$.

This representation was given by Pinsky [22] and by Birkhoff and Lynch [1].

It was noticed in [7, 8] that it can be rewritten in the simpler form

$$(9) \qquad U_i(x,t) = E_i(f_{v(t)}(x + \sum v_i \gamma_i(t)))$$

where the random variable $\gamma_i(t)$ is the amount of time up to the epoch t in which the particle possesses velocity v_i.

Formula (9) was presented in [7, 8] as a particular case of a more abstract formula. If $A_1, \ldots, A_n$ are operators in a Banach space L, each of which generates a semigroup $\exp(tA_j)$, then one can pose an abstract Cauchy problem, where the initial data f and the solution $U(t)$ are each an n-tuple of vectors in L:

$$(10) \qquad \begin{cases} \dfrac{\partial U}{\partial t} = \text{diag} \langle A_i \rangle U + QU \\ U(0) = f = \langle f_j \rangle \quad j = 1, \ldots, n. \end{cases}$$

This is solved stochastically in terms of a Markov chain $j(t)$, with n states and generator Q. We use this chain to index the operator $A_{j(t)}$, defining thereby an operator-valued Markov chain.

Let $\tau_0 = 0$, and let τ_1 be the epoch of the first jump of $j(t)$, when it leaves its initial state $j(0)$. Let τ_2 be the epoch of the second jump, and so on. The number of jumps performed up to a given time t is an integer-valued random variable, which we call $N(t)$. Then (10) is solved in terms of the following random product of operators:

$$(11)\quad \begin{cases} M(t) = \\ \exp(\tau_1 A_{j(0)})\exp((\tau_2-\tau_1)A_{j(\tau_1)})\ldots\exp((t-\tau_N)A_{j(\tau_N)}) \end{cases}$$

$$(12)\qquad U_i(t) = E_i[M(t)f_{j(t)}].$$

Here again E_i means expected value conditioned on $j(0) = i$.

If the A_i commute with each other, then (12) reduces to

$$(13)\qquad U_i(t) = E_i\left[\exp\left[\sum_{k=1}^{n} \gamma_k(t)A_k\right]f_{j(t)}\right]$$

and if $A_k = c_k\, d/dx$, then $e^{tA_k} f(x) = f(x + c_k t)$, and (13) specializes to (9).

Of course, (10) is a system of partial differential equations if the A_i are chosen to be differential operators. They need not be hyperbolic; they could be parabolic or of Schrodinger type. If they have constant coefficients, they commute and (13) is valid. If they have variable (space-dependent) coefficients, we must be satisfied with (11) - (12).

If $n = 2$, $Q = \begin{pmatrix} -a & a \\ a & -a \end{pmatrix}$, and $A_1 = A$, $A_2 = -A$,

then $N(t)$ turns out to be a Poisson process with intensity a. In this case we have a system of two equations which is equivalent to a single equation of second-order--namely the abstract telegraph equation

$$(14)\qquad U_{tt} + 2aU_t = A^2U.$$

We can think of the random product (11) in connection with the following intuitive picture: an abstract system is capable of evolving in any of n modes; a random mechanism switches its mode of operation from one channel to another; $M(t)$ is a typical outcome of such a "random evolution".

If in particular $A = \sqrt{\Delta}$, (14) is the n-dimensional telegrapher's equation. We recover Kac's formula (4), along with a stochastic model to motivate it (random evolution forward and backward, according to the generator $\sqrt{\Delta}$.)

As a formula for solving hyperbolic systems, (9) or even (12) is somewhat special, because the matrix Q is independent of x, and is restricted to have zero row-sums and non-negative off-diagonal elements. In 1969 David Heath obtained a stochastic solution for a general first-order linear hyperbolic system:

$$(15)\quad \begin{cases} \dfrac{\partial u_i}{\partial t} = v_i(x)\,\dfrac{\partial u_i}{\partial x} + \sum g_{ij}(x)\,u_j \\[2ex] u_i = f(x) \quad \text{at} \quad t = 0. \end{cases}$$

This equation still looks special, because the derivatives of u_i appear only in the i'th equation; but this is no real restriction, since it is well known that any first-order hyperbolic system in two independent variables can be put into such a form.

To allow a general zero-order term $g_{ij}(x)$ instead of just the restricted constants q_{ij}, Heath used a method of "piecing out" to construct a process in which a particle moving on the x-axis according to one of n

different position-dependent velocities $v_i(x)$ undergoes a random jump in velocity from $v_i(x)$ to $v_j(x)$ with a probability $|g_{ij}(x)|$. Then he introduces a multiplicative functional, $m(t,\omega)$ which depends on the signs of the off-diagonal elements g_{ij}, and on the sign and magnitude of g_{ii}:

$$m(t,\omega) = \exp\left[\int_0^t d(x(s,\omega))ds\right] \cdot \prod_{\ell=1}^{k} \operatorname{sign} g_{i_\ell j_\ell}(x(\tau_\ell)).$$

Here $d(x) = g_{ii}(x) + \sum_{j\neq i} |g_{ij}(x)|$ and $\tau_k \leq t \leq \tau_{k+1}$.

The jumps of the process are at time τ_ℓ, $\ell = 1,\ldots,k$ and at the ℓ'th jump it goes from velocity $v_{i_\ell}(x)$ to velocity $v_{j_\ell}(x)$, if the particle at the jump time has position x.

With this functional, Heath is able to show that

$$E_{(x,i)}\{m(t,\omega)f(x,(t,\omega))\}$$

satisfies (15).

By constructing an equivalent first-order system, he can solve second-order equations of the form

$$u_{tt} = u_{xx} + 2r(x)u_x.$$

He also gives a theory of generalized solutions, analogous to Doob's theory of "parabolic" functions.

It would be interesting to see if Heath's technique could be extended to more general cases.

For arbitrary second-order elliptic operators V_i one should be able to solve parabolic systems of the form

$$\frac{\partial u_i}{\partial t} = V_i u_i + \sum g_{ij}(x) u_j$$

by piecing together diffusions and using the same multiplicative functional $m(t,\omega)$.

4. SINGULAR PERTURBATIONS

The perturbation theorem for $\varepsilon u_{tt} + 2au_t = u_{xx}$, which was mentioned in the discussion on second-order equations, has been generalized to systems of n equations, with either constant or variable coefficients. This was accomplished in stages, by a series of papers [22,7,8,13,12,19,4,20]. In the most recent and general formulations, instead of finite-valued index i in $U_i(x,t)$, we can permit $U_v(x,t)$ to depend on a parameter v which ranges over a set in R^n. The matrix Q is replaced by an integral operator and the velocity of our particle is allowed to be, for example, any vector in R^3, as is called for by the physical problems of transport theory. For more discussion of this generalization, see [10,20]. This model leads to a differential-integral equation. Here we consider only systems of finitely many differential equations.

It turns out that there are two types of limit theorems. First consider

$$(16) \qquad \frac{\partial U}{\partial t} = \operatorname{diag} \langle A_i \rangle U + \frac{1}{\varepsilon} QU \qquad U_i(0) = f_i .$$

In particular we could have $A_i = c_i\, d/dx$ or $c_i(x)\, d/dx$.

Now we assume also that Q is <u>ergodic</u>; i.e., it has zero as a simple eigenvalue. In that case Q has a unique left unit null vector $<\Pi_i>$, $i = 1,\ldots,n$. $\sum \Pi_i = 1$, $\Pi Q = 0$, and it is known that e^{tQ} converges as $t \to \infty$, to a matrix all of whose rows are equal to Π. The i'th component of Π is the proportion of time that the chain $j(t)$ spends in the i'th state, asymptotically as $t \to \infty$.

With this probabilistic interpretation in mind, it is not hard to guess that speeding up the chain (replacing Q by $\frac{1}{\varepsilon} Q$, $\varepsilon \to 0$) has the effect of averaging out the generators A_i, according to the weighted averages Π_i. In fact, one has the theorem:

THEOREM. *As* $\varepsilon \to 0$, *each component* $U_i(t)$ *of the solution of* (16) *converges to the solution of*

$$\left\{\frac{\partial U}{\partial t} = \sum \Pi_i A_i U, \quad U(0) = \sum \Pi_i f_i\right\}. \tag{17}$$

The initial data as well as the coefficients are averaged out. Convergence holds only for $t > 0$; near $t = 0$ there is an initial-layer discontinuity.

For a proof, see [23].

If $\sum \Pi_i A_i = 0$, the theorem just stated simply says that $U(t)$ converges to $\sum \Pi_i f_i$, a constant independent of t.

In this case it is more interesting to consider a more precise limit theorem. We now write

$$\left\{\begin{aligned} &\frac{\partial U}{\partial t} = \frac{1}{\varepsilon} \operatorname{diag} < A_i > + \frac{1}{\varepsilon^2} QU \\ &u_i(0) = f_i. \end{aligned}\right. \tag{18}$$

THEOREM. *If* Q *is ergodic,* $\Pi Q = 0, \sum \Pi_i A_i = 0,$ *then as* $\varepsilon \to 0$ *each component of the solution of* (18) *converges to the solution of*

$$\frac{\partial u}{\partial t} = \bar{V}u, \quad u(0) = \sum \Pi_i f_i \tag{19}$$

where

$$\bar{V} = \sum_{\alpha,\beta=1}^{n} \Pi_\alpha A_\alpha A_\beta \lim_{\lambda \to 0} \left[(\lambda - Q)^{-1}_{\alpha,\beta} - \frac{\Pi_\beta}{\lambda} \right]$$

$$= \sum_{\alpha,\beta=1}^{n} \Pi_\alpha A_\alpha A_\beta \frac{(d/d\lambda)\ \mathrm{cof}_{\beta,\alpha}(\lambda - Q)}{(d/d\lambda)\ \det(\lambda - Q)} \Bigg|_{\lambda=0}$$

where $\mathrm{cof}_{\beta,\alpha}(\lambda - Q)$ *is the cofactor of the element in the* β*'th row and* α*'th column of* $\lambda - Q$.

This theorem was proved in [12]. There is a simplified discussion in [23], and a more general theorem, proved without reliance on probabilistic arguments, in [19].

Notice that $\bar{V}$ is a non-symmetric quadratic form in non-commuting products. It is symmetric only for special Q ("reversible chains"). If the A_i commute with each other, simpler proofs and formulas can be given (see [13]). For the special case, $A_i = c_i\, d/dx$, f independent of i, it was proved in [22].

The significance of (19) is clearest if the A_i are first-order differential operators (constant or variable coefficients). Then $\bar{V}$ is second-order elliptic, and

u^ε converges to the solution of a parabolic equation--we have in fact the "diffusion approximation" to our linear transport model.

For in this case the groups e^{tA_i} are groups of translations, and we are back to our model of a particle undergoing random changes of speed. In replacing A_i by A_i/ε in (10), we are speeding up the motion by a factor $1/\varepsilon$. In replacing Q by Q/ε^2, we are shortening the time between "collisions" (changes of speed) by a factor ε^2. The combined effect of these two replacements is to multiply the average distance traveled between collisions ("mean free path") by a factor of ε. The hypothesis $\sum \Pi_i A_i = 0$ means that the average distance traveled to the right is equal to the average distance traveled to the left--i.e., the total "drift" is zero. Our limit theorem then says that in the limit of small mean free path, the linear transport model behaves like a diffusion.

By a simple change of parameters and of the time-scale, the more general equation

$$\frac{\partial U}{\partial t} = \varepsilon_1 \operatorname{diag} \langle A_i \rangle U + \varepsilon_2 QU$$

may be reduced to (18). The parameters ε_1 and ε_2 may be small or large. Provided only that $\varepsilon_1/\varepsilon_2$ is small, U is again approximated by a solution of (19) (see [13]).

In [13] it is also shown how, in the case that $A_i A_j \equiv A_j A_i$, the convergence theorem can be modified to permit $\sum \Pi_i A_i \neq 0$, and to allow Q to have several null vectors or to have non-zero row-sums.

The rate of convergence was studied by Pinsky [22] and, in much greater generality, by Papanicolaou and Varadhan [20]. Under reasonable restrictions, convergence is $O(\varepsilon)$.

5. OPEN QUESTIONS

In all our discussion except on the work of Kaplan, we had coefficients independent of time. Both the representation theory and the asymptotics should be extended to permit time-dependent coefficients.

The representation theory was extended by Heath to equations of general form; can any asymptotic results be proved in such generality?

For physical applications, it is desirable to consider equations such as (6) or (15) on a bounded interval, with suitable boundary conditions. Here there will be a boundary-layer effect as well as an initial-layer effect. No rigorous results of an asymptotic nature have been published as yet in this case. Recently formal expansions have been calculated by Matkowski and Habetler and by Larsen and Keller [17].

REFERENCES

1. BIRKHOFF, G. and LYNCH, R.E., *Numerical solution of the telegraph and related equations*, in Numerical Solutions of Partial Differential Equations, Proc. Symp. Univ. Md., 1965; New York, Academic Press, 1966.

2. BOBISUD, L. and HERSH, R., *Perturbation and approximation theory for higher-order abstract Cauchy problems*, Rocky Mt. J. of Math. 2 (1972), 57-73.

3. CHABROWSKI, J., *Les solutions non negatives d'un systeme parabolique d'equations*, Ann. Polon. Math. 19 (1967), 193-197.

4. COGBURN, R. and HERSH, R., *Two limit theorems for random differential equations*, Indiana Univ. Math. J. 22 (1973), 1067-1089.

5. ELLIS, R. S., and PINSKY, M. A., *Limit theorems for model Boltzmann equations with several conserved quantities*, preprint.

6. GOLDSTEIN, S., *On diffusion by discontinuous movements and on the telegraph equation*, Quart. J. Mech. Appl. Math. 4 (1951), 129-156.

7. GRIEGO, R. J., and HERSH, R., *Random evolutions, Markov chains, and systems of partial differential equations*, Proc. of National Academy of Sciences 62 (1969), 305-308.

8. GRIEGO, R. J., and HERSH, R., *Theory of random evolutions with applications to partial differential equations*, Trans. Amer. Math. Soc. 156 (1971), 405-418.

9. HEATH, D. C., *Probabilistic Analysis of Hyperbolic Systems of Partial Differential Equations*, Doctoral dissertation, University of Illinois, 1969.

10. HERSH, R., *Random evolutions: a survey of results and problems*, Rocky Mt. J. of Math., Summer 1974.

11. HERSH, R., *The method of transmutations*, this volume.

12. HERSH, R., and PAPANICOLAOU, G., *Non-commuting random evolutions, and an operator-valued Feynman-Kac formula*, Comm. Pure and Appl. Math. 25 (1972), 337-367.

13. HERSH, R. and PINSKY, M., *Random evolutions are asymptotically Gaussian*, Comm. Pure and Appl. Math., 25 (1972), 33-44.

14. KAC, M., *A stochastic model related to the telegrapher's equation*, Magnolia Petroleum Co. Colloquium Lectures, 2 (1956); reprinted in Rocky Mt. J. of Math., Summer 1974.

15. KAPLAN, S., *Differential equations in which the Poisson process plays a role*, Bull. Amer. Math. Soc. 70 (1964), 264-268.

16. KEEPLER, M., *Backward and Forward Equations for Random Evolutions*, Doctoral dissertation, University of New Mexico, 1973.

17. KELLER, J. and LARSEN, E., *Asymptotic solution of neutron transport problems for small mean free paths*, J. Math. Phys. 15 (1974), 75-81.

18. KERTZ, R., *Limit Theorems for Discontinuous Random Evolutions*, Doctoral dissertation, Northwestern University, 1972.

19. KURTZ, T. G., *A limit theorem for perturbed operator semigroups with applications to random evolutions*, J. Func. Anal. 12 (1973), 55-67.

20. PAPANICOLAOU, G. C. and VARADHAN, S. R. S., *A limit theorem with strong mixing in Banach space and two applications to stochastic differential equations*, Comm. Pure and Appl. Math. 26 (1973), 497-523.

21. PINSKY, M., *Multiplicative operator functionals of a Markov process*, Bull. Amer. Math. Soc. 77 (1971), 377-380.

22. PINSKY, M., *Differential equations with a small parameter and the central limit theorem for functions defined on a finite Markov chain*, Z. Wahrscheinlichkeitstheorie verw. Geb. 9 (1968), 101-111.

23. PINSKY, M., *Multiplicative operator functionals and their asymptotic properties*, Advances in Probability, Vol. 3, Marcel Dekker, New York, 1974, 1-100.

24. SCHOENE, A., *Semi-groups and a class of singular perturbation problems*, Indiana U. Math. J. 20 (1970), 247-263.

REMARKS ON SOME NEW NONLINEAR BOUNDARY VALUE PROBLEMS

by

J. L. LIONS
Collège de France
and
I. R. I. A.
Domaine de Voluceau - Rocquencourt
B. P. 5 - 78150 Le Chesnay, France

INTRODUCTION

In a paper of A. Bensoussan and the author (cf. Bensoussan-Lions [1]) it was observed how problems of optimal stopping times lead to free-surface problems which can be solved by the technique of Variational Inequalities (V. I.), in the terminology of Stampacchia and the author.

In a series of papers of Bensoussan and the author (see the Bibliography) it was shown how problems of impulse control lead to a large number of new nonlinear problems, again of the type of "free surface" problems, but with new aspects which make it necessary to introduce new tools. It is to that effect that Bensoussan and the author introduced the Quasi-Variational Inequalities (Q. V. I.).

In Sections 1, 2, 3 below we briefly recall the motivations for impulse control problems and the study of Q. V. I., and we give, without proofs, two results of Bensoussan and the author which are proved elsewhere; we also use in Section 2 a result of L. Tartar [1].

The Q. V. I. can be of <u>elliptic</u> or of <u>parabolic</u> type for <u>second order operators</u> (or special systems of such operators), and they can also be of hyperbolic type for first order operators (or special systems).

In the note [3] 3) of Bensoussan-Lions, we introduced (this time without motivation from optimal control) Q. V. I. for <u>second order hyperbolic</u> operators.

In Section 4 below, we introduce a Q. V. I. for a system which is not of Cauchy-Kowaleskaya type; in this section the proofs are given; they also give <u>some</u> ideas of the proofs which can be used in general for solving Q. V. I.

We do not enter here into the question of the numerical approximation of the solutions of Q. V. I.; we refer to Bensoussan-Lions [5], [6].

It turns out that the Q. V. I. are useful in different contexts; they can be used for the solution of classical free surface problems arising in infiltration theory (cf. C. Baiocchi [1]) and in other problems of Mechanics, as we shall report elsewhere; the Q. V. I. are also of some interest in the theory of equilibrium points in Economics; we refer to Bensoussan-Lions [7] and also to the work of Joly and Mosco [1].

The plan is as follows:

1. Motivation.
2. Q. V. I. of elliptic type.

1. MOTIVATION

Let us consider the management of n goods, with a stochastic demand of Gaussian type (for deterministic demands or for demands of Poisson type, cf. Bensoussan and the author, loc. cit.) More precisely, the demand during the period $[t, t+\Delta t]$ of time is given by

(1.1) $$D(t, t+\Delta t) = \mu\Delta t + \sigma\Delta b(t)$$

where $\mu \in \mathbb{R}^n$, σ is an $n \times n$ matrix and where $\Delta b(t)$ is a Gaussian random variable, with zero mean and with covariance $= (\Delta t)$ Identity.

Let T denote the horizon of the problem; we suppose $T < \infty$ to fix the ideas. We suppose that at time t our stock is $x \in \mathbb{R}^n$. A <u>control</u> (or a <u>policy</u>) will consist of a sequence of times θ^i where we place orders:

(1.2) $$t \leq \theta^1 < \theta^2 < \ldots < \theta^N \leq T,$$

<u>where</u> N <u>is not given</u>, and of a sequence of vectors

(1.3) $$\xi^1, \xi^2, \ldots, \xi^N,$$

where $\xi^j \in \mathbb{R}^n$ denotes the order we place at time θ^j. The constraints on ξ^j are

(1.4) $$\begin{cases} \xi_i^j \geq 0 \quad \text{for each } i, \text{ and also an} \\ \qquad \text{upper bound, say } \xi_i^j \leq C_i. \end{cases}$$

In short:

$$v = \{\theta^1, \xi^1; \theta^2, \xi^2; \ldots ; \theta^N, \xi^N\}. \tag{1.5}$$

Of course the variables $\theta^1, \xi^1, \ldots$ are random variables. If we assume that the time of delivery is zero*, <u>the state</u> $y(s;v)$ <u>of our system</u> (i.e. the amount of goods we have at time s, applying the policy v) is given (as a random variable) by

$$y(s;v) = x - D(t,s), \quad t \le s < \theta^1,$$
$$y(\theta^1;v) = x - D(t,\theta^1) + \xi^1, \quad \text{etc.}$$

Let us now introduce the <u>cost function</u> (or payoff) $J(v)$. We suppose that there is a <u>fixed price</u>, say $k > 0$, we have to pay for placing an order (k = set up cost)**; let $f(y,s)$ be a <u>positive</u> function which takes care of the <u>storage cost</u> and of the <u>shortage cost</u>; then

$$J(v) = E[\int_t^T e^{-\alpha(t-s)} f(y(s;v), s)ds + kN(v)], \tag{1.6}$$

where E denotes the expectation and $N(v)$ denotes the number of orders in the policy v as given by (1.5), and where α denotes a discount rate.

We define next

$$u(x,t) = \inf_v J(v). \tag{1.7}$$

The problem is now to characterize, if possible, the function u defined by (1.7) and to construct the op-

---

* Not a realistic hypothesis. But the case where we have delays in the deliveries can be treated along similar lines. Cf. Bensoussan-Lions in the Bibliography.

** We can also consider cases when the set up cost has a more complicated structure; the fundamental point is that this set up cost is $\ge k > 0$.

timal policy (if it exists) assuming u to be known.

A formal computation shows (cf. Bensoussan-Lions, loc. cit.) that u satisfies:

$$(1.8)\quad \begin{cases} -\dfrac{\partial u}{\partial t} + Au - f \le 0, \\ u - M(u) \le 0, \\ \left(-\dfrac{\partial u}{\partial t} + Au - f\right)\left(u - M(u)\right) = 0, \quad x \in \mathbb{R}^n,\ t < T, \end{cases}$$

where A is the second order elliptic operator given by

$$(1.9)\qquad A\phi = -\tfrac{1}{2}\,\mathrm{tr}\,\frac{\partial^2\phi}{\partial x^2}\,\sigma\sigma^* + \sum_{i=1}^{n} \mu_i \frac{\partial\phi}{\partial x_i} + \alpha\phi$$

and where Mu is given by

$$(1.10)\qquad Mu(x,t) = k + \inf_{0\le\xi_i\le c_i} u(x+\xi,t).$$

To the set of inequalities (1.8) we add the Cauchy condition

$$(1.11)\qquad u(x,T) = 0, \quad x \in \mathbb{R}^n.$$

In what follows we intend to give some indications of the methods which can be used to study the problem (1.8)-(1.11) and problems of somewhat similar types.

Before we proceed, some remarks are in order.

REMARK 1.1. It is clear that in (1.7) we can restrict ourselves to the policies with a finite number of orders; otherwise $kN(v) = +\infty$. This remark explains the terminology of "impulse controls".

REMARK 1.2. The operator M is nonlinear and of non local type.

REMARK 1.3. The problem (1.8)-(1.11) is of the type

of a free boundary problem. Indeed there will be two regions in the space (x,t):

(i) the saturated region where $u = M(u)$,

(ii) the continuation region where $u < M(u)$ and where $-\frac{\partial u}{\partial t} + Au = f$.

The interface S is unknown, and actually the study of S is one of the most important problems of the theory.

REMARK 1.4. We can consider similar problems with constraints on the state y. This leads to the following problem. Let $\mathcal{O}$ be a bounded open set in $\mathbb{R}^n$, with a smooth boundary Γ. Then we are looking for a function $u = u(x,t)$ defined for $x \in \mathcal{O}$, $t < T$, such that

$$(1.12) \quad \begin{cases} -\frac{\partial u}{\partial t} + Au - f \leq 0, \quad u - M(u) \leq 0, \\ \left(-\frac{\partial u}{\partial t} + Au - f\right)\left(u - M(u)\right) = 0 \\ \qquad \text{for } x \in \mathcal{O}, \quad t < T, \end{cases}$$

where M is (for instance) given by

$$(1.13) \quad \begin{cases} Mu(x,t) = k + \inf u(x+\xi,t), \\ \xi_i \geq 0, \quad x + \xi \in \mathcal{O}; \end{cases}$$

the boundary conditions are

$$(1.14) \quad \begin{cases} \frac{\partial u}{\partial \nu_A} \leq 0, \quad u - M(u) \leq 0, \\ \frac{\partial u}{\partial \nu_A}(u - M(u)) = 0 \quad \text{on } \Sigma = \Gamma \times]-\infty, T] \end{cases}$$

(where $\frac{\partial}{\partial \nu_A}$ denotes the conormal derivative associated with A, and directed toward the exterior of $\mathcal{O}$), and the <u>initial condition</u> is

$$u(x,T) = 0, \quad x \in \mathcal{O}. \tag{1.15}$$

REMARK 1.5. The <u>stationary problem</u> which corresponds to (1.12) is

$$\begin{cases} Au - f \leq 0, \quad u - M(u) \leq 0, \\ (Au - f)(u - M(u)) = 0 \quad \text{in } \mathcal{O} \end{cases} \tag{1.16}$$

with boundary conditions (1.14) on Γ.

REMARK 1.6. If the demand is <u>deterministic</u> ($\sigma = 0$), then the operator which appears in (1.8) becomes <u>hyperbolic</u>, instead of being <u>parabolic</u> when $\sigma \neq 0$.

REMARK 1.7. One can meet in the applications a wide variety of different operators M. The main property which is needed for M is:

$$u \leq v \text{ implies } M(u) \leq M(v). \tag{1.17}$$

In Bensoussan-Lions, loc. cit., we used (1.17) together with some (weak) continuity properties of M. But L. Tartar [1] observed that <u>no continuity at all is needed on</u> M (provided we have (1.17)); but then the proofs of existence of a solution are <u>not constructive</u> (one uses Zorn's lemma).

REMARK 1.8. One also meets in applications similar problems for <u>systems</u> of operators. We shall study such a situation in Section 4 below.

2. Q. V. I. OF ELLIPTIC TYPE

We consider the problem (1.16) subject to the boundary conditions (1.14). We write A in the form

$$
(2.1) \quad \begin{aligned} A\phi = & -\sum_{i,j=1}^{n} \frac{\partial}{\partial x_i}\left(a_{ij}(x)\frac{\partial \phi}{\partial x_j}\right) \\ & + \sum_{j=1}^{n} a_j(x)\frac{\partial \phi}{\partial x_j} + a_0(x)\phi \end{aligned}
$$

and we suppose that

$$
(2.2) \quad \begin{cases} a_{ij},\ a_j,\ a_0 \in L^{\infty}(\mathcal{O}), \\ \sum a_{ij}(x)\xi_i\xi_j \geq \alpha \sum \xi_i^2, \quad a_0(x) \geq \alpha > 0. \end{cases}
$$

Let $H^1(\mathcal{O})$ denote the Sobolev space of order 1; for $u,v \in H^1(\mathcal{O})$, we define

$$
(2.3) \quad \begin{aligned} a(u,v) = & \sum \int_{\mathcal{O}} a_{ij} \frac{\partial u}{\partial x_j}\frac{\partial v}{\partial x_i}\,dx \\ & + \sum \int_{\mathcal{O}} a_j \frac{\partial u}{\partial x_j} v\,dx \\ & + \int_{\mathcal{O}} a_0\, uv\,dx. \end{aligned}
$$

For f and g in $L^2(\mathcal{O})$, we set

$$
(2.4) \quad (f,g) = \int_{\mathcal{O}} fg\,dx.
$$

It is now a simple matter, at least <u>formally</u>, to check that problem (1.16) is <u>equivalent</u> to the following one: find $u \in H^1(\mathcal{O})$ such that

$$
(2.5) \quad \begin{cases} a(u,v-u) \geq (f,v-u) \\ \text{for all } v \in H^1(\mathcal{O}), \quad v \leq M(u), \quad u \leq M(u); \end{cases}
$$

(2.5) is what we call a <u>Quasi Variational Inequality</u>

(Q. V. I.).

REMARK 2.1. Let us recall what an (ordinary) Variational Inequality (V. I.) looks like: let K be a (non empty) closed convex subset of $H^1(\mathcal{O})$; then the V. I. associated to $a(u,v)$ and K consists in finding u such that

$$\begin{cases} a(u,v-u) \geq (f,v-u) & \text{for all } v \in K, \\ u \in K. \end{cases} \tag{2.6}$$

For instance if we take

$$K = \{v \mid v \in H^1(\mathcal{O}), \quad v \leq \psi \text{ on } \mathcal{O}\}, \tag{2.7}$$

where ψ is a given function, say in $L^2(\mathcal{O})$, we can see the difference between (2.6)-(2.7) and (2.5): in (2.5), $\psi = M(u)$ is not known.

REMARK 2.2. One can of course extend the formulation (2.5). See, in particular, L. Tartar [1]; see also Joly and Mosco [1]. One can prove the following (cf. Bensoussan-Lions, loc. cit., Bensoussan-Goursat-Lions [1], Th. Laetsch [1], L. Tartar [1]*).

THEOREM 2.1. *Let us assume that* (2.2) *holds true and that* f *is given satisfying*

$$f \in L^\infty(\mathcal{O}), \quad f \geq 0. \tag{2.8}$$

Then there exists a unique solution $u \geq 0$, $u \in L^\infty(\mathcal{O})$, *which satisfies* (2.5).

---

* The uniqueness is proved in Laetsch and in Tartar.

We do not give the complete proof here. We just give below some brief indications on the construction of a solution.

REMARK 2.3. The V. I. (2.6) <u>admits a unique solution</u> (cf. Lions-Stampacchia [1]) if we suppose that

$$(2.9)\quad \begin{cases} a(v,v) \geq \beta \|v\|^2, & \beta > 0, \\ \|v\| = \text{norm of } v \text{ in } H^1(\mathcal{O}), \end{cases}$$

which is a <u>stronger hypothesis</u> than (2.2).

REMARK 2.4. The V. I. can be solved for elliptic operators of <u>any order</u>. For the time being the theory of Q. V. I. is (essentially) restricted to the case of <u>second order operators</u>; this is due to the fact that we heavily rely (for proving the existence of a solution) on the maximum principle*.

<u>Construction of a solution of (2.5)</u>.

Let u^0 be a constant $\geq (\sup f)(\inf a_0)^{-1}$. Then

$$(2.10)\quad \begin{cases} a(u^0,v) = (\tilde{f},v) \\ \text{for all } v \in H^1(\mathcal{O}), \quad \tilde{f} \geq f. \end{cases}$$

We introduce λ such that

$$(2.11)\quad \begin{cases} a(v,v) + \lambda |v|^2 \geq \alpha_0 \|v\|^2, & \alpha_0 > 0, \\ \text{for all } v \in H^1(\mathcal{O}), \end{cases}$$

where $|v|^2 = (v,v)$; such a λ exists, by virtue of (2.2). We then inductively define $u^1, u^2, \ldots, u^n, \ldots$ by the solution of the V. I.

---

* But we have <u>examples</u> where we can solve Q. V. I. for higher order operators.

$$(2.12)\quad \begin{cases} a(u^n,v-u^n) + \lambda(u^n,v-u^n) \\ \quad \geq (f+\lambda u^{n-1},v-u^n), \\ \text{for all} \quad v \leq M(u^{n-1}), \quad u^n \leq M(u^{n-1}). \end{cases}$$

We remark that (2.12) is the V. I. (2.6)-(2.7) with $a(u,v)$ replaced by $a(u,v) + \lambda(u,v)$ and with $\psi = M(u^{n-1})$; therefore according to Remark 2.3, (2.12) uniquely defines u^n.

One then proves that

$$(2.13)\quad u^0 \geq u^1 \geq \ldots \geq u^{n-1} \geq u^n \geq \ldots \geq 0.$$

It is possible to take $v = 0$ in (2.12) (since $u^{n-1} \geq 0$ implies $M(u^{n-1}) \geq 0$), hence the following estimate holds:

$$(2.14)\quad \|u^n\| \leq C.$$

It is then possible to pass to the limit: $u^n \to u$ in $H^1(\mathcal{O})$ weakly and in $L^p(\mathcal{O})$ strongly for all finite p, and u is a solution of the Q. V. I. (2.5).

3. Q. V. I. OF PARABOLIC TYPE

With the notations of Section 2, the problem (1.12), (1.14), (1.15) (when we change t into $T - t$) can be formulated as follows: find a function

$$t \longrightarrow u(t) = \text{"}x \longrightarrow u(x,t)\text{"},$$

defined for $t > 0$, with values in $H^1(\mathcal{O})$, such that

$$(3.1)\quad \begin{cases} \left(\dfrac{\partial u}{\partial t}, v-u\right) + a(u,v-u) \geq (f,v-u) \\ \text{for all} \quad v \leq M(u), \quad u \leq M(u) \end{cases}$$

subject to the initial condition

$$u(0) = 0. \tag{3.2}$$

This is what is called a <u>Q. V. I. of evolution</u>. One can prove (Bensoussan-Lions, loc. cit.) the following:

THEOREM 3.1. *We assume that* (2.2) *holds true.* We suppose that*

$$(3.3)\quad \begin{cases} f \in L^{\infty}(\mathcal{O}\times]0,T[), \quad f \geq 0, \quad \dfrac{\partial f}{\partial t} \geq 0, \\ \dfrac{\partial f}{\partial t} \in L^2(\mathcal{O}\times]0,T[). \end{cases}$$

Then there exists a function u *which satisfies:*

$$(3.4)\quad \begin{cases} u \in L^{\infty}(0,T;H^1(\mathcal{O})) \cap L^{\infty}(\mathcal{O}\times]0,T[), \\ \dfrac{\partial u}{\partial t} \in L^2(\mathcal{O}\times]0,T[), \quad \dfrac{\partial u}{\partial t} \geq 0, \end{cases}$$

and which is a solution of (3.1), (3.2).

REMARK 3.1. The uniqueness is extremely likely but it is, in general, an open problem. If u is smooth enough (and there are indeed cases where one can prove regularity theorems which are sufficient for our purposes), then one can prove the uniqueness of u by showing that u is of the form $\inf_v J(v)$, in the problem we started from.

For the proof of Theorem 3.1 we refer to the papers mentioned in the Bibliography.

---

* But now a_0 can be <u>arbitrary</u> in $L^{\infty}(\mathcal{O})$.

4. Q. V. I. FOR A SYSTEM WHICH IS NOT OF CAUCHY-KOWALESKAYA TYPE

4.1 Statement of the problem

Let $\mathcal{O}$ be a bounded open set in $\mathbb{R}^n$, with smooth boundary Γ. Let A_0 be defined by

$$A_0\phi = -\sum \frac{\partial}{\partial x_i}\left(a_{ij}\frac{\partial\phi}{\partial x_j}\right), \tag{4.1}$$

where

$$\begin{cases} a_{ij} \in C^1(\bar{\mathcal{O}}), \quad \sum\limits_{i,j} a_{ij}(x)\xi_i\xi_j \geq \alpha_0 \sum\limits_i \xi_i^2, \quad \alpha_0 > 0, \\ a_{ij} = a_{ji} \quad \text{for all } i,j, \end{cases} \tag{4.2}$$

and let B_1, B_2 be defined by

$$\begin{cases} B_1\phi = \sum a_{j1}(x)\dfrac{\partial\phi}{\partial x_j}, \quad a_{j1} \in L^\infty(\mathcal{O}), \\ B_2\phi = \sum a_{j2}(x)\dfrac{\partial\phi}{\partial x_j} + a_2(x)\phi, \quad a_{j2} \in L^\infty(\mathcal{O}), \\ \qquad a_2 \geq \alpha > 0. \end{cases} \tag{4.3}$$

We are looking for functions $u_1 = u_1(x,t)$, $u_2 = u_2(x,t)$ which satisfy

$$\begin{cases} \dfrac{\partial u_1}{\partial t} + A_0u_1 + B_1u_1 - f_1 \leq 0, \\ u_1 - k_2 - u_2 \leq 0, \\ \left(\dfrac{\partial u_1}{\partial t} + A_0u_1 + B_1u_1 - f_1\right)\left(u_1 - k_2 - u_2\right) = 0 \\ \qquad \text{in } \mathcal{O}\times]0,+\infty[, \end{cases} \tag{4.4}$$

$$(4.5)\quad \begin{cases} A_0u_2 + B_2u_2 - f_2 \le 0, \quad u_2 - k_1 - u_1 \le 0, \\ (A_0u_2 + B_2u_2 - f_2)\ (u_2 - k_1 - u_1) = 0 \\ \qquad \text{in} \quad \mathcal{O} \times]0,\infty[\end{cases}$$

subject to the boundary conditions

$$(4.6)\qquad \frac{\partial u_1}{\partial \nu_{A_0}} = \frac{\partial u_2}{\partial \nu_{A_0}} = 0 \quad \text{on} \quad \Gamma \times]0,+\infty[,$$

and subject to the initial condition

$$(4.7)\qquad u_1(x,0) = 0 \quad \text{in} \quad \mathcal{O}.$$

In (4.4), (4.5) f_i and k_i are given. We shall assume

$$(4.8)\quad \begin{cases} f_i \in L^\infty(\mathcal{O} \times]0,T[), & f_i \ge 0, \\ \dfrac{\partial f_i}{\partial t} \in L^2(\mathcal{O} \times]0,T[), & \dfrac{\partial f_i}{\partial t} \ge 0, \quad i = 1,2 \end{cases}$$

and

$$(4.9)\qquad k_i = \text{constant}, \quad k_i > 0, \quad i = 1,2.$$

4.2 Problem (4.4)-(4.7) as a Q. V. I. of a non Cauchy-Kowaleskaya type

Let us introduce the following notations:

$$a_0(u,v) = \sum \int_{\mathcal{O}} a_{ij} \frac{\partial u}{\partial x_j} \frac{\partial v}{\partial x_i}\, dx,$$

$$a_1(u,v) = a_0(u,v) + (B_1u,v),$$

$$a_2(u,v) = a_0(u,v) + (B_2u,v).$$

Then the problem can be stated as follows:

$$(4.10)\quad \begin{cases} \left(\dfrac{\partial u_1}{\partial t}, v_1-u_1\right) + a_1(u_1, v_1-u_1) \geq (f_1, v_1-u_1), \\ a_2(u_2, v_2-u_2) \geq (f_2, v_2-u_2), \end{cases}$$

for all $v = \{v_1, v_2\}$ satisfying

$$(4.11)\quad v_1 \leq k_2 + u_2, \quad v_2 \leq k_1 + u_1,$$

where u_1 should satisfy (4.7), and $\{u_1, u_2\}$ should satisfy

$$(4.12)\quad u_1 \leq k_2 + u_2, \quad u_2 \leq k_1 + u_1.$$

If we set

$$a(u,v) = a_1(u_1, v_1) + a_2(u_2, v_2),$$

$$(f,v) = (f_1, v_1) + (f_2, v_2),$$

$$\Lambda u = \left\{\frac{\partial u_1}{\partial t}, 0\right\}, \quad Mu = \{k_2+u_2, k_1+u_1\},$$

then the problem can be stated as a Q. V. I. of evolution:

$$(4.13)\quad \begin{cases} (\Lambda u, v-u) + a(u,v-u) \geq (f,v-u), \text{ for all } v \leq M(u),^* \\ u \leq M(u), \\ \Lambda u(0) = 0. \end{cases}$$

<u>The operator</u> Λ <u>is degenerate</u>. This is why the Q. V. I. (4.13) is <u>not</u> of Cauchy-Kowaleskaya type.

REMARK 4.1. The boundary conditions which are (formally) related to the formulation (4.10) are:

---

* By this we mean, as it is natural, $v_1 \leq (M(u))_1$, $v_2 \leq (M(u))_2$.

$$
(4.14)\quad \begin{cases} \dfrac{\partial u_1}{\partial \nu_{A_0}} \le 0, \quad u_1 - k_2 - u_2 \le 0, \\ \qquad \dfrac{\partial u_1}{\partial \nu_{A_0}} (u_1 - k_2 - u_2) = 0, \\ \dfrac{\partial u_2}{\partial \nu_{A_0}} \le 0, \quad u_2 - k_1 - u_1 \le 0, \\ \qquad \dfrac{\partial u_2}{\partial \nu_{A_0}} (u_2 - k_1 - u_1) = 0; \end{cases}
$$

but we shall actually <u>prove</u> that (4.14) is satisfied in the particular sense of (4.6) (together with the inequalities $u_1 - k_2 - u_2 \le 0$, $u_2 - k_1 - u_1 \le 0$ which are satisfied not only on Γ but also in $\mathcal{O}$).

REMARK 4.2. We consider in what follows only the case when Λ is given by $\Lambda u = \left\{\frac{\partial u_1}{\partial t}, 0\right\}$; we shall return to a systematic study of (4.13) for more general operators Λ.

4.3 Statement of an existence theorem

In what follows, we denote by $H^2(\mathcal{O})$ the Sobolev space of order 2 on $\mathcal{O}$.

THEOREM 4.1. *Let* A_0, B_1, B_2 *be given satisfying* (4.1), (4.2), (4.3). *We assume that* (4.8), (4.9) *hold true. Then there exist functions* u_1, u_2 *which satisfy*

(4.15) $$u_i \ge 0, \quad i = 1,2,$$

(4.16) $$u_i \in L^2(0,T;H^2(\mathcal{O})) \cap L^\infty(0,T;H^1(\mathcal{O})), \quad i = 1,2,$$

(4.17) $$\frac{\partial u_1}{\partial t} \in L^2(\mathcal{O} \times]0,T[), \quad \frac{\partial u_1}{\partial t} \ge 0,$$

and which satisfy (4.10), (4.11), (4.12), (4.6), (4.7).

REMARK 4.3. The uniqueness is an open problem. One can prove that the solution we are going to construct in the following sub-sections is the maximum solution among all (possible only one!) nonnegative solutions.

We prove Theorem 4.1 in several steps. We begin with an "approximate system", using penalty terms.

4.4. The penalized system

We are going to solve first the system:

$$(4.18)\quad \begin{cases} \dfrac{\partial u_{1\varepsilon}}{\partial t} + (A_0 + B_1)u_{1\varepsilon} + \dfrac{1}{\varepsilon}\beta_{1\varepsilon}^{+} = f_1 , \\ (A_0 + B_2)u_{2\varepsilon} + \dfrac{1}{\varepsilon}\beta_{2\varepsilon}^{+} = f_2 , \end{cases}$$

where $\varepsilon > 0$,

$$(4.19)\quad \beta_{1\varepsilon} = u_{1\varepsilon} - k_2 - u_{2\varepsilon} , \quad \beta_{2\varepsilon} = u_{2\varepsilon} - k_1 - u_{1\varepsilon} ,$$

and where as usual

$$\phi^{+} = \sup \{\phi, 0\} .$$

To the equations (4.18) we add the boundary and initial conditions

$$(4.20)\quad \begin{cases} \dfrac{\partial u_{i\varepsilon}}{\partial \nu_{A_0}} = 0, \quad i = 1,2, \\ u_{1\varepsilon}(0) = 0 . \end{cases}$$

REMARK 4.4. The problem (4.18)-(4.20) is a nonlinear problem which is not of Cauchy-Kowaleskaya type.

REMARK 4.5. The terms $\frac{1}{\varepsilon}\beta_{i\varepsilon}^{+}$ are the "penalty terms" ; they are intended to take care, when $\varepsilon \to 0$, of the constraints $u_1 - k_2 - u_2 \le 0$, $u_2 - k_1 - u_1 \le 0$.

We now show how to solve the nonlinear problem (4.18), (4.20). We use an iteration procedure. We define first u_1^0, u_2^0 by

$$(4.21)\quad \begin{cases} \dfrac{\partial u_1^0}{\partial t} + (A_0 + B_1)u_1^0 = f_1 \quad \text{in } \mathcal{O} \times]0,T[, \\ u_1^0(0) = 0, \quad \dfrac{\partial u_1^0}{\partial \nu_{A_0}} = 0 \quad \text{on } \Gamma \times]0,T[, \end{cases}$$

$$(4.22)\quad \begin{cases} (A_0 + B_2)u_2^0 = f_2, \\ \dfrac{\partial u_2^0}{\partial \nu_{A_0}} = 0 \quad \text{on } \Gamma. \end{cases}$$

We remark that (4.22) is an elliptic problem where t plays the role of a parameter; since f_2 depends on t, then u_2^0 depends on t. We then define in an inductive manner u_1^n, u_2^n by

$$(4.23)\quad \begin{cases} \dfrac{\partial u_1^n}{\partial t} + (A_0 + B_1)u_1^n + \dfrac{1}{\varepsilon}(u_1^n - k_2 - u_2^{n-1})^+ = f_1, \\ \qquad (A_0 + B_2)u_2^n + \dfrac{1}{\varepsilon}(u_2^n - k_1 - u_1^{n-1})^+ = f_2,^* \\ \dfrac{\partial u_i^n}{\partial \nu_{A_0}} = 0, \quad i = 1,2, \\ u_1^n(0) = 0. \end{cases}$$

---

* To simplify a little bit, we shall assume that $a_2(v,v) \geq \alpha_2 \|v\|^2$, $\alpha_2 > 0$. If not, we should add λu_2^n (resp. λu_2^{n-1}) to the right (resp. left) hand side of this equation and choose λ large enough.

Of course this problem is uncoupled; we can compute u_1^n and u_2^n separately, once u_1^{n-1}, u_2^{n-1} are known. The functions u_i^n defined in this way depend on ε; we write u_i^n instead of $u_{i\varepsilon}^n$ to simplify the notations. Each problem in (4.23) is nonlinear of monotonic type so that the existence and uniqueness of a solution is standard. We now proceed to obtain a number of estimates.

LEMMA 4.1. *One has*

$$u_1^0 \geq u_1^1 \geq 0, \quad u_2^0 \geq u_2^1 \geq 0.$$

PROOF. We take the scalar product of (4.21) with $(u_1^0 - u_1^1)^-$ * and we take the scalar product of the first equation in (4.23) (for $n = 1$) with $-(u_1^0 - u_1^1)^-$. Adding up, we obtain

$$(4.24) \quad \begin{cases} \left(\dfrac{\partial}{\partial t}(u_1^0 - u_1^1), (u_1^0 - u_1^1)^-\right) \\ + a_1(u_1^0 - u_1^1, (u_1^0 - u_1^1)^-) \\ - \dfrac{1}{\varepsilon}((u_1^1 - k_2 - u_2^0)^+, (u_1^0 - u_1^1)^-) = 0. \end{cases}$$

We shall write $a_i(\phi)$ instead of $a_i(\phi,\phi)$ and we observe that $a_1(\phi,\phi^-) = -a_1(\phi^-)$. Then (4.24) can be written

$$\begin{cases} \dfrac{1}{2}\dfrac{d}{dt}|(u_1^0 - u_1^1)^-|^2 + a((u_1^0 - u_1^1)^-) + \dfrac{1}{\varepsilon}X = 0, \\ X = ((u_1^1 - k_2 - u_2^0)^+, (u_1^0 - u_1^1)^-) \geq 0, \end{cases}$$

---

* $\phi^- = \sup\{-\phi, 0\}$.

hence

$$(4.25)\qquad \frac{1}{2}\frac{d}{dt}\,|(u_1^0 - u_1^1)^-|^2 + a((u_1^0 - u_1^1)^-) \le 0.$$

But $(u_1^0 - u_1^1)^- = 0$ for $t = 0$ so that (4.25) and Gronwall's Lemma imply $(u_1^0 - u_1^1)^- = 0$. Therefore $u_1^0 \ge u_1^1$. One proves $u_2^0 \ge u_2^1$ in a similar manner. To prove that $u_1^1 \ge 0$, we multiply the first equation in (4.23) (for $n = 1$) by $(u_1^1)^-$. We obtain

$$(4.26)\qquad \begin{cases} \frac{1}{2}\frac{d}{dt}\,|(u_1^1)^-|^2 + a((u_1^1)^-) + \frac{1}{\varepsilon}Y + (f_1, (u_1^1)^-) = 0 \\ \text{where } \; Y = -((u_1^1 - k_2 - u_2^0)^+, (u_1^1)^-). \end{cases}$$

But $Y = 0$; indeed to compute Y we have to integrate over the set $u_1^1 \le 0$ <u>and</u> $u_1^1 - k_2 - u_2^0 \ge 0$ which is empty (since $k_2 > 0$ and $u_2^0 \ge 0$). Then, since $f_1 \ge 0$, (4.26) implies

$$(4.27)\qquad \frac{1}{2}\frac{d}{dt}\,|(u_1^1)^-|^2 + a((u_1^1)^-) \le 0$$

hence $(u_1^1)^- = 0$ follows. One proves that $u_2^1 \ge 0$ in a similar manner.

LEMMA 4.2. *One has*

$$(4.28)\qquad \begin{cases} u_1^0 \ge u_1^1 \ge \dots \ge u_1^{n-1} \ge u_1^n \ge \dots \ge 0, \\ u_2^0 \ge u_2^1 \ge \dots \ge u_2^{n-1} \ge u_2^n \ge \dots \ge 0. \end{cases}$$

PROOF. We assume by induction that

$$(4.29)\qquad u_i^{n-2} \ge u_i^{n-1} \ge 0, \quad i = 1,2,$$

and we prove that

$$(4.30) \qquad u_i^{n-1} \geq u_i^n \geq 0, \quad i = 1,2.$$

We multiply the first equation (4.23), for $n-1$, by $(u_1^{n-1} - u_1^n)^-$, and for n by $-(u_1^{n-1} - u_1^n)^-$. Adding up, we obtain:

$$(4.31) \quad \begin{cases} -\frac{1}{2}\frac{d}{dt}|(u_1^{n-1} - u_1^n)^-|^2 - a_1((u_1^{n-1} - u_1^n)^-) \\ \qquad\qquad + \frac{1}{\varepsilon} X^n = 0, \end{cases}$$

where

$$(4.32) \quad \begin{cases} X^n = ((u_1^{n-1} - k_2 - u_2^{n-2})^+ \\ \qquad - (u_1^n - k_2 - u_2^{n-1})^+, (u_1^{n-1} - u_1^n)^-). \end{cases}$$

In X^n we integrate over the subset, say E^n, of $\mathcal{O}$ where $u_1^{n-1} \leq u_1^n$; but since $-u_2^{n-1} \geq -u_2^{n-2}$ a.e. by (4.29), we have $u_1^n - k_2 - u_2^{n-1} \geq u_1^{n-1} - k_2 - u_2^{n-2}$ over E^n, and therefore $(u_1^n - k_2 - u_2^{n-1})^+ \geq (u_1^{n-1} - k_2 - u_2^{n-2})^+$ over E^n, hence it follows that $X^n \leq 0$. Then (4.31) and $(u_1^{n-1} - u_1^n)^- = 0$ for $t = 0$ imply that $(u_1^{n-1} - u_1^n)^- = 0$.

To prove that $u_1^n \geq 0$, we multiply the first equation (4.23) by $(u_1^n)^-$; we observe that (since $u_2^{n-1} \geq 0$) $((u_1^n - k_2 - u_2^{n-1})^+, (u_1^n)^-) = 0$ so that $-\frac{1}{2}\frac{d}{dt}|(u_1^n)^-|^2 - a((u_1^n)^-) = (f_1, (u_1^n)^-)$ hence $u_1^n \geq 0$ follows. One proves in a similar manner (4.30) for $i = 2$.

LEMMA 4.3. *One has*

$$(4.33) \qquad \|u_i^n\|_{L^2(0,T;H^1(\mathcal{O}))} \leq C, \quad i = 1,2,$$

where C *is a constant which does not depend on* n *or on* ε.

PROOF. We multiply the first (resp. second) equation (4.23) by u_1^n (resp. u_2^n). Since $u_i^n \geq 0$, the penalty terms give a positive contribution, and therefore

$$\text{(4.34)} \quad \begin{cases} \left(\dfrac{\partial u_1^n}{\partial t}, u_1^n\right) + a_1(u_1^n) \leq (f_1, u_1^n), \\ \qquad\qquad a_2(u_2^n) \leq (f_2, u_2^n), \end{cases}$$

hence (4.33) follows by standard arguments. Actually we obtain more for u_2^n, namely

$$\text{(4.35)} \quad \|u_2^n\|_{L^\infty(0,T;H^1(\mathcal{O}))} \leq C.$$

LEMMA 4.4. *One has*

$$\text{(4.36)} \quad \frac{\partial u_i^n}{\partial t} \geq 0, \quad i = 1,2.$$

PROOF. Let us set $\dfrac{\partial u_i^n}{\partial t} = w_i^n$. Taking the t-derivative of the equations (4.23) we obtain

$$\text{(4.37)} \quad \begin{cases} \dfrac{\partial w_1^n}{\partial t} + (A_0 + B_1)w_1^n + \dfrac{1}{\varepsilon}\dfrac{\partial}{\partial t}(u_1^n - k_2 - u_2^{n-1})^+ = \dfrac{\partial f_1}{\partial t}, \\ (A_0 + B_2)w_2^n + \dfrac{1}{\varepsilon}\dfrac{\partial}{\partial t}(u_2^n - k_1 - u_1^{n-1})^+ = \dfrac{\partial f_2}{\partial t}. \end{cases}$$

We notice that

$$\text{(4.38)} \quad w_1^n(0) = f_1(0) \geq 0.$$

We prove (4.36) by induction on n. The result is straightforward for $n = 0$. We assume it until $n - 1$ and we prove it for n. We multiply the first equation

(4.37) by $(w_1^n)^-$; we obtain

$$(4.39)\quad -\frac{1}{2}\frac{d}{dt}\,|(w_1^n)^-|^2 - a_1((w_1^n)^-) + \frac{1}{\varepsilon}\,Y^n = \left(\frac{\partial f_1}{\partial t}, (w_1^n)^-\right),$$

where

$$Y^n = \left(\frac{\partial}{\partial t}\,(u_1^n - k_2 - u_2^{n-1})^+, (w_1^n)^-\right)$$

$$= \int_{\mathcal{O}\cap[u_1^n \geq k_2+u_2^{n-1}]} (w_1^n - w_2^{n-1})\,(w_1^n)^-\,dx$$

$$= -\int_{[u_1^n \geq k_2+u_2^{n-1}]} [((w_1^n)^-)^2 + w_2^{n-1}(w_1^n)^-]\,dx \leq 0.$$

Then (4.39) gives

$$(4.40)\quad \frac{1}{2}\frac{d}{dt}\,|(w_1^n)^-|^2 + a_1((w_1^n)^-) + \left(\frac{\partial f_1}{\partial t}, (w_1^n)^-\right) \leq 0.$$

By virtue of (4.38), $(w_1^n)^- = 0$ for $t = 0$, hence $(w_1^n)^- = 0$ follows. We prove similarly that $(w_2^n)^- = 0$.

LEMMA 4.5. *One has*

$$(4.41)\quad \left\|\frac{\partial u_1^n}{\partial t}\right\|_{L^2(0,T;L^2(\mathcal{O}))} + \|u_1^n\|_{L^\infty(0,T;H^1(\mathcal{O}))} \leq C,$$

$$(4.42)\quad \|u_2^n\|_{L^\infty(0,T;H^1(\mathcal{O}))} \leq C.$$

PROOF. We take the scalar product of the first equation (4.23) by $\frac{\partial u_1^n}{\partial t}$ and of the second equation by $\frac{\partial u_2^n}{\partial t}$. Since the contribution of the penalty terms is ≥ 0 (by virtue of Lemma 4.4), we obtain:

$$(4.43) \qquad \left|\frac{\partial u_1^n}{\partial t}\right|^2 + a_1\left(u_1^n, \frac{\partial u_1^n}{\partial t}\right) \le \left(f_1, \frac{\partial u_1^n}{\partial t}\right),$$

$$(4.44) \qquad a_2\left(u_2^n, \frac{\partial u_2^n}{\partial t}\right) \le \left(f_2, \frac{\partial u_2^n}{\partial t}\right).$$

Since a_0 is symmetric, we obtain

$$(4.45) \qquad \begin{cases} \left|\frac{\partial u_1^n}{\partial t}\right|^2 + \frac{1}{2}\frac{d}{dt} a_0(u_1^n) \le \left(f_1 - B_1 u_1^n, \frac{\partial u_1^n}{\partial t}\right), \\ \frac{1}{2}\frac{d}{dt} a_0(u_2^n) \le \left(f_2 - B_2 u_2^n, \frac{\partial u_2^n}{\partial t}\right). \end{cases}$$

But by Lemma 4.3, $B_1 u_1^n$ and $B_2 u_2^n$ are bounded in $L^2(\mathcal{O} \times]0,T[)$ hence (4.41), (4.42) follow.

We can now pass to the limit in n. By virtue of the preceding estimates, we have

$$(4.46) \qquad \begin{cases} u_i^n \downarrow u_{i\varepsilon} \text{ in } L^p(\mathcal{O} \times]0,T[) \\ \text{strongly for every finite } p, \quad i = 1,2, \end{cases}$$

$$(4.47) \qquad \begin{cases} u_i^n \to u_{i\varepsilon} \text{ in } L^\infty(0,T;H^1(\mathcal{O})) \\ \text{weak star}, \quad i = 1,2, \end{cases}$$

$$(4.48) \qquad \frac{\partial u_1^n}{\partial t} \to \frac{\partial u_{1\varepsilon}}{\partial t} \text{ in } L^2(\mathcal{O} \times]0,T[) \text{ weakly.}$$

It is a simple matter to pass to the limit, at least in the variational form. We write (4.23) in the form

$$\left(\frac{\partial u_1^n}{\partial t}, v_1\right) + a_1(u_1^n, v_1) +$$

$$+ \frac{1}{\varepsilon}((u_1^n - k_2 - u_2^{n-1})^+, v_1) = (f_1, v_1),$$

$$a_2(u_2^n, v_2) + \frac{1}{\varepsilon}((u_2^n - k_1 - u_1^{n-1})^+, v_2) = (f_2, v_2)$$

for all $v_1, v_2 \in H^1(\mathcal{O})$ and we obtain in the limit the variational form of (4.18), (4.20). It is then a standard matter to check that $u_{i\varepsilon}$ has values in $H^2(\mathcal{O})$ and the non-variational form (4.18), (4.20) follows. From the estimates for u_i^n we obtain:

(4.49) $$u_{i\varepsilon} \geq 0, \quad \frac{\partial u_{i\varepsilon}}{\partial t} \geq 0,$$

(4.50) $$\|u_{i\varepsilon}\|_{L^\infty(0,T;H^1(\mathcal{O}))} \leq C, \quad i = 1,2,$$

(4.51) $$\left\|\frac{\partial u_{i\varepsilon}}{\partial t}\right\|_{L^2(\mathcal{O} \times]0,T[)} \leq C.$$

4.5. Another estimate for the penalized system

We multiply the first (resp. second) equation (4.18) by $A_0(u_{1\varepsilon} - u_{2\varepsilon})$ (resp. by $A_0(u_{2\varepsilon} - u_{1\varepsilon})$). We observe that

$$(\beta_{1\varepsilon}^+, A_0(u_{1\varepsilon} - u_{2\varepsilon})) = a_0(\beta_{1\varepsilon}^+, \beta_{1\varepsilon}) = a_0(\beta_{1\varepsilon}^+) \geq 0$$

and similarly $(\beta_{2\varepsilon}^+, A_0(u_{2\varepsilon} - u_{1\varepsilon})) \geq 0$. Therefore, by adding up, we obtain

(4.52) $$\left\{ \begin{aligned} |A_0(u_{1\varepsilon} - u_{2\varepsilon})|^2 \leq \Big(&f_1 - f_2 - \frac{\partial u_{i\varepsilon}}{\partial t} \\ &- B_1 u_{1\varepsilon} + B_2 u_{2\varepsilon}, A_0(u_{1\varepsilon} - u_{2\varepsilon})\Big). \end{aligned} \right.$$

But by virtue of (4.50), (4.51), we have that

$$g_\varepsilon = f_1 - f_2 - \frac{\partial u_{1\varepsilon}}{\partial t} - B_1 u_{1\varepsilon} + B_2 u_{2\varepsilon}$$

satisfies

$$\|g_\varepsilon\|_{L^2(\mathcal{O}\times]0,T[)} \le C \tag{4.53}$$

so that (4.52) implies

$$\|A_0(u_{1\varepsilon}-u_{2\varepsilon})\|_{L^2(\mathcal{O}\times]0,T[)} \le C. \tag{4.54}$$

But (4.18) implies

$$\begin{cases} \frac{1}{\varepsilon}(\beta^+_{1\varepsilon}-\beta^+_{2\varepsilon}) = h_\varepsilon, \\ h_\varepsilon = g_\varepsilon - A_0(u_{1\varepsilon}-u_{2\varepsilon}). \end{cases} \tag{4.55}$$

By virtue of (4.53), (4.54) we have

$$\|h_\varepsilon\|_{L^2(\mathcal{O}\times]0,T[)} \le C. \tag{4.56}$$

But one easily checks that $\beta^+_{1\varepsilon}\,\beta^+_{2\varepsilon} = 0$ a.e., so that (4.55), (4.56) imply

$$\begin{cases} \|\frac{1}{\varepsilon}(\beta^+_{1\varepsilon}-\beta^+_{2\varepsilon})\|^2_{L^2(\mathcal{O}\times]0,T[)} = \|\frac{1}{\varepsilon}\beta^+_{1\varepsilon}\|^2_{L^2(\mathcal{O}\times]0,T[)} \\ \qquad + \|\frac{1}{\varepsilon}\beta^+_{2\varepsilon}\|^2_{L^2(\mathcal{O}\times]0,T[)} \le C. \end{cases} \tag{4.57}$$

Then (4.18), (4.57) imply:

$$\begin{cases} \frac{\partial u_{1\varepsilon}}{\partial t} + A_0 u_{1\varepsilon} + B_1 u_{1\varepsilon} = F_{1\varepsilon}, F_{1\varepsilon} \\ \text{bounded in } L^2(\mathcal{O}\times]0,T[), \\ \qquad A_0 u_{2\varepsilon} + B_2 u_{2\varepsilon} = F_{2\varepsilon}, F_{2\varepsilon} \\ \text{bounded in } L^2(\mathcal{O}\times]0,T[). \end{cases} \tag{4.58}$$

It follows that

LEMMA 4.6. *One has*

(4.59) $$\|u_{i\varepsilon}\|_{L^2(0,T;H^2(\mathcal{O}))} \le C, \quad i = 1,2.$$

4.6. Proof of Theorem 4.1

One can show that

(4.60) $u_{i\varepsilon}$ decreases as ε decreases,

as in the case of Q. V. I. of Cauchy-Kowaleskaya type (cf. Bensoussan-Lions, loc. cit.). Therefore as $\varepsilon \downarrow 0$,

(4.61) $$\begin{cases} u_{i\varepsilon} \downarrow u_i \text{ in } L^p(\mathcal{O} \times]0,T[) \text{ strongly,} \\ \text{for each finite } p, \end{cases}$$

(4.62) $$\begin{cases} u_{i\varepsilon} \to u_i \text{ in } L^2(0,T;H^2(\mathcal{O})) \text{ weakly,} \\ \text{in } L^\infty(0,T;H^1(\mathcal{O})) \text{ weak star,} \end{cases}$$

(4.63) $$\frac{\partial u_{1\varepsilon}}{\partial t} \to \frac{\partial u_1}{\partial t} \text{ in } L^2(\mathcal{O} \times]0,T[) \text{ weakly,}$$

and $u_i \ge 0$, $\frac{\partial u_i}{\partial t} \ge 0$. We obtain from (4.18) that

(4.64) $$\begin{cases} \left(\frac{\partial u_{1\varepsilon}}{\partial t} + (A_0 + B_1)u_{1\varepsilon} - f_1\right)\beta^-_{1\varepsilon} = 0, \\ ((A_0 + B_2)u_{2\varepsilon} - f_2)\beta^-_{2\varepsilon} = 0, \end{cases}$$

and since we have (4.61), then $\beta^-_{i\varepsilon} \to \beta^-_i$ in $L^2(\mathcal{O} \times]0,T[)$ strongly so that

(4.65) $$\begin{cases} \left(\frac{\partial u_1}{\partial t} + (A_0 + B_1)u_1 - f_1\right)\beta^-_1 = 0, \\ ((A_0 + B_2)u_2 - f_2)\beta^-_2 = 0. \end{cases}$$

But by (4.57), $\beta^+_1 = \beta^+_2 = 0$ so that $\beta^-_i = \beta_i$ and the theorem is proved.

BIBLIOGRAPHY

BAIOCCHI, C.

[1] *Problemes à frontière libre en hydraulique.* C. R. A. S. Paris, April (1974).

BENSOUSSAN, A., and LIONS, J. L.

[1] *Problemes de temp d'ârret optimal et Inequations Variationelles paraboliques.* Applicable Analysis (1974).

[2] C. R. A. S. Paris, 276 (1973)
1.) pp. 1189-1192; 2.) pp. 1333-1338.

[3] C. R. A. S. Paris, (1974)
1.) March; 2.) March; 3.) April.

[4] Inter. J. of Optimization and Applied Math., (1974).

[5] Symposium Versailles, December (1973).

[6] I. F. I. P. Symposium, Novosibirsk, July (1974).

[7] Q. V. I. and Nash points, to appear.

BENSOUSSAN, A., GOURSAT, M., and LIONS, J. L.

[1] C. R. A. S. Paris, 276 (1973), pp. 1333-1336.

JOLY, J. L., and MOSCO, U.

[1] C. R. A. S. Paris, (1974).

LAETSH, TH.

[1] to appear.

LIONS, J.L., and STAMPACCHIA, G.

[1] *Variational Inequalities.* C. P. A. M., XX (1967), pp. 493-519.

TARTAR, L.

[1] C. R. A. S. Paris, (1974).

SEMILINEAR WAVE EQUATIONS

by

L. A. MEDEIROS

Department of Mathematics
Federal University of Rio de Janeiro
Rio de Janeiro, G. B., Brasil

INTRODUCTION

In 1951, Schiff [16] published a paper which motivated an existence theorem for an important class of nonlinear wave equations, like equation (*), proved by Jörgens [8]. This paper of Jörgens gave origin to a set of papers on this class of equations using different methods and boundary conditions. The first abstract approach was done by Browder [1], using the spectral theory of self adjoint operations in Hilbert spaces. Also Segal [17] studied the same type of problems as an application of the Hille-Yosida theorem for linear semigroups. For abstract second order equations, the reader can also look Carroll-Cooper [5], Goldstein [6], Carroll [4] and Medeiros [13]. For the Cauchy problem in a noncylindrical domain for the equation

$$(*) \qquad \frac{\partial^2 u}{\partial t^2} - \Delta u + m^2 u - u^{2k+1} = f,$$

$k \geq 2$, $m > 0$, the first results were obtained by Lions

[11] by the penalty method. This result was generalized in two different directions. One by Cooper-Bardos [3], considering a more general class of noncylindrical domains, another by Medeiros [13] considering a general nonlinear term in (*), proving his results with the help of a convergence theorem of Strauss [18]. Finally in Cooper-Medeiros [2] can be found a general result which generalizes all those above mentioned for the noncylindrical case, including also one obtained by Rogak-Kazarinoff [15].

This is an expository paper that contains the lectures I delivered in the Department of Mathematics of Tulane University, during the winter of 1974. We divide the paper into two sections. Section 1 contains certain results about Sobolev spaces and the proof of the theorem of Strauss on convergence of measurable functions. In Section 2 is given the proof of the existence of weak solutions for the Cauchy problem for the equation (*) with a general nonlinear term.

1. NOTATIONS AND TERMINOLOGY

Let us represent by $\mathbb{R}^n$ the euclidean space of dimension n. Its vectors we represent by $x = (x_1, x_2, \ldots, x_n)$. Let Ω be an open set of $\mathbb{R}^n$ and $L^2(\Omega)$ be the Hilbert space of real square integrable functions u in Ω. By $H^m(\Omega)$ we represent the Sobolev space of functions u of $L^2(\Omega)$ such that the derivative $D^\alpha u$, in the sense of distributions, belongs to $L^2(\Omega)$ for all $|\alpha| \le m$ with the inner product

(1) $$(u, v)_m = \sum_{|\alpha| \le m} \int_\Omega D^\alpha u \, D^\alpha v \, dx.$$

The space $\mathcal{D}(\Omega)$ of infinitely continuously differentiable functions with compact support in Ω, is contained in $H^m(\Omega)$. The closure of $\mathcal{D}(\Omega)$ in $H^m(\Omega)$ we represent by $H_0^m(\Omega)$. It is true that $\mathcal{D}(\mathbb{R}^n)$ is dense in $H^m(\mathbb{R}^n)$, but $H_0^m(\Omega)$ is usually properly contained in $H^m(\Omega)$ when $\Omega \neq \mathbb{R}^n$. For all the results mentioned in this paragraph on Sobolev space, the reader can look at Lions [11]. Suppose we have $\Omega = \{(x', x_n) \mid x_n > 0\}$, where $x' = (x_1, x_2, \ldots, x_{n-1})$. It is true that $\mathcal{D}(\overline{\Omega})$ is dense in $H^m(\Omega)$. Let us represent by Γ the set of (x', x_n) such that $x_n = 0$, that is, Γ is a hyperplane of $\mathbb{R}^n$. If $u \in \mathcal{D}(\overline{\Omega})$ it makes sense to talk of the restriction of u to Γ, which is called the <u>trace</u> of u on Γ represented by $u|\Gamma$. Let γ be the mapping from $\mathcal{D}(\overline{\Omega})$ into $\mathcal{D}(\Gamma)$ defined by $\gamma u = u|\Gamma$. It is possible to define $H^\alpha(\Omega)$ for $\alpha > 0$ real, for example $H^{1/2}(\Omega)$ (Lions [11]). It is known that γ is continuous. The extension of γ to $H^1(\Omega)$ by closure is represented by γ_0 and it is known that γ_0 maps $H^1(\Omega)$ into $H^{1/2}(\Gamma)$ continuously. When Ω is a bounded open set of Ω with smooth boundary Γ, one can establish the above results for Γ and that the kernel of γ_0 is $H_0^1(\Omega)$; that is, the set of u in $H^1(\Omega)$ such that $\gamma_0 u = u|_\Gamma = 0$ is equal to $H_0^1(\Omega)$. Therefore, we say that if u belongs to $H_0^1(\Omega)$, it is zero on the boundary of Ω, when Ω has a smooth boundary. This is a very elementary formulation of the trace theorem (cf. Lions [11]).

In the next section we need a theorem on convergence proved by Strauss [18].

THEOREM 1. *Let* Ω *be an open set of* $\mathbb{R}^n$ *with finite measure and* $(U_\nu)_{\nu\in\mathbb{N}}$ *a sequence of real measurable functions on* Ω. *Let us consider two sequences* $(F_\nu)_{\nu\in\mathbb{N}}$ *and* $(G_\nu)_{\nu\in\mathbb{N}}$ *of functions from* $\mathbb{R}$ *into* $\mathbb{R}$, *such that* $F_\nu \circ U_\nu$, $G_\nu \circ U_\nu$ *are measurable on* Ω *for each* $\nu \in \mathbb{N}$. *We suppose:*

a) $F_\nu \circ U_\nu$ *converges to* v *almost everywhere on* Ω.

b) $\int_\Omega |F_\nu(U_\nu(x))G_\nu(U_\nu(x))|dx < C$, C *independent of* $\nu \in \mathbb{N}$.

c) $G_\nu \to \infty$ *as* $F_\nu \to \infty$.

Then

d) The function v *is in* $L^1(\Omega)$,

e) $F_\nu \circ U_\nu \to v$ *in* $L^1(\Omega)$.

REMARK 1. By $F \circ v$ we represent the function defined by $(F\circ v)(x) = F(v(x))$. The hypothesis c) is equivalent to saying that for each $M > 0$ there exists $N > 0$ independent of ν, such that if $|F_\nu(s)| \geq N$ then $|G_\nu(s)| \geq M$ for all s in $\mathbb{R}$.

PROOF OF THEOREM 1. For each ν in $\mathbb{N}$ let us represent by Ω_ν the set

$$\Omega_\nu = \{x \in \Omega \mid |G_\nu(U_\nu(x))| \leq 1\}$$

and put $\Omega'_\nu = \Omega - \Omega_\nu$. For all x in Ω'_ν we have $|G_\nu(U_\nu(x))| > 1$, therefore

$$|F_\nu(U_\nu(x))| \le |G_\nu(U_\nu(x))|\,|F_\nu(U_\nu(x))|,$$

that is

$$(2) \qquad \int_{\Omega'_\nu} |F_\nu(U_\nu(x))|dx < C$$

by b), for all $\nu \in \mathbb{N}$.

For each x in Ω_ν it follows from hypothesis c) and from Remark 1, that since $|G_\nu(U_\nu(x))| \le 1$ in Ω_ν, we have $|F_\nu(U_\nu(x))| \le N$ for some constant N, for all x in Ω_ν and all $\nu \in \mathbb{N}$. Therefore,

$$(3) \qquad \int_{\Omega_\nu} |F_\nu(U_\nu(x))|dx < N\mu(\Omega_\nu) \le N\mu(\Omega).$$

From (2) and (3) it follows that

$$\int_\Omega |F_\nu(U_\nu(x))|dx \le C + N\mu(\Omega),$$

except for a finite number of ν, that is, the sequence of integrals of $(F_\nu \circ U_\nu)_{\nu\in\mathbb{N}}$ is bounded. This implies that $(F_\nu \circ U_\nu)_{\nu\in\mathbb{N}}$ satisfies the conditions of the Fatou lemma and we can say that $|v| = \lim_\nu |F_\nu \circ U_\nu|$ is integrable on Ω, that is, v is integrable on Ω, which proves the first part of Theorem 1.

To complete the proof, we observe that, from Egoroff's theorem, for each $\delta > 0$ there exists a measurable set $\Omega_0 \subset \Omega$, such that $\mu(\Omega_0) < \delta$ and $F_\nu \circ U_\nu$ converges to v uniformly on $\Omega-\Omega_0$. It follows that in $\Omega-\Omega_0$ we have:

$$\lim_{\nu\to\infty} \int_{\Omega-\Omega_0} |F_\nu(U_\nu(x)) - v(x)|dx = 0.$$

We need only to prove that

$$\lim_{\nu\to\infty} \int_{\Omega_0} |F_\nu(U_\nu(x)) - v(x)|dx = 0.$$

In fact, we have

$$\int_{\Omega_0} |F_\nu(U_\nu(x)) - v(x)|dx \le \int_{\Omega_0} |F_\nu(U_\nu(x))|dx + \int_{\Omega_0} |v(x)|dx.$$

Since v is integrable and $\mu(\Omega_0) < \delta$, it follows that $\int_{\Omega_0} |v(x)|dx < \varepsilon$. It is sufficient to prove that

$$\sup_\nu \int_{\Omega_0} |F_\nu(U_\nu(x))|dx < \varepsilon$$

when $\mu(\Omega_0) < \delta$. Let $M = 2C/\varepsilon$, C being the constant of hypothesis b) and let N be the constant of Remark 1. Consider the set

$$\Omega_{0\nu} = \{x \in \Omega_0 \mid |G_\nu(u_\nu(x))| < M\}$$

and $\Omega'_{0\nu} = \Omega_0 - \Omega_{0\nu}$. On $\Omega'_{0\nu}$ we have:

$$(4) \quad \int_{\Omega'_{0\nu}} |F_\nu(u_\nu(x))|dx \le M^{-1} \int_{\Omega'_{0\nu}} |G_\nu(u_\nu(x))|\,|F_\nu(u_\nu(x))|dx <$$

$$< \frac{C}{M} = \frac{\varepsilon}{2}.$$

On $\Omega_{0\nu}$ we have, by Remark 1, that $|F_\nu(u_\nu(x))| < N$ because $|G_\nu(u_\nu(x))| < M$.

It follows that:

$$(5) \qquad \int_{\Omega_{0\nu}} |F_\nu(u_\nu(x))|dx \le N\mu(\Omega_{0\nu}) < N\delta = \frac{\varepsilon}{2},$$

when we take $\delta = 2N/\varepsilon$.

Combining (4) and (5) yields the desired result. This completes the proof of Theorem 1.

REMARK 2. Suppose $G_\nu(s) = s$ for all $\nu \in \mathbb{N}$ and $s \in \mathbb{R}$, that is, G_ν is the identity mapping. We have:

a) $F_\nu \circ u_\nu \to v$ almost everywhere on Ω.

b) $\int_\Omega |F_\nu(u_\nu(x))| |u_\nu(x)| dx < C$ for all ν.

c) $s \to \infty$ when $F_\nu \to \infty$, that is, for each $M > 0$, there exists $N > 0$ such that $|s| \geq M$ when $|F_\nu(s)| \geq N$, which is equivalent to say that $(F_\nu)_{\nu \in \mathbb{N}}$ is uniformly bounded on bounded sets of $\mathbb{R}$.

COROLLARY 1. *Let* $(u_\nu)_{\nu \in \mathbb{N}}$ *be a sequence of real measurable functions, bounded in* $L^p(\Omega)$ *and convergent to* u *almost everywhere in* Ω. *Then it converges strongly in* $L^q(\Omega)$ *to the same limit* u, *for all* $1 \leq q < p$ *and hence weakly in* $L^p(\Omega)$.

2. THE CAUCHY PROBLEM IN NONCYLINDRICAL DOMAINS

Let B be a strip of the euclidean space $\mathbb{R}^{n+1}$ defined by

$$B = \{(x,t) \mid x = (x_1, x_2, \ldots, x_n) \in \mathbb{R}^n, \ 0 < t < T < \infty\},$$

and suppose that Q is a bounded open set contained in B. We represent by Ω_s the intersection of Q with the plane $t = s$ of $\mathbb{R}^{n+1}$. By Ω_0 and Ω_T we represent the intersections of Q with the planes $t = 0$ and $t = T$. Represent by $\partial Q = \Gamma$ the boundary of Q and by $\Gamma_s = \partial Q \cap [t = s]$ the boundary of Ω_s. By

$\Sigma = \bigcup_{0 \le s \le T} \Gamma_s$ we denote the lateral boundary of Q. therefore, Q is the subset of B whose boundary ∂Q is $\Omega_0 \cup \Sigma \cup \Omega_T$.

In this paragraph we study the following problem: Given u_0, u_1 in appropriate functions spaces and $F : \mathbb{R} \to \mathbb{R}$ continuous and $sF(s) \ge 0$, find a function $u : Q \to \mathbb{R}$ such that

$$u(x,0) = u_0(x) \ , \ u_t(x,0) = u_1(x) \quad \text{on} \ \Omega_0$$

and

$$\frac{\partial^2 u}{\partial t^2} - \Delta u + F(u) = 0 \quad \text{on} \ Q.$$

In this paper this problem is solved under the following hypotheses on the lateral boundary of Q:

ASSUMPTION 1. *Let* Ω_t^* *the projection of* Ω_t *on the hyperplane* $t = 0$. *We suppose that* Ω_t^* *is increasing with* t, *that is,* $\Omega_t^* \subseteq \Omega_s^*$ *if* $t \le s$.

ASSUMPTION 2. *For each* t *in* $[0,T]$, Ω_t *has the following regularity property: If* $u \in H^1(\mathbb{R}^n)$ *and* $u = 0$ *on* $\mathbb{R}^n - \Omega_t^*$ *a.e., then the restriction of* u *to* Ω_t *belongs to* $H_0^1(\Omega_t)$.

The following theorem generalizes Lions [10] to the case of a general nonlinear term $F(s)$, Strauss [18] to the case of noncylindrical domains, and also includes the results contained in Rogak-Kazarinoff [15].

THEOREM 1. *Suppose that Assumptions 1, 2 hold and suppose* f, u_0, u_1 *are given such that*

(1) $f \in L^2(Q)$,

(2) $u_0 \in H_0^1(\Omega_0)$,

(3) $u_1 \in L^2(\Omega_0)$.

Then there exists a function $u : Q \to \mathbb{R}$ *with the following properties:*

(4) $u \in L^\infty(0,T;H_0^1(\Omega_t))$,

(5) $u_t \in L^\infty(0,T;L^2(\Omega_t))$,

(6) $\dfrac{\partial^2 u}{\partial t^2} - \Delta u + m^2 u + F(u) = f$ *in the weak sense on* Q,

(7) $u(x,0) = u_0(x)$, $u_t(x,0) = u_1(x)$ *on* Ω_0.

REMARK 1. The conditions *(4,5,6)* and the continuity of F implies that u and u_t are continuous in $0 \le t \le T$, so that the initial values $u(x,0)$ and $u_t(x,0)$ make sense. Here m is a nonnegative real number, that could be zero.

The proof of Theorem 1 will be done by the penalty method. Let $M : B \to \mathbb{R}$ be defined by $M(x,t) = 0$ on Q and $M(x,t) = 1$ outside Q. Let us call U_0, U_1 the extensions of u_0, u_1 to the $\mathbb{R}^n$, defined by $U_0 = U_1 = 0$ outside Ω_0. Let us prove the theorem in the case $f = 0$, but the same argument can be used in the case $f \neq 0$. The proof of Theorem 1 will be done as a consequence of the following theorem:

THEOREM 2. *For each* $\varepsilon > 0$ *there exists a function* $U^\varepsilon : B \mapsto \mathbb{R}$ *satisfying the following conditions:*

(8) $U^\varepsilon \in L^\infty(0,T;H^1(\mathbb{R}^n))$ *and* U^ε *belongs to a bounded set in this space when* $\varepsilon \to 0$.

(9) $U_t^\varepsilon \in L^\infty(0,T;L^2(\mathbb{R}^n))$ *and belongs to a bounded set in this space when* $\varepsilon \to 0$.

(10) $\frac{1}{\sqrt{\varepsilon}} M \frac{\partial U^\varepsilon}{\partial t}$ *belongs to a bounded set of* $L^2(B)$ *when* $\varepsilon \to 0$.

(11) $\frac{\partial^2 U^\varepsilon}{\partial t^2} - \Delta U^\varepsilon + m^2 U^\varepsilon + \frac{1}{\varepsilon} M \frac{\partial U^\varepsilon}{\partial t} + F(U^\varepsilon) = 0$ *on* B *in the weak sense.*

(12) $U^\varepsilon(x,0) = U_0(x)$, $U_t(x,0) = U_1(x)$.

REMARK 2. To prove Theorem 1 it is sufficient to prove that the solution U^ε of Theorem 2 converges to U in the weak sense, when $\varepsilon \to 0$, and that the restriction of U to Q satisfies all the conclusions of Theorem 1.

We shall prove Theorem 2 in three steps. First, we study the linear case and we use the Galerkin method; second, we suppose F Lipschitzian and we use the Picard method; finally, we suppose $F : \mathbb{R} \to \mathbb{R}$ continuous plus $sF(s) \geq 0$ and we approximate F by a sequence $(F_\nu)_{\nu \in \mathbb{N}}$ of Lipschitzian functions, of the same type as F, and we use Theorem 1 of Section 1.

PROPOSITION 1. *The linear problem*

(13) $\frac{\partial^2 U^\varepsilon}{\partial t^2} - \Delta U^\varepsilon + m^2 U^\varepsilon + \frac{1}{\varepsilon} M \frac{\partial U^\varepsilon}{\partial t} = h$ *weakly on* B,

with B *in* $L^2(B)$,

(14) $U^\varepsilon(x,0) = U_0(x)$, $U_t^\varepsilon(x,0) = U_1(x)$ *on* Ω_0,

has a solution U^ε *satisfying the regularity conditions of Theorem 2.*

PROOF OF PROPOSITION 1. We use the Galerkin method. Let $w_1, w_2, \dots, w_\nu, \dots$ be a family of test functions in $H^1(\mathbb{R}^n)$ such that for each natural number m the vectors $w_1, w_2, \dots, w_m$ are linearly independent and the finite linear combinations $\sum_{i=1}^{m} \alpha_i w_i$ (with m varying) are dense in $H^1(\mathbb{R}^n) \cap L^1(\mathbb{R}^n)$. For each m, let $U_m^\varepsilon(x,t) = \sum_{i=1}^{m} g_{im}(t) w_i(x)$ be defined by

$$\left(\frac{\partial^2 U_m^\varepsilon}{\partial t^2}, w_j\right) + m^2\left(U_m^\varepsilon, w_j\right) + \sum_{i=1}^{m}\left(\frac{\partial U_m^\varepsilon}{\partial x_i}, \frac{\partial w_j}{\partial x_i}\right) + \\ + \frac{1}{\varepsilon}\left(M\frac{\partial U_m}{\partial t}, w_j\right) = \left(h, w_j\right),$$

$$U_m^\varepsilon(x,0) = U_{0m}^\varepsilon(x) = \sum_{i=1}^{m} \alpha_{im} w_i \to U_0 \quad \text{strongly in } H^1(\mathbb{R}^n),$$

$$\frac{\partial U_m}{\partial t}(x,0) = U_{1m}(x) = \sum_{i=1}^{m} \beta_{im} w_i \to U_1 \quad \text{strongly in } L^2(\mathbb{R}^n).$$

We have to solve a system of ordinary differential equations to find the functions $g_{im}(t)$. Since $\det(w_1, w_j) \neq 0$, it follows that the system has a solution. The above system can be written in the following form:

$$(15)\quad \left(\frac{\partial^2 U_m^\varepsilon}{\partial t^2}, w_j\right) + \left(U_m^\varepsilon, w_j\right)_1 + \frac{1}{\varepsilon}\left(M\frac{\partial U_m}{\partial t}, w_j\right) = \left(h, w_j\right)$$

$$(16)\quad U_m^\varepsilon(x,0) = U_{0m}^\varepsilon \to U_0 \quad \text{strongly in} \quad H^1(\mathbb{R}^n),$$

$$(17)\quad \frac{\partial U_m^\varepsilon}{\partial t}(x,0) = U_{1m} \to U_1 \quad \text{strongly in} \quad L^2(\mathbb{R}^n), \quad \text{where}$$

$$(u,v)_1 = m^2(u,v) + \sum_{i=1}^{n}\left(\frac{\partial u}{\partial x_i}, \frac{\partial v}{\partial x_i}\right)$$

and (u,v) is the inner product in $L^2(\mathbb{R}^n)$.

Multiply the equation *(15)* by w_j and add in j; we obtain

$$\frac{d}{dt}\left\{\frac{1}{2}\left\|U_m^\varepsilon(t)\right\|_1^2 + \frac{1}{2}\left\|\frac{\partial U_m^\varepsilon}{\partial t}\right\|^2\right\} + \frac{1}{\varepsilon}\left(M\frac{\partial U_m^\varepsilon}{\partial t}, \frac{\partial U_m^\varepsilon}{\partial t}\right) = \left(h, \frac{\partial U_m^\varepsilon}{\partial t}\right).$$

Let us write U_m in the place of U_m^ε. It follows that

$$(18)\quad \frac{d}{dt}\left\{\frac{1}{2}\left\|U_m(t)\right\|_1^2 + \frac{1}{2}\left\|U_m'(t)\right\|^2\right\} + \frac{1}{\varepsilon}\left(MU_m', U_m'\right) \le \le \frac{1}{2}\|h(t)\|^2 + \frac{1}{2}\|U_m'(t)\|^2 .$$

REMARK 3. If we write $E(t) = \frac{1}{2}\left\|U_m(t)\right\|_1^2 + \frac{1}{2}\left\|U_m'(t)\right\|^2$, $H(t) = \frac{1}{\varepsilon}(MU_m', U_m') \ge 0$, $C(t) = \frac{1}{2}\|h(t)\|^2_{L^2(\mathbb{R}^n)}$, the inequality *(18)* can be written in the following form:

$$E'(t) - E(t) + H(t) \le C(t).$$

Multiplying both sides of the last inequality by e^{-t}, we obtain

$$\frac{d}{dt}\left[e^{-t} E(t)\right] + e^{-t} H(t) \le e^{-t} C(t).$$

Integrating from 0 to $t \le T$, we have:

$$E(t) + \int_0^t e^{t-s} H(s)ds \le \int_0^t e^{t-s} C(s)ds + e^T E(0).$$

From the Remark 3, the inequality *(18)* can be written in the following form:

$$(19) \quad \frac{1}{2}\left\|U_m(t)\right\|_1^2 + \frac{1}{2}\left\|U'_m(t)\right\|^2 + \frac{1}{\varepsilon}\int_0^T (MU'_m \,,\, U'_m)dt \le$$
$$\le K_T\left(1 + \left\|U_m(0)\right\|_1^2 + \left\|U'_m(0)\right\|^2\right).$$

By the initial conditions *(16)* and *(17)*, the second member of *(19)* can be bounded by a constant $K(T)$, that is:

$$(20) \quad \frac{1}{2}\left\|U_m^\varepsilon(t)\right\|_1^2 + \frac{1}{2}\left\|\frac{\partial U_m^\varepsilon}{\partial t}(t)\right\|^2 + \int_0^T\left(\frac{1}{\sqrt{\varepsilon}}\frac{\partial U_m^\varepsilon}{\partial t}, \frac{1}{\sqrt{\varepsilon}}\frac{\partial U_m^\varepsilon}{\partial t}\right)dt$$
$$\le K(T),$$

for each $0 \le t < T < \infty$, all $\varepsilon > 0$ and $m \in \mathbb{N}$. Therefore, from *(20)* we can conclude:

(21) U_m^ε belongs to a bounded set B_1 of $L^\infty(0,T;H^1(\mathbb{R}^n))$ independent of ε and m.

(22) $\dfrac{\partial U_m^\varepsilon}{\partial t}$ belongs to a bounded set B_2 of $L^\infty(0, T; L^2(\mathbb{R}^n))$ independent of ε and m.

(23) $\frac{1}{\sqrt{\varepsilon}} M \frac{\partial U_m^\varepsilon}{\partial t}$ belongs to a bounded set of $L^2(B)$ when $\varepsilon \to 0$ and $m \to \infty$. From *(21)*, *(22)* and *(23)*, there exists a subsequence U_ν^ε of U_m^ε satisfying the following conditions:

(24) $U_\nu^\varepsilon \to U^\varepsilon$ in $L^\infty(0,T;H^1(\mathbb{R}^n))$ in the weak star topology.

(25) $\frac{\partial U_\nu^\varepsilon}{\partial t} \to \frac{\partial U^\varepsilon}{\partial t}$ in $L^\infty(0,T;L^2(\mathbb{R}^n))$ in the weak star topology.

(26) $\frac{1}{\sqrt{\varepsilon}} M \frac{\partial U_\nu^\varepsilon}{\partial t} \to \frac{1}{\sqrt{\varepsilon}} M \frac{\partial U^\varepsilon}{\partial t}$ in $L^2(B)$ weakly.

The limit U^ε is a weak solution of *(11)*. To check the initial value, we have that *(16)* plus *(24)* imply $U^\varepsilon(0) = U_0$ and that *(17)* plus *(25)* imply $\frac{\partial U^\varepsilon}{\partial t}(0) = U_1$.

PROPOSITION 2. *The Cauchy problem*

$$(13') \qquad \frac{\partial^2 U^\varepsilon}{\partial t^2} - \Delta U^\varepsilon + m^2 U^\varepsilon + \frac{1}{\varepsilon}\frac{\partial U^\varepsilon}{\partial t} + F(U^\varepsilon) = f$$

weakly on B, *with* F *Lipschitzian and* f *in* $L^2(B)$,

$$(14') \qquad U^\varepsilon(x,0) = U_0(x),\ U_t^\varepsilon(x,0) = U_1(x)$$

has a solution U^ε *satisfying the regularity conditions of Theorem 2.*

PROOF OF PROPOSITION 2. Let $U_0^\varepsilon, U_1^\varepsilon, \ldots, U_\nu^\varepsilon, \ldots$ be a sequence of approximations for the solution of equation *(11)* on B, defined by the following conditions:

$$U_0^\varepsilon(x,t) = 0 \quad \text{on} \quad B$$

and U_n^ε, for $n = 1,2,\ldots$ are defined as the weak solution of the following linear problem:

$$\text{(27)} \qquad \frac{\partial^2 U_\nu^\varepsilon}{\partial t^2} - \Delta U_\nu^\varepsilon + m^2 U_\nu^\varepsilon + \frac{1}{\varepsilon} M \frac{\partial U_\nu^\varepsilon}{\partial t} + F(U_{\nu-1}^\varepsilon) = h,$$

$$\text{(28)} \qquad U_\nu^\varepsilon(x,0) = U_0(x),$$

$$\text{(29)} \qquad \frac{\partial U_\nu^\varepsilon}{\partial t}(x,0) = U_1(x).$$

By Proposition 1 we know that the system *(27)-(29)* has a solution for each $\nu = 1,2,\ldots,$ satisfying the hypotheses *(8)-(10)* of Theorem 2. Therefore, to prove Proposition 2 we need only to verify that the sequence of successive approximations converges in an appropriate topology. To prove the convergence, let U_n^ε and U_{n-1}^ε be two solutions of *(27)-(29)* and put $W_n^\varepsilon = U_n^\varepsilon - U_{n-1}^\varepsilon$. It follows that W_n^ε is a solution of

$$\text{(30)} \qquad \frac{\partial^2 W_n}{\partial t^2} - \Delta W_n^\varepsilon + m^2 W_n^\varepsilon + \frac{1}{\varepsilon} M \frac{\partial W_n^\varepsilon}{\partial t} + \{F(U_n^\varepsilon) - F(U_{n-1}^\varepsilon)\} = 0$$

with zero initial conditions. Taking the inner product of both sides of *(30)* with $\frac{\partial W_n^\varepsilon}{\partial t}$, we obtain

$$(31)\quad \frac{d}{dt}\frac{1}{2}\left\{\left\|\frac{\partial W_n^\varepsilon}{\partial t}\right\|^2 + \left\|W_n\right\|_1^2\right\} + \frac{1}{\varepsilon}\left(M\frac{\partial W_n^\varepsilon}{\partial t}, \frac{\partial W_n^\varepsilon}{\partial t}\right) =$$

$$= -\left(F\left(U_n^\varepsilon\right) - F\left(U_{n-1}^\varepsilon\right), \frac{\partial W_n^\varepsilon}{\partial t}\right)$$

Let us write $W_n^\varepsilon = w^n$, $U_n^\varepsilon = u^n$ and

$$\|w^n\|_E^2 = \frac{1}{2}\|w_t^n\|^2 + \frac{1}{2}\|w^n\|_1^2 .$$

Since $\frac{1}{\varepsilon}(Mw_t^n, w_t^n) \geq 0$, $w^n(0) = 0$, we obtain, after integrating *(31)* from 0 to $t < T < \infty$:

$$(32)\qquad \|w^n(t)\|_E^2 \leq \int_0^t \left|(F(u^{n-1}) - F(u^{n-2}), w_t^n)\right| ds$$

for all $\varepsilon > 0$, $n = 1,2,\ldots$. By the Schwarz inequality and by the Lipschitz condition we can write *(32)* in the form:

$$(33)\qquad \|w^n(t)\|_E^2 \leq C\int_0^t \|u^{n-1} - u^{n-2}\|\,\|w_t^n\| ds .$$

Let $e_n(t)$ defined by

$$(34)\qquad e_n(t) = \operatorname*{ess\,sup}_{0\leq s\leq t} \|w^n(s)\|_E^2$$

for $0 < t < T < \infty$, $n = 1,2,\ldots$. From *(33)* we obtain:

$$\|w^n(t)\|_E^2 \leq C^2 t\int_0^t \|w^{n-1}(s)\|^2\, ds + \frac{1}{4t}\int_0^t \|w_t^n(s)\|^2 ds,$$

and by the definition *(34)* it follows that

$$(35)\qquad e_n(t) \leq C^2 t\int_0^t \|w^{n-1}(s)\|^2 ds + \frac{1}{4}e_n(t).$$

Now, let us estimate $\|w^{n-1}(s)\|^2$ by $e_{m-1}(t)$. Since $w^{n-1}(0) = 0$, we obtain:

$$w^{n-1}(t) = \int_0^t w_t^{n-1}(s)ds \ , \quad \text{so}$$

$$|w^{n-1}(t)| \leq \int_0^t |w_t^{n-1}(s)|ds \leq \sqrt{t}\left(\int_0^t |w_t^{n-1}(s)|^2 \, ds\right)^{\frac{1}{2}}.$$

Square both sides of the last inequality and integrate on $\mathbb{R}^n$; we obtain:

$$\|w^{n-1}(\xi)\|^2 \leq \xi \int_0^{\xi} \|w_t^{n-1}(s)\|^2 \, ds \ , \ 0 < \xi \leq t < \infty,$$

or

$$(36) \quad \int_0^t \|w^{n-1}(s)\|^2 \, ds \leq \frac{T^2}{2} \int_0^t \|w^{n-1}(s)\|_E^2 \, ds.$$

From *(35)* and *(36)* we obtain:

$$e_n(t) \leq K(T)t \int_0^t \|w^{n-1}(s)\|_E^2 \, ds,$$

that is,

$$e_n(t) \leq K(T)t \int_0^t e_{n-1}(s)ds.$$

It follows from *(37)* that for $n = 1,2,\ldots$, we obtain:

$$e_n(t) \leq K^{n-1} \frac{t^{n-1}}{(n-1)!} e_1(t)$$

and we have the convergence of the series $\sum_{n=1}^{\infty}(u_n - u_{n-1})$ in the norm $\sqrt{e_m(t)}$ and therefore, the convergence of the sequence $(u_n)_{n\in\mathbb{N}}$ in the same norm. We can say that $(U_n^\varepsilon)_{u\in\mathbb{N}}$ is convergent in the norm $\sqrt{e_m(t)}$, therefore in $H^1(\mathbb{R}^n)$ and almost everywhere, and the limit U^ε satisfies the conditions:

$$U^\varepsilon \in L^\infty(0,T;H^1(\mathbb{R}^n)) \ , \ U_t^\varepsilon \in L^\infty(0,T;L^2(\mathbb{R}^n))$$

and is a solution of *(13')* with the initial condition *(14')*, which proves Proposition 2.

PROOF OF THEOREM 2. Suppose F is continuous and $sF(s) \geq 0$ for all s in $\mathbb{R}$. Let $(F_\nu)_{\nu\in\mathbb{N}}$ be a sequence of Lipschitzian functions such that $sF_\nu(s) \geq 0$ and $F_\nu \to F$ uniformly on bounded sets of $\mathbb{R}$ (cf. Strauss [18]). It follows by Proposition 2, that for each $\nu \in \mathbb{N}$ there exists a solution U_ν^ε of the following Cauchy problem:

$$\text{(38)} \qquad \frac{\partial^2 U_\nu}{\partial t^2} - \Delta U_\nu^\varepsilon + m^2 U_\nu^\varepsilon + \frac{1}{\varepsilon} M \frac{\partial U_\nu^\varepsilon}{\partial t} + F_\nu(U_\nu^\varepsilon) = 0$$

weakly on B,

$$\text{(39)} \qquad U_\nu^\varepsilon(x,0) = U_0(x) \ , \ \frac{\partial U_\nu^\varepsilon}{\partial t}(x,0) = U_1(x).$$

Taking the inner product of both sides of *(38)* with $\frac{\partial U_\nu^\varepsilon}{\partial t}$, we obtain

$$\frac{d}{dt}\left\{\frac{1}{2}\|U_\nu^\varepsilon(t)\|_1^2 + \frac{1}{2}\left\|\frac{\partial U_\nu^\varepsilon}{\partial t}\right\|^2\right\} + \frac{1}{\varepsilon}\left(M\frac{\partial U_\nu^\varepsilon}{\partial t}, \frac{\partial U_\nu^\varepsilon}{\partial t}\right) + \int_{\mathbb{R}^n} F_\nu(U_\nu^\varepsilon)\frac{\partial U_\nu^\varepsilon}{\partial t}\,dx = 0.$$

Let G_ν be a function such that $\frac{d}{ds} G_\nu = F_\nu$ and $G_\nu(0) = 0$. It follows that the last equation can be written as:

$$\frac{d}{dt}\left\{\frac{1}{2}\|U_\nu^\varepsilon(t)\|_1^2 + \frac{1}{2}\left\|\frac{\partial U_\nu^\varepsilon}{\partial t}\right\|^2 + \int_{\mathbb{R}^n} G_\nu(U_\nu^\varepsilon)dx\right\} + \\ + \frac{1}{\varepsilon}\left(M\frac{\partial U_\nu^\varepsilon}{\partial t}, \frac{\partial U_\nu^\varepsilon}{\partial t}\right) = 0.$$

Integrating from 0 to t, we obtain

$$(40) \qquad \frac{1}{2}\left\|\frac{\partial U_\nu^\varepsilon}{\partial t}\right\|^2 + \frac{1}{2}\|U_\nu^\varepsilon(t)\|_1^2 + \int_{\mathbb{R}^n} G_\nu(U_\nu^\varepsilon)dx + \\ + \frac{1}{\varepsilon}\int_0^T\left(M\frac{\partial U_\nu^\varepsilon}{\partial t}, \frac{\partial U_\nu^\varepsilon}{\partial t}\right) = \frac{1}{2}\left\|\frac{\partial U_\nu^\varepsilon}{\partial t}(0)\right\|^2 + \frac{1}{2}\|U_\nu^\varepsilon(0)\| + \int_{\mathbb{R}^n} G_\nu(U_\nu^\varepsilon(0))dx.$$

REMARK 4. Let us suppose U_0 is bounded. For the unbounded case look at Strauss [18].

By the hypothesis on U_0, U_1 and G_ν, it follows that the right hand side of *(40)* is bounded because

$$\int_{\mathbb{R}^n} G_\nu(U_0)dx \longrightarrow \int_{\mathbb{R}^n} G(U_0)dx$$

where G is such that $\frac{dG}{ds} = F$, $G(0) = 0$. Therefore the right hand side of *(40)* is bounded by a constant which does not depend on ε or ν, that is,

$$(41) \qquad \frac{1}{2}\left\|\frac{\partial U_\nu^\varepsilon}{\partial t}\right\|^2 + \frac{1}{2}\|U_\nu^\varepsilon\|_1^2 + \int_{\mathbb{R}^n} G_\nu(U_\nu^\varepsilon)dx + \\ + \frac{1}{\varepsilon}\int_0^T\left(M\frac{\partial U_\nu^\varepsilon}{\partial t}, \frac{\partial U_\nu^\varepsilon}{\partial t}\right) \le C_T ,$$

for all $\nu \in \mathbb{N}$, $\varepsilon > 0$, $0 \le t \le T$. From *(41)* we have:

(42) U_ν^ε is in a bounded set of $L^\infty(0,T;H^1(\mathbb{R}^n))$, independent of ν and ε.

(43) $\dfrac{\partial U_\nu^\varepsilon}{\partial t}$ belongs to a bounded set of $L^\infty(0,T;L^2(\mathbb{R}^n))$ independent of ν and $\varepsilon > 0$.

(44) $\dfrac{1}{\sqrt{\varepsilon}} M \dfrac{\partial U_\nu^\varepsilon}{\partial t}$ belongs to a bounded set of $L^2(B)$ independent of ν, when $\varepsilon \to 0$.

From *(42)*, *(43)* and *(44)*, we can select a subsequence U_i^ε of U_ν^ε such that $U_i^\varepsilon \to U^\varepsilon$ in $L^\infty(0,T;H^1(\mathbb{R}^n))$ in the weak star topology; $\dfrac{\partial U_i^\varepsilon}{\partial t} \to \dfrac{\partial U^\varepsilon}{\partial t}$ in $L^\infty(0,T;L^2(\mathbb{R}^n))$ in the weak star topology and $\dfrac{1}{\sqrt{\varepsilon}} M \dfrac{\partial U_i^\varepsilon}{\partial t} \to \dfrac{1}{\sqrt{\varepsilon}} M \dfrac{\partial U^\varepsilon}{\partial t}$ weakly in $L^2(B)$.

Now we need to control the term $\int_{\mathbb{R}^n} G_i(U_i^\varepsilon)dx$ in the inequality *(41)*. By *(42)* and *(43)* it follows that U_i^ε and its first derivatives with respect to x and t are bounded in $L^2(\mathbb{R}^n)$, for all $0 \le t \le T$. Therefore, by Rellich theorem, there exists a subsequence $(U_k^\varepsilon)_{k \in \mathbb{N}}$ such that $U_k^\varepsilon \to U^\varepsilon$ almost everywhere on B. It follows that if we fix (x,t) in B, $U_k^\varepsilon(x,t) \to U^\varepsilon(x,t)$ a.e. and $F_k(U_k^\varepsilon(x,t)) \to F_k(U^\varepsilon(x,t))$ a.e., because the F_k are continuous. Since $F_k \to F$ uniformly on bounded sets, we have

$$F_k(U_k^\varepsilon(x,t)) \to F(U^\varepsilon(x,t)) \quad \text{a.e.}$$

To obtain what we need, we must prove that

$F_k(U_k^\varepsilon) \to F(U^\varepsilon)$ in $L^1(B)$. For this, we need only to verify the conditions of Theorem 1 of Section 1, that is, we need to verify that for each bounded set $B' \subset B$, we have

$$\int_{B'} |F_k(U_k^\varepsilon)||U_k^\varepsilon \, dx| < C$$

for all k. In fact, for each k we have

$$\frac{\partial^2 U_k^\varepsilon}{\partial t^2} - \Delta U_k^\varepsilon + m^2 U_k^\varepsilon + \frac{1}{\varepsilon} M \frac{\partial U_k^\varepsilon}{\partial t} + F_k(U_k^\varepsilon) = 0$$

and taking the inner product of both sides with U_k^ε we obtain:

$$(45) \quad \int_B \frac{\partial^2 U_k^\varepsilon}{\partial t^2} U_k^\varepsilon \, dx\, dt - \int_B \Delta U_k^\varepsilon U_k^\varepsilon \, dx\, dt +$$

$$+ \frac{1}{\varepsilon}\int_B M \frac{\partial U_k^\varepsilon}{\partial t} U_k^\varepsilon \, dx\, dt + \int_B F_k(U_k^\varepsilon) U_k^\varepsilon \, dx\, dt = 0$$

The first and the second term in *(45)* are bounded. To bound the last term, is sufficient to prove the following lemma:

LEMMA 1. *If* $\frac{1}{\sqrt{\varepsilon}} MU_t^\varepsilon$ *is bounded in* $L^2(B)$ *when* $\varepsilon \to 0$, *the same is true for* $\frac{1}{\varepsilon} MU_t^\varepsilon$.

PROOF OF LEMMA 1. In fact, we have:

$$(46) \quad \left|\frac{1}{\varepsilon}\int_B MU_t^\varepsilon U^\varepsilon \, dx\, dt\right| \le \frac{1}{\varepsilon}\int_B |MU_t^\varepsilon||U^\varepsilon| dx\, dt \le$$

$$\le \frac{1}{\varepsilon}\int_0^T \|MU_t^\varepsilon\| \, \|MU^\varepsilon\| dt.$$

Since the domain Q is increasing and $U^{\varepsilon}(0) = U_0$ is zero outside Ω_0, we obtain:

$$|MU^{\varepsilon}(t)| \leq \int_0^T |MU_t^{\varepsilon}(t)|dt.$$

By Schwarz inequality, taking the square of both sides and integrating on $\mathbb{R}^n$, we obtain:

$$\text{(47)} \qquad \int_{\mathbb{R}^n} |MU^{\varepsilon}(t)|^2 \, dx \leq T \int_B |MU_t^{\varepsilon}|^2 \, dx\, dt.$$

From *(46)* and *(47)* we obtain the proof of Lemma 1.

It follows from Lemma 1 and *(45)* that $\int_B F_k(U_k^{\varepsilon})U_k^{\varepsilon}\, dx\, dt$ is bounded independent of k and ε. Therefore, by Theorem 1 of Section 1, we have $F_\nu(U_\nu^{\varepsilon}) \to F(U^{\varepsilon})$ in $L^1(B)$, and it is possible to prove that U^{ε} is, in fact, the solution claimed in Theorem 2.

PROOF OF THEOREM 1. To prove Theorem 1, we take U^{ε} to be the solution given by Theorem 2 of the penalized equation and we shall prove that there exists a subsequence of U^{ε} that converges weakly to U on B; the restriction of U to Q gives the solution claimed in Theorem 1. In fact, by Theorem 2 we obtain a subsequence U^{η} of U^{ε} satisfying the following conditions:

(48) $U^{\eta} \to U$ in $L^{\infty}(0,T;H^1(\mathbb{R}^n))$ in the weak star topology.

(49) $U_t^{\eta} \to U_t$ in $L^{\infty}(0,T;L^2(\mathbb{R}^n))$ in the weak star topology.

Now, we have to prove that $F(U^{\eta}) \to F(U)$ in $L^1(B)$.

But by the Rellich theorem, we obtain a subsequence U^η such that $U^\eta \to U$ a.e. on B. To apply Theorem 1 of Section 1, it is sufficient to prove that $\int_B |F(U^\eta)||U^\eta| dx\,dt$ is bounded by a constant independent of η. In fact, take the inner product in $L^2(B)$ with U^η; we obtain:

$$\left\langle \frac{\partial^2 U}{\partial t^2}, U^\eta \right\rangle + \left\langle -\Delta U^\eta, U^\eta \right\rangle + \left\langle m^2 U^\eta, U^\eta \right\rangle + \\ + \frac{1}{\eta}\left\langle M \frac{\partial U^\eta}{\partial t}, U^\eta \right\rangle + \left\langle F(U^\eta), U^\eta \right\rangle = 0.$$

It follows in $L^2(\mathbb{R}^n)$ the following equality:

$$(50)\quad (U_t^\eta(T), U_t^\eta(T)) - (U_t^\eta(0), U_t^\eta(0)) - \langle U_t^\eta, U_t^\eta \rangle + \\ + \langle \nabla U^\eta, \nabla U^\eta \rangle + \langle mU^\eta, mU^\eta \rangle + \\ + \frac{1}{\varepsilon}\int_0^T \int_{\mathbb{R}^n} M \frac{\partial U^\eta}{\partial t} U^\eta \, dx\, dt + \langle F(U^\eta), U^\eta \rangle = 0.$$

The first five terms of *(50)* are bounded and to bound the sixth one is sufficient to prove that $\frac{1}{\eta}\langle MU_t^\eta, U^\eta \rangle$ is bounded when $\eta \to 0$, which is a consequence of Lemma 1. Therefore, $\langle F(U^\eta), U^\eta \rangle$ is bounded and we have $F(U^\eta) \to F(U)$ in $L^1(B)$.

We can suppose that $\frac{1}{\sqrt{\eta}} M \frac{\partial U^\eta}{\partial t} \to \psi$ weakly on $L^2(B)$ when $\eta \to 0$. Therefore $M \frac{\partial U^\eta}{\partial t} \to 0$ in the same sense. Since $U_t^\eta \to U_t$ weakly in $L^\infty(0,T;L^2(\mathbb{R}^n))$, we have $M \frac{\partial U}{\partial t} = 0$ on B, that is, $\frac{\partial U}{\partial t} = 0$ on $B - \overline{Q}$ by the definition of M. Let us represent by u^η the restriction of U^η to Q. We have for u^η

$$\frac{\partial^2 u^\eta}{\partial t^2} - \Delta u^\eta + m^2 u^\eta + F(u^\eta) = 0.$$

Taking the limit in the sense of the distributions on B and representing by u the restriction of U to Q, we obtain:

$$\frac{\partial^2 u}{\partial t^2} - \Delta u + m^2 u + F(u) = 0$$

and $u(0) = u_0$, $u_t(0) = u_1$. By *(48)* and *(49)* we have $u \in L^\infty(0,T;H^1(\Omega_t))$ and $u_t \in L^\infty(0,T;L^2(\Omega_t))$. To complete the proof we need only to verify that $u \in L^\infty(0,T;H_0^1(\Omega_t))$. We have $\frac{\partial U}{\partial t} = 0$ on $B - \overline{Q}$, thus $U(x,t) = w(x)$ a.e. on $B - \overline{Q}$ with w independent of t. We have $x \to U(x,t)$ is in $H^1(\mathbb{R}^n)$ a.e. in $t \in (0,T)$. Then $w \in H^1(\mathbb{R}^n - \overline{\Omega}_0)$ and $U(x,0) = w$ on $\mathbb{R}^n - \overline{\Omega}_0$. Since $U(x,0) = U_0 = 0$ on $\mathbb{R}^n - \overline{\Omega}_0$, by definition of U_0, it follows that $w = 0$ on $\mathbb{R}^n - \Omega_0$. We supposed Ω_t increasing, therefore $U = 0$ on $\mathbb{R}^n - \Omega_t^*$ a.e. We have $U(t) \in H^1(\mathbb{R}^n)$, $U(t) = 0$ on $\mathbb{R}^n - \Omega_t^*$ a.e. in t, then by the regularity assumption on Σ, the restriction u of U to Q is in $H_0^1(\Omega_t)$, which completes the proof of Theorem 1.

BIBLIOGRAPHY

1. BROWDER, F. E., *On nonlinear wave equations*, Math. Zeit. 80 (1962), 249-264.

2. COOPER, J. and MEDEIROS, L. A., *The Cauchy problem for nonlinear wave equations in domains with moving boundary,* Annali della Scuola Normale Superiori di Pisa 26 (1972), 829-838.

3. COOPER, J. and BARDOS, C., *A nonlinear wave equation in time dependent domain,* Journal of Mathematical Analysis and Applications 42 (1973), 29-60.

4. CARROLL, R. W., *Abstract Methods in Partial Differential Equations,* Harper-Row, New York, 1969.

5. CARROLL, R. W. and COOPER, J., *Remarks on some variable domain problems in abstract evolution equations,* Math. Ann. 188 (1970), 143-164.

6. GOLDSTEIN, J. A., *Semi-groups and second order differential equations,* Journal of Functional Analysis 4 (1969), 50-70.

7. GOLDSTEIN, J. A., *Time dependent hyperbolic equations,* Journal of Functional Analysis, 4 (1969), 31-49; 6 (1970), 347.

8. JÖRGENS, K., *Über die Anfangswertprobleme im Grossen für eine Klasse nichtlinearer Wellengleichungen,* Math. Zeit. 77 (1961), 295-308.

9. LIONS, J. L., *Une remarque sur les problémes d'evolution nonlinéaires dans les domaines noncylindriques,* Rev. Romaine de Math. 9 (1964), 11-18.

10. LIONS, J. L., *Quelques Methodes de Resolution des Problẽmes aux Limites Nonlinéaires,* Dunod, Paris, 1969.

11. LIONS, J. L., *Problẽmes aux Limites dans les Equations aux Derivées Partielles,* L'Université de Montréal, Canada, 1965.

12. MEDEIROS, L. A., *The initial value problem for nonlinear wave equations in Hilbert space,* Trans. Amer. Math. Soc. 136 (1969), 305-327.

13. MEDEIROS, L. A., *Solutions faibles de l'equation des ondes semi-lineaires dans les domaines non-cylindriques*, Comptes Rendus Acad. Sci. Paris 272 (1971), 1457-1458.

14. RIVERA, P. H., *El problema de Cauchy para una clase de ecuaciones no lineales de la onda*, Anais Acad. Bras. Ciencias 44 (1973), 393-397.

15. ROGAK, E. D. and KAZARINOFF, N. D., *Exterior initial value problem for quasi linear hyperbolic equations in time dependent domains*, Journal of Math. Analysis and Appl. 27 (1969), 116-126.

16. SCHIFF, L. I., *Nonlinear meson theory of nuclear forces*, Phys. Reviews 84 (1951), 1-9.

17. SEGAL, I. E., *Nonlinear semigroups*, Ann. of Math. 78 (1963), 339-364.

18. STRAUSS, W. A., *On weak solutions of semilinear hyperbolic equations*, Anais Acad. Bras. Ciencias 42 (1970), 645-651.

LECTURE #1. FIVE PROBLEMS: AN INTRODUCTION TO THE QUALITATIVE THEORY OF PARTIAL DIFFERENTIAL EQUATIONS

by

JEFFREY RAUCH
Department of Mathematics
University of Michigan
Ann Arbor, Michigan 48104

INTRODUCTION

The three lectures presented here have several goals. For emphasis we list them.

(I) It is often remarked that one of the justifications for proving existence and uniqueness theorems in partial differential equations is that the methods and ideas developed are also useful in the more interesting qualitative questions concerning solutions. The problems discussed illustrate this point.

(II) The five problems of the first lecture are quite simple and in my opinion physically interesting. Despite this fact they do not seem to be in any of the standard elementary P.D.E. treatises. It is my hope that they will find their way into introductory courses.

(III) It is an important fact that interesting physical questions often take a qualitative rather than quantitative form and as such can be treated by the methods mentioned in (I).

(IV) Physical intuition is a very valuable aid in formulating questions; however, it is often unreliable, especially when it comes to answering these questions. We give several examples where reasonable physical intuition is dead wrong. The instances where mathematics tells us our intuition is faulty are clearly of great value. Formulas are smarter than people.

(V) By studying the qualitative implications of mathematical models our intuition can be improved. After all, intuition only reflects our belief based on accumulated experience.

(VI) The last two lectures are intended to serve as an introduction to work with M. Taylor on scattering by unusual obstacles. The original paper is not easy reading. I hope these are.

1. A PROBLEM IN HEAT CONDUCTION

Let $\Omega \subset \mathbb{R}^n$ be a connected, bounded, open set with smooth boundary (of course $n = 2,3$ are of special importance). We assume that the region Ω is filled with a homogeneous medium and is insulated at $\partial\Omega$. For example we could be considering the flow of heat in a metal plate or in jello inside a thermos bottle. The mathematical model for the temperature $u(t,x)$ at time t and place x is

(1) $$u_t = c\,\Delta u \quad \text{in } \Omega ,$$

(2) $$\frac{\partial u}{\partial \nu} = 0 \qquad \text{on } \partial\Omega.$$

Here c is a physical constant depending on the medium, $\Delta = \Sigma \frac{\partial^2}{\partial x_i^2}$, and $\frac{\partial}{\partial \nu}$ is differentiation in the direction of the normal to $\partial\Omega$. Given the initial temperature, $u(0,x)$, (1) and (2) suffice to determine the time evolution.

QUESTION. *If initially the medium is strongly heated in a neighborhood of a point* x_0 *(for example with a torch) so that* $u(0,x)$ *has a sharp maximum at* x_0, *is it true that for each* $t > 0$ *the maximum of* $u(t,x)$ *occurs at or near* x_0*?*

Intuitively, one feels that heat will flow from the hot spot to the cooler ones with the result that the temperature gradually decreases near x_0 and increases at other points. This vague idea which even sounds like the second law of thermodynamics indicates that the answer is yes.

ANALYSIS. We solve (1), (2) by eigenfunction expansion. It is well known that there are eigenvalues $0 = \lambda_0 > \lambda_1 \geq \dots$ converging to $-\infty$ and eigenfunctions $\Phi_0, \Phi_1, \dots$ such that

$$\Delta\Phi_j = \lambda_j \Phi_j \ , \ j = 0,1,2,\dots \ ,$$

$$\frac{\partial}{\partial \nu} \Phi_j = 0 \quad \text{on} \quad \partial\Omega \ ,$$

$\{\Phi_j\}$ is an orthonormal basis for $L_2(\Omega)$.

The solution $u(t,x)$ is then given by

$$u = \sum_j a_j e^{c\lambda_j t} \Phi_j$$

where $a_j = (u(0,x), \Phi_j)_{L_2(\Omega)}$. We choose $\Phi_0 = \frac{1}{|\Omega|^{1/2}} > 0$. Then since $\lambda_j < 0$ for $j > 0$ we have

$$u = \frac{1}{|\Omega|} \int_\Omega u + o(1).$$

To be precise, for any k

$$\left\| u(t) - \frac{1}{|\Omega|} \int_\Omega u \right\|_{H_k(\Omega)} = O(e^{c\lambda_1 t}) \quad \text{as} \quad t \to \infty.$$

EXERCISE. Prove this.

The conclusion is that u converges rapidly to its average value. This is a rigorous version of the "approach to equilibrium".

HYPOTHESIS. λ_1 *is a simple eigenvalue, that is,* $\lambda_1 > \lambda_2$.

This hypothesis is satisfied by "most" domains. Its failure is usually due to some sort of symmetry. For example, in $\mathbb{R}^2$ it is valid for all ellipses except the circle and all rectangles except the square.

EXERCISE. Verify the assertions of the last sentence.

With this assumption we now look at the next term in the expression for u.

$$u = a_0\Phi_0 + a_1 e^{c\lambda_1 t} \Phi_1 + O(e^{c\lambda_2 t}) .$$

EXERCISE. Formulate and prove a precise sense for $O(e^{c\lambda_2 t})$.

HYPOTHESIS. *The eigenfunction* Φ_1 *has a unique maximum (resp. minimum) at* P_+ *(resp.* P_-*).*

Again this behavior is expected in unexceptional cases. Since Φ_1 is orthogonal to constants we have $P_+ > 0 > P_-$.

ANSWER TO QUESTION. *If* $a_1 > 0$ *(resp.* < 0*) the point where* $u(t)$ *achieves its maximum approaches* P_+ *(resp.* P_-*) as* $t \to \infty$.

Thus, rather than staying in the neighborhood of the original hot spot the maximum always moves to one of two points. It is very instructive to consider the case $\Omega = [0,\pi] \subset \mathbb{R}^1$, $\Phi_1(x) = \sqrt{2/\pi}\cos x$. If the initial hot spot is to the left (resp. right) of $\pi/2$ then the maximum moves to 0 (resp. π).

EXERCISE. Verify the last assertion.

One can even get a feeling for why this happens. If the initially hot region is to the left of $\pi/2$ then there is more medium to the right that must be heated. Thus the heat is drawn off more rapidly on the right and the region of high temperature moves to the left because of "erosion on the right".

2. A PROBLEM FROM ELECTROSTATICS

Let Ω_1, Ω_2 in $\mathbb{R}^3$ be disjoint bounded open sets with smooth boundaries. We suppose that perfect conductors occupy these regions. Suppose a positive charge q is placed on Ω_1. The charges on the conductors distribute themselves rapidly and equilibrium is reached. What happens is that the positive charge on Ω_1 attracts negatives on Ω_2 and a charge distribution as in Figure 1 is established.

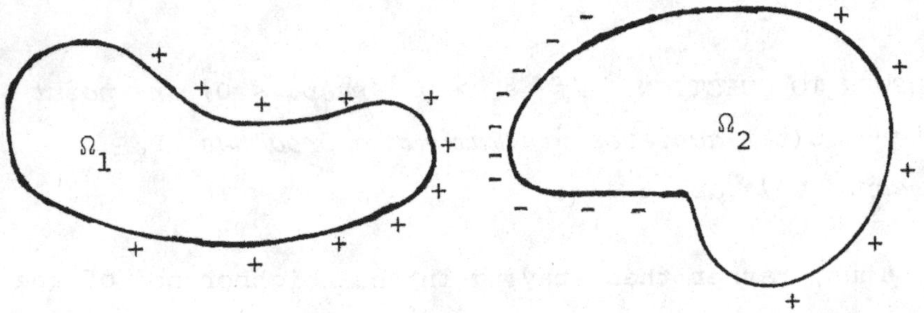

Figure 1

QUESTION. *Is it possible that the attractive force between charges is so strong that so much plus charge is drawn to one side of* Ω_1 *that a net negative charge is present at some point of* Ω_1*?*

ANALYSIS. We suppose that units are chosen so that the electrostatic potential due to a unit positive charge at the origin is $(4\pi|x|)^{-1}$. Recall that if ϕ is the electrostatic potential then the electric field is given by $E = -\text{grad}\,\phi$ and the force on a point charge q at x is qE. In the exterior of the conductors Ω_1, Ω_2 the potential $\phi(x)$ satisfies

$$(3) \qquad \Delta\phi = 0, \quad \phi = O(1/|x|) \quad \text{as} \quad |x| \to \infty.$$

In addition the charge on the conductors is located entirely on the surface and

(4) $\frac{\partial\phi}{\partial\nu}$ = charge density per unit area at $\partial\Omega_1$, $\partial\Omega_2$.

Here ν is the normal direction pointing into Ω_i, i = 1,2. Thus the condition that Ω_1 carry charge q and Ω_2 be neutral is

(5) $$\int_{\partial\Omega_1}\frac{\partial\phi}{\partial\nu} = q, \quad \int_{\partial\Omega_2}\frac{\partial\phi}{\partial\nu} = 0.$$

Since charge is free to move in the tangential directions at $\partial\Omega_i$ the tangential force must vanish there. Thus we have $E_{tan} = 0$ at $\partial\Omega_i$, or in terms of ϕ

(6) $$\phi = \text{constant on } \partial\Omega_i, \ i = 1,2,.$$

There is a unique function ϕ satisfying (3)-(6).

PROOF OF UNIQUENESS. Suppose there were two solutions ϕ_1, ϕ_2. Let $\psi = \phi_1 - \phi_2$. If $\psi \neq 0$ it must assume a positive maximum or negative minimum since $\psi = O(1/|x|)$ as $|x| \to \infty$. By the maximum principle this extremum must occur on one of the conductors, which we call Ω_i. Since ψ is constant on this conductor, each point of Ω_i is an extremum of the same type. By the Hopf maximum principle $\frac{\partial\psi}{\partial\nu} \neq 0$ at every point of Ω_i so $\int_{\Omega_i}\frac{\partial\psi}{\partial\nu} \neq 0$, which cannot be. Therefore $\psi \equiv 0$. □

PROOF OF EXISTENCE. Let V_1 be the solution of the standard exterior Dirichlet problem

$$\Delta V_1 = 0 \quad \text{in} \quad \mathbb{R} \setminus (\Omega_1 \cup \Omega_2) \,,$$

$$V_1 = O(1/|x|) \quad \text{as} \quad |x| \to \infty,$$

$$V_1 = 1 \quad \text{on} \quad \partial\Omega_1 \,, \; V_1 = 0 \quad \text{on} \quad \partial\Omega_2 \,.$$

Let V_2 be defined similarly except that the role of the conductors is reversed. If ϕ is a solution and $c_i = \phi|_{\Omega_i}$, $i = 1,2$ then $\phi = c_1V_1 + c_2V_2$. We will try to find numbers c_1, c_2 such that $c_1V_1 + c_2V_2$ solves (3)-(6). Conditions (3), (4), (6) are automatic. On the other hand (5) yields a pair of simultaneous linear equations for c_1 and c_2. To prove existence it suffices to show that the corresponding homogeneous equations have only the trivial solution. This follows from uniqueness. □

PROPOSITION. $\phi > 0$ *on* $\mathbb{R}^3 \setminus (\Omega_1 \cup \Omega_2)$ *and attains its maximum on* $\partial\Omega_1$ *where* $\frac{\partial\phi}{\partial\nu} > 0$.

PROOF. First we show that $\phi \geq 0$. If not ϕ would attain a negative minimum which by the maximum principle would occur on $\partial\Omega_i$ for $i = 1$ or 2. Also since ϕ is constant on $\partial\Omega_i$ each point is a minimum so $\frac{\partial\phi}{\partial\nu} < 0$ at each point of $\partial\Omega_i$, violating (5). Once we know $\phi \geq 0$ the maximum principle and (5) imply that $\phi > 0$.

EXERCISE. Prove the last assertion.

Since $\phi > 0$ it must attain a positive maximum at all points of $\partial\Omega_i$ for $i = 1$ or 2. Then $\frac{\partial\phi}{\partial\nu} > 0$ at $\partial\Omega_i$ so by (5), $i = 1$. □

ANSWER TO QUESTION. *Since* $\frac{\partial\phi}{\partial\nu} > 0$ *on* $\partial\Omega_1$ *we see that the answer is* *<u>NO</u>.*

3. ANOTHER PROBLEM FROM ELECTROSTATICS

Since we have developed these nice tools we'll do another problem. We have essentially the same situation as before except that conductor number 2 is grounded. Mathematically this means that the boundary condition on Ω_2 becomes

$$\phi = 0 \quad \text{on} \quad \partial\Omega_2 .$$

Practically this means that Ω_2 is connected to some very large object, for example, Lake Michigan. Physically a new phenomenon occurs. The attractive force of the positive charges on Ω_1 causes negative charge to flow from the "large object" to Ω_2 so that Ω_2 becomes negatively charged. The negative charge is called induced charge. As before we could ask whether any point of Ω_1 has a net negative charge or if any point of Ω_2 has a net positive charge. The answer to both questions is NO.

QUESTION. *Is it possible that the total negative charge induced on* Ω_2 *is greater (in absolute value) than the positive charge on* Ω_1*?*

If c is the value of ϕ on Ω_1 then it is easy to see that $\phi = cV_1$. It is also a simple matter to show that $0 \le V_1 \le 1$ so that $\frac{\partial V_1}{\partial\nu} > 0$ at all points of Ω_1.

It follows that $c > 0$ since $\int_{\partial\Omega_1} \frac{\partial\phi}{\partial\nu} = q > 0$. To get more information about ϕ we apply Green's identity in the region $R = \{x \in \mathbb{R}^3 \setminus (\Omega_1 \cup \Omega_2) \mid |x| \le r\}$ to obtain

$$0 = \int_R \Delta\phi = \int_{|x|=r} \frac{\partial\phi}{\partial r} + \int_{\partial\Omega_1} \frac{\partial\phi}{\partial\nu} + \int_{\partial\Omega_2} \frac{\partial\phi}{\partial\nu} .$$

Therefore

$$\int_{|x|=r} \frac{\partial\phi}{\partial r} = -q - \int_{\partial\Omega_2} \frac{\partial\phi}{\partial\nu} . \tag{7}$$

This proves that the left hand side is independent of r for r large. In addition, using the standard multipole expansion of harmonic functions (see [4; Ch. 5, §7, 8]*) one shows that (3) and $\phi \ge 0$ imply

$$\frac{\partial\phi}{\partial r} = \frac{\text{negative constant}}{|x|^2} + O(1/|x|^3). \tag{8}$$

Thus the left hand side of (7) is negative for r large, and hence it is negative. Therefore

$$|\text{Total charge on } \Omega_2| = -\int_{\Omega_2} \frac{\partial\phi}{\partial\nu} = q + \int_{|x|=r} \frac{\partial\phi}{\partial r} < q.$$

ANSWER TO QUESTION. *NO.*

For a heuristic proof using lines of force see [6; §89c]. In fact Chapter III of [6] is one of the nicest treatments of electrostatics available. A more

---

*References are to the Bibliography at the end of Lecture #3.

mathematical approach can be found in [3] or [4; Ch. 9, §12].

4. A PROBLEM ABOUT WAVE MOTION WITH FRICTION

We consider damped wave motion in $\mathbb{R}^3$. A typical mathematical model is

$$u_{tt} = \Delta u - a(x)u_t$$

where $a(x) \geq 0$ represents a frictional resistance. Given the initial position and velocity $u(0,x)$, $u_t(0,x)$, the motion is uniquely determined. It is a simple matter to show that the energy

$$E(t) = \int_{\mathbb{R}^3} (u_t^2 + |\nabla u|^2)dx$$

is a decreasing function of time. In fact,

$$\frac{dE}{dt} = -\int_{\mathbb{R}^3} a(x)u^2(t,x)dx \leq 0.$$

It appears from this formula and from physical intuition that if the friction coefficient $a(x)$ is increased the energy dissipation is enhanced. For fixed initial data we can consider the solution as depending parametrically on $a(x)$, that is, we have $u(t,x;\, a(x))$ and $E(t;\, a(x))$.

QUESTION. *If* $\tilde{a}(x) \geq a(x)$ *for all* $x \in \mathbb{R}^3$ *it is true that* $E(t;\, a(x)) \geq E(t;\, \tilde{a}(x))$ *for* $t \geq 0$ *and all initial data?*

ANALYSIS. We attack the problem in case the functions

$a, \tilde{a}$ are constant. In this case we can effectively use the Fourier transform. Let

$$\hat{u}(t,\xi;\ a) = (2\pi)^{-3/2} \int u(t,x;\ a) e^{ix\cdot\xi}\, dx$$

be the partial Fourier transform of u. For $\hat{u}$ we have the ordinary differential equation in t depending parametrically on ξ,

$$\hat{u}_{tt} = -|\xi|^2 \hat{u} - a\hat{u}_t .$$

For fixed ξ this is just the equation of a damped spring. Furthermore, the energy is given by

$$E = \int (|\hat{u}_t|^2 + |\xi|^2 |\hat{u}|^2) d\xi$$

which is merely the "sum" of the separate spring energies. Let us concentrate on the spring equation

$$\ddot{y} + a\dot{y} + y = 0$$

with energy

$$\dot{y}^2 + y^2 = e(t) = e(t;\ a).$$

REDUCED QUESTION. *For the spring equation is* $e(t;\ a)$ *a monotone decreasing function of* a *for arbitrary initial data?*

EXERCISE. Show that a yes or a no answer for the reduced question implies the same answer for the original question.

ANSWER TO QUESTIONS. *NO.*

Since any question about constant coefficient second order ordinary differential equations must be trivial this is left as an <u>exercise</u>. However, as a hint I suggest you investigate the overdamped case $a \gg 1$.

Once the answer has been found and the root cause identified as overdamping several observations can be made. First the answer is not entirely unreasonable since in the case of extreme damping <u>everything</u> happens very slowly including energy decay. Second, looking at the Fourier transform solution we see that overdamping corresponds to $|\xi|$ small compared to a. Thus after a while only the slowly decaying modes will be noticeable. That is, $\hat{u}(t,\xi)$ will tend to become concentrated near $\xi = 0$. This corresponds to a flattening out of u. Not only does dissipation decrease energy but it tends to iron out the "wrinkles" in u.

5. A PROBLEM ABOUT PERFECT SHADOWS

In this section we will study wave motion in the presence of a periodic (in time) driving force. This falls into the class of problems called <u>radiation problems</u>. We will investigate whether obstacles can form perfect shadows, that is, whether there can be an open set which is unaffected by the radiation. A typical mathematical model is

$$(9) \qquad u_{tt} = c^2 \Delta u + F(x) e^{i\alpha t}, \quad (t,x) \in \mathbb{R} \times (\mathbb{R}^3 \backslash \Omega),$$

where Ω represents an obstacle. In addition we have some condition which prescribes how the wave interacts with the obstacle. This usually takes the form of a boundary condition, for example, a Dirichlet or Neumann condition at $\partial\Omega$. In many situations it can be shown that the solution can be written as

(10) $$v(x)e^{i\alpha t} + \text{transient wave motion.}$$

Here $v(x)e^{i\alpha t}$ is a motion at the same frequency as the driving term and the transient term tends to zero at each point $x \in \mathbb{R}^3 \setminus \Omega$ as $t \to \infty$. Thus after an initial adjustment an observer sees the steady state $ve^{i\alpha t}$, for the transient has died away. This is called the principle of limiting amplitude (see [5; Thm. 4.4] for a proof).

We suppose that the radiating term $Fe^{i\alpha t}$ is spatially localized in a region $R \equiv \text{supp } F$.

QUESTION. *Is it possible for there to be a perfect shadow? Precisely, can there be an open set* ω *in the exterior of* $\Omega \cup R$ *such that* $v = 0$ *on* ω*?*

Physically one might try to construct such a set as in Figure 2.

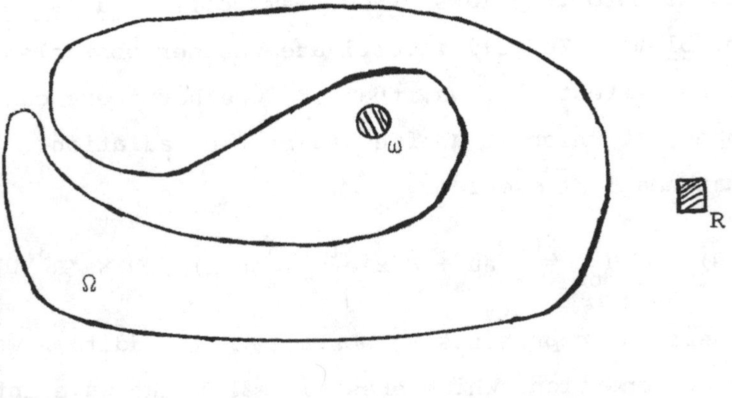

Figure 2

ANALYSIS. Plugging the expression (10) into the differential equation (9), we see that v must satisfy

$$(c^2\Delta + \alpha^2)v = -F.$$

Thus exterior to R we have $(c^2\Delta + \alpha^2)v = 0$, so v is real analytic outside R. Therefore if $v = 0$ on an open set ω exterior to R, then $v \equiv 0$ outside R.

ANSWER. *The only way for there to be any region that is not affected by the light is for the radiation to be confined entirely to* R, *that is, there is no radiation at all.*

There is one possibility that is overlooked here. This is illustrated by Figure 3.

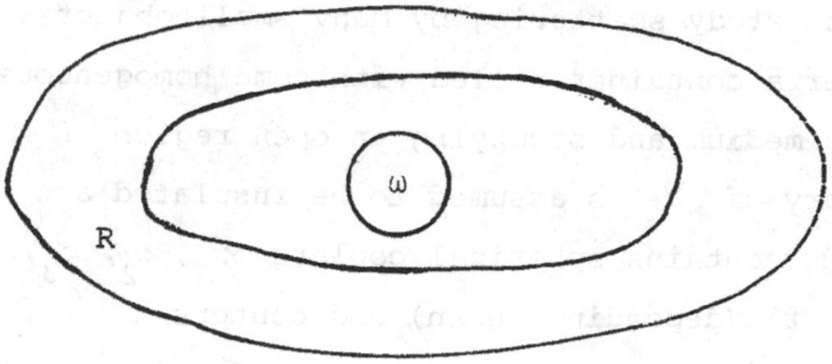

Figure 3

It is possible to have a perfect shadow inside the source. For example the source can radiate outgoing spherical waves which never affect the inside of the "antenna". For more detailed information see [8].

LECTURE #2. THE MATHEMATICAL THEORY OF CRUSHED ICE

by

JEFFREY RAUCH
Department of Mathematics
University of Michigan
Ann Arbor, Michigan 48104

In this lecture we will investigate the cooling efficiency of crushed ice. With certain idealizations this becomes a problem of estimating the smallest eigenvalue for an elliptic boundary value problem. These eigenvalue estimates will be needed in the next lecture in order to study scattering by many small objects.

Consider a container filled with some homogeneous continuous medium and occupying an open region $\Omega \subset \mathbb{R}^3$. The boundary of Ω is assumed to be insulated and smooth. Ω contains spherical coolers $\kappa_1, \kappa_2, \kappa_3, \ldots, \kappa_n$ of radius r (depending on n) and centers $x_1, x_2, \ldots, x_n$.* Overlap of coolers is permitted. We view $K_n = \bigcup_{i=1}^{n} \kappa_i$ as one large cooler for the medium in $\Omega_n \equiv \Omega \setminus K_n$. The coolers are assumed to be stationary and to have internal mechanisms which maintain their boundaries at temperature zero. If $u(t,x)$ is the

---

*Other shapes can easily be treated. See the remark after the statement of the theorem.

temperature at time t and place x, then heat flow in Ω is modeled by the boundary value problem

$$u_t = c\,\Delta u \quad \text{for} \quad t \geq 0,\ x \in \Omega,$$

$$\frac{\partial u}{\partial \nu} = 0 \quad \text{on} \quad \partial\Omega,$$

$$u = 0 \quad \text{on} \quad \partial K_n .$$

Here c is a constant depending on the conductivity and specific heat of the medium and Δ is the Laplacian in $\mathbb{R}^3$. Notice that only heat flow by conduction is considered here - the medium is stationary. To completely describe u, the initial temperature distribution $u(0,x)$ must be given. The boundary value problem is solved by eigenfunction expansion

$$u = \Sigma a_j e^{c\lambda_j t} \Phi_j$$

where $a_j = (u(0,x), \Phi_j(x))_{L^2(\Omega_n)}$ and the Φ_j are normalized eigenfunctions of Δ:

$$\Delta\Phi_j = \lambda_j \Phi_j ,$$

$$\Phi_j = 0 \quad \text{on} \quad \partial K_n ,$$

$$\frac{\partial \Phi_j}{\partial \nu} = 0 \quad \text{on} \quad \partial\Omega .$$

The eigenvalues are ordered so that $0 > \lambda_1 > \lambda_2 \geq \ldots$. As in Lecture 1, u goes to zero like $e^{c\lambda_1 t}$, so λ_1 gives the rate of cooling. We are interested in the behavior of λ_1 as $n \to \infty$, $r \to 0$.

This problem is related to common experience with

crushed ice. It is well known that a given volume of ice is more efficient as a cooler if it is divided into many small pieces. The idealization we have made is that the ice neither moves nor melts.

ORTHODOX EXPLANATION. The reason that crushed ice is an efficient coolant is that the surface area of the ice is large. After all, the cooling takes place by contact of the medium with the coolers, and increasing the surface area increases the amount of contact. Furthermore the total volume of coolers in our case goes like nr^3 and the total surface area like nr^2 so for fixed volume, the surface area $nr^2 = \frac{1}{r} \cdot (nr^3)$ goes to infinity as $r \to 0$. This analysis leads to

GUESS #1. If $n \to \infty$, $r \to 0$ in such a way that the total volume of the coolers stays constant then the cooling becomes infinitely efficient, that is $\lambda_1 \to -\infty$, provided the coolers are evenly spaced throughout Ω.

In addition there is a companion conjecture.

GUESS #2. If $n \to \infty$ and $r \to 0$ in such a way that the total surface area goes to zero, then the cooling disappears in the limit, i.e. $\lambda_1 \to 0$.

Reasonable as all this is, <u>one</u> of the above guesses is incorrect. If you think that is not surprising try to figure out which one without reading further. The correct answer is provided by the following

THEOREM. *For* nr *small we have* $-\lambda_1 \leq \frac{2nr}{|\Omega|}(1 + O(nr))$.

If the coolers are evenly spaced in Ω *then there are positive constants* c_1, c_2 *such that* $-\lambda_1 \geq c_1 nr - c_2$ *for all* n.

The precise sense of even spacing will emerge in the proof. In addition the constants c_1, c_2 and the term $O(nr)$ can be crudely estimated from the data of the problem if necessary. If one is interested in coolers which are not spherical, one can get lower bounds by replacing the coolers by spherical ones contained inside them and upper bounds by larger spherical coolers. In this way the estimates can be carried over to more or less arbitrary shapes.

Applying the theorem we see that Guess #1 is correct, for if $nr^2 \to \infty$ then $nr = \frac{1}{r}(nr^2) \to \infty$, so $\lambda_1 \to -\infty$. Notice, however, that for r small $nr >> nr^2$ so the cooling efficiency is greater than predicted by surface area considerations. In particular, nr may grow infinitely large even though $nr^2 \to 0$ so that Guess #2 is wrong. The correct results replace nr^2 by nr in both guesses. I do not have any intuitive idea why this should be but in the next lecture we will gain some insight into the failure of our intuition.

PROOF OF THE UPPER BOUND FOR $-\lambda_1$

We use the variational characterization of λ_1.

$$-\lambda_1 = \inf \frac{\int_{\Omega_n} |\nabla\psi|^2}{\int_{\Omega_n} \psi^2} \quad ,$$

the infimum over all $\psi \in C^\infty(\overline{\Omega_n})$ such that $\psi \not\equiv 0$ and

$\psi = 0$ on ∂K_n. The infimum is attained for multiples of Φ_1. It is important to notice that the condition $\frac{\partial\psi}{\partial\nu} = 0$ on $\partial\Omega$ is <u>not</u> imposed; this is a natural boundary condition (see [1; Ch. 6, §1]). To get an upper bound we will plug in a good trial function. To describe the function let me review the notion of <u>capacity</u>. For a reasonably well-behaved set $\Gamma \subset \mathbb{R}^3$ there is a unique solution to the Dirichlet problem

$$\Delta\phi = 0 \text{ in } \mathbb{R}^3 \setminus \Gamma,$$
$$\phi = 1 \text{ on } \partial\Gamma,$$
$$\phi = O(1/|x|) \text{ as } |x| \to \infty.$$

ϕ is called the <u>capacitary potential</u> of Γ. The capacity of Γ, $\operatorname{cap}(\Gamma)$, is defined as

$$\int_{\partial\Gamma} \frac{\partial\phi}{\partial\nu}$$

where ν is the outward normal to $\mathbb{R}^3 \setminus \Gamma$. Physically this is the amount of charge which must be placed on a conductor occupying the region Γ in order to raise its potential to 1. Thus capacity measures the ability to hold charge. For ϕ we have $|\nabla\phi| = O(1/|x|^2)$ so a straightforward application of Green's identity shows that

$$0 = \int_{\mathbb{R}^3\setminus\Gamma} \phi\,\Delta\,\phi = -\int_{\mathbb{R}^3\setminus\Gamma} |\nabla\phi|^2 + \int_{\partial\Gamma} \phi\frac{\partial\phi}{\partial\nu} ,$$

so

$$\int_{\partial\Gamma} \frac{\partial\phi}{\partial\nu} = \int_{\mathbb{R}^3\setminus\Gamma} |\nabla\phi|^2 ,$$

an alternate expression for the capacity. We need two facts (see [7; p.125 ff.]):

(α) The capacity of a sphere of radius r is r.

(β) Capacity is a subadditive set function.

Using these we have

$$\mathrm{cap}(K_n) = \mathrm{cap}(\cup \kappa_i) \le \Sigma\, \mathrm{cap}(\kappa_i) = nr.$$

Let ϕ_n be the capacitary potential of K_n. The trial function will be $(1-\phi_n)|\Omega_n$. We immediately have that

(i) $\phi_n = 1$ on ∂K_n,

(ii) $\int_{\Omega_n} |\nabla\phi_n|^2 < \int_{\mathbb{R}^3\setminus K_n} |\nabla\phi_n|^2 \le nr.$

Furthermore, for any $x \notin \Gamma$ we can perform the "ball of radius ϵ argument" on

$$\int_{\mathbb{R}^3\setminus K_n} \phi_n \cdot \Delta_y \frac{1}{|x-y|}$$

to conclude that

$$\phi_n(x) = \frac{1}{4\pi} \int_{\partial K_n} \frac{\partial\phi_n}{\partial\nu}(y) \frac{1}{|x-y|}\, dy.$$

Therefore

$$4\pi|\phi_n(x)| \le \mathrm{cap}(K_n) \cdot \mathrm{dist}(x, K_n) \le nr \cdot \mathrm{dist}(x, K_n).$$

Let $\mathcal{O}$ be a bounded open set at positive distance from Ω. Then there is a constant c such that

$$\|\Phi\|_{L^2(\Omega)} \le c(\|\Phi\|_{L^2(\mathcal{O})} + \|\nabla\Phi\|_{L^2(\mathbb{R}^3)}),$$

the inequality holding for all $\Phi \in H_1(\mathbb{R}^3)$. Applying this to

$$\Phi = \begin{cases} \phi_n & \text{in } \mathbb{R}^3 \setminus K_n \\ 1 & \text{in } K_n \end{cases}$$

we get

$$\|\phi_n\|_{L^2(\Omega_n)} \le c\,n\,r$$

with c independent of n. The function $\psi = 1 - \phi_n$ then satisfies

$$\int_{\Omega_n} |\nabla\psi|^2 = \int_{\Omega_n} |\nabla\phi_n|^2 \le c\,n\,r,$$

$$\begin{aligned}\int_{\Omega_n} \psi^2 &= |\Omega_n| - 2\int_{\Omega_n} \phi_n + \int_{\Omega_n} \phi_n^2 \\ &\ge |\Omega_n| - 2|\Omega_n|^{\frac{1}{2}}\|\phi_n\| + \|\phi_n\|^2 \\ &\ge |\Omega_n| - (\tfrac{1}{2}|\Omega_n| + 2\|\phi_n\|^2) + \|\phi_n\|^2 \\ &\ge \tfrac{1}{2}|\Omega_n| - O(nr)\end{aligned}$$

provided nr is small. Therefore

$$-\lambda_1 \le \int_{\Omega_n} |\nabla\psi|^2 \Big/ \int_{\Omega_n} \psi^2 \le \frac{c\,n\,r}{\frac{1}{2}|\Omega| - O(nr)}$$

which completes the proof of the upper bound.

PROOF OF THE LOWER BOUND FOR $-\lambda_1$. It is a well known general principle that lower bounds are harder to obtain than upper bounds. Ours is no exception to the rule. The notion of evenly spaced is that one can cover Ω by balls with centers at $x_1, x_2, x_3, \ldots, x_n$ with the property that there is not a great deal of overlap. To be precise we assume that there are numbers $R(n) > 2r$ such that the balls $B_i = \{x \mid |x-x_i| < R(n)\}$, $i=1,2,\ldots,n$ satisfy

(i) $\bigcup_{i=1}^{n} B_i \supset \Omega$,

(ii) there is a number M independent of n such that each point of Ω is in at most M of the B_i.

As a consequence of (ii) we see that $nR^3 \le c|\Omega|$ for some constant c independent of n. The number R serves as a measure of distance between adjacent coolers.

Suppose $\psi \in H_1(\Omega_n)$, $\psi = 0$ on ∂K_n. We extend ψ to $\psi_{ext} \in H_1(\mathbb{R}^3)$ by extending it as zero in K_n and then doing a Lions type reflection across $\partial\Omega$. This can be done so that

$$\|\psi_{ext}\|_{H_1(\mathbb{R}^3)} \le c\|\psi\|_{H_1(\Omega_n)}$$

provided $\frac{|K_n|}{|\Omega|} \le \bar{c} < 1$ with $\bar{c}$ independent of n. When this condition fails the problem becomes uninteresting. Let $\Omega_{ext} = \bigcup_{i=1}^{n} B_i$. First we get a lower bound for $\int_{\Omega_{ext}} |\nabla\psi_{ext}|^2$ by estimating $\int_{B_i} |\nabla\psi_{ext}|^2$ from below. We use the

LEMMA. *If* $r < \frac{1}{2}R$ *and* $A = \{x \mid r < |x| < R\}$ *then* $\int_A |\nabla\phi|^2 \geq \frac{cr}{R^3} \int_A \phi^2$ *for all* $\phi \in H_1(A)$ *with* $\phi = 0$ *on* $|x| = r$.

PROOF. The minimum of $\left(\int_A |\nabla\phi|^2\right) / \int_A \phi^2$ occurs for the eigenfunction $\Delta\phi = \lambda\phi$, $\phi = 0$ for $|x| = r$, $\phi_r = 0$ for $|x| = R$ corresponding to the largest eigenvalue λ. This eigenfunction must be positive. (The proof of this well-known fact is not trivial.) It follows (exercise) that the eigenfunction is rotationally symmetric so $\phi(x) = f(|x|)$. Using this fact the Lemma is reduced to showing

SUBLEMMA. $\int_r^R f'(t)^2 t^2 \, dt \geq \frac{cr}{R^3} \int_r^R f^2(t) t^2 \, dt$ *for all* $f \in C^1[r,R]$ *satisfying* $f(r) = 0$.

The sublemma is a consequence of the inequality

$$(\dagger) \quad \int_r^\rho f^2(t)\phi(t)dt \leq \int_r^\rho \phi(t)dt \int_r^\rho \frac{1}{\phi(t)}dt \int_r^\rho (f'(t))^2\phi(t)dt$$

with $\phi(t) = t^2$. One proof of this inequality can be found in [9; Lemma 4.5]. We present an argument due to Jim Ralston. Write $f(t) = \int_r^t g(s)ds$ with $g = f'$. Then

$$f^2(t) = \left(\int_r^t g(s)\sqrt{\phi(s)} \cdot \frac{1}{\sqrt{\phi(s)}}ds\right)^2 \leq \int_r^\rho g^2(s)\phi(s)ds \int_r^\rho \frac{1}{\phi(s)}ds$$

and (†) follows immediately. □

We then have

$$\int_{\Omega_{ext}} |\nabla\psi_{ext}|^2 \geq \frac{1}{M}\sum\int_{B_i} |\nabla\psi_{ext}|^2 \geq \frac{cr}{MR^3}\sum\int_{B_i}\psi^2$$

$$\geq \frac{c}{M}\frac{nr}{nR^3}\int_{\Omega_{ext}}\psi^2 \geq \frac{c'}{M|\Omega|}\, nr\int_{\Omega}\psi^2 .$$

From the inequality for the Lions reflection defining ψ_{ext} we have

$$\int_{\Omega_n} |\nabla\psi|^2 + c\int_{\Omega_n}\psi^2 \geq \int_{\mathbb{R}^3} |\nabla\psi_{ext}|^2$$

which yields the estimate

$$\int_{\Omega_n} |\nabla\psi|^2 \geq \left(\frac{c'}{M|\Omega|}\, nr - c\right)\int_{\Omega_n}\psi^2 ,$$

proving the lower bound for $-\lambda_1$. □

LECTURE #3. SCATTERING BY MANY TINY OBSTACLES

by

JEFFREY RAUCH

Department of Mathematics
University of Michigan
Ann Arbor, Michigan 48104

The subject of scattering theory, as treated by mathematicians, is usually concerned with interactions with one target object. For example we have potential scattering in quantum mechanics and the problems of acoustic scattering by macroscopic obstacles.

However, one of the most commonly encountered situations is the scattering by many similar small targets. For example in the classical Rutherford experiment alpha particles (Helium nuclei) are scattered by a thin foil containing an enormous number of individual targets (atoms). In this case a very satisfactory treatment can be given by treating the scattering by a single atom and just adding up the results assuming that the various scattering events are independent (incoherent scattering). An adequate mathematical explanation for this success is still lacking. In this paper we focus attention on the opposite situation where the presence of many obstacles leads to qualitatively new phenomena. The most striking examples are when the many obstacles behave as if they were a solid object or in the opposite extreme case where because of their small size many tiny obstacles may have negligible effect on incident waves. We call the first <u>solidification</u> and the second <u>fading</u>.

EXAMPLE 1. Dust particles in the atmosphere have a negligible effect on the propagation of sound waves.

EXAMPLE 2. A cloud of small conductors sprayed into the air appears solid on a radar screen.

EXAMPLE 3. The water droplets in a cloud give a solid appearance while those in the atmosphere are essentially invisible. Here you can clearly see that it is a balance between number and size that determines whether there is fading or solidification.

EXAMPLE 4. The atomic nature of crystalline matter is not apparent in its interaction with macroscopic objects. This is perhaps the most common example of solidification.

EXAMPLE 5. It is well-known that a region enclosed by walls made of conductors will have no electric field on the inside. In practice it is observed that a screen made of conductors has essentially the same effect as a solid conductor. On the other hand if the wire is sufficiently thin it is clear that this screening effect will not be present. The conventional explanation (see, for example [2; Ch. 7, §10]) does not take this into account. The problem of electrostatic screening is discussed in a forthcoming paper by M. Taylor and myself.

BASIC PROBLEM. *What values of the relevant physical parameters correspond to solidifying obstacles and which correspond to fading?*

We will study one situation in which a substantial first step toward solving the basic problem can be made. The obstacles are assumed to lie in a nice bounded open set Ω. The obstacles will be n spheres $\kappa_1, \kappa_2, \ldots, \kappa_n$ of radius r (depending on n) and centers $x_1, x_2, \ldots, x_n$; and as before $K_n = \bigcup_{i=1}^{n} \kappa_i$ is the effective obstacle. We will consider two problems simultaneously:

$$(1) \qquad u_{tt} = \Delta u \quad \text{in } \mathbb{R}^3 \setminus K_n , \ u = 0 \quad \text{on} \quad \partial K_n ,$$

$$(2) \qquad u_t = i\Delta u \quad \text{in } \mathbb{R}^3 \setminus K_n , \ u = 0 \quad \text{on} \quad \partial K_n .$$

The first corresponds to acoustic scattering by soft spheres and the second quantum mechanical scattering by impenetrable spheres. The solutions are defined for $(t,x) \in \mathbb{R} \times (\mathbb{R}^3 \setminus K_n)$ and we consider them to be extended as zero inside K_n. We will consider the limit as $n \to \infty$, $r \to 0$. Solidifying obstacles means that the solutions (for fixed initial data) converge to solutions of

$$(1)_\infty \qquad u_{tt} = \Delta u \quad \text{in } \mathbb{R}^3 \setminus \Omega , \ u = 0 \quad \text{on} \quad \partial\Omega ,$$

$$(2)_\infty \qquad u_t = i\Delta u \quad \text{in } \mathbb{R}^3 \setminus \Omega , \ u = 0 \quad \text{on} \quad \partial\Omega .$$

Disappearance would mean that the solutions converge to the solutions of the wave equation or Schrödinger equations on all of $\mathbb{R}^3$, that is, without obstacles. If we view the solutions as functions of t with values in $L^2(\mathbb{R}^3)$ the notion of convergence that is natural is uniform convergence on compact time intervals, that is convergence in $C((-\infty,\infty)\,;\,L^2(\mathbb{R}^3))$.

We have the following result.

If $nr \to 0$ *the obstacles fade.*
If $nr \to \infty$ *and the spheres are evenly spaced in* Ω *the obstacles solidify.*

Thus the critical combination of parameters is nr. In order to show the flavor of our method and the connection with crushed ice a proof of the solidification half of the above statement will be given in detail. This proof is extracted from [9] where one may also find a proof of the fading assertion (Theorem 4.2) and many other results in the spirit of this lecture.

NOTATIONS:

(i) Δ_n is the selfadjoint operator on $L^2(\mathbb{R}^3 \setminus K_n)$ defined by Dirichlet conditions on ∂K_n. That is, $D(\Delta_n) = \{u \in H_2(\Omega_n) \mid u = 0 \text{ on } \partial K_n\}$, and $\Delta_n u = \sum_{i=1}^{3} \frac{\partial^2 u}{\partial x_i^2}$

(ii) Δ_∞ is the selfadjoint operator on $L^2(\mathbb{R}^3 \setminus \Omega)$ with Dirichlet conditions on $\partial\Omega$.

(iii) If $v \in L^2(U)$ for some set $U \subset \mathbb{R}^3$ we consider $v \in L^2(\mathbb{R}^3)$ by setting v equal to zero on the complement of U. Thus for $v \in L^2(U)$, $\tilde{v} \in L^2(\tilde{U})$ we may consider $\|v-\tilde{v}\|_{L^2(\mathbb{R}^3)}$.

The solution u to problem (1) is given by

$$u(t) = \frac{\sin\sqrt{-\Delta_n}\, t}{\sqrt{-\Delta_n}} u_t(0) + (\cos\sqrt{-\Delta_n}\, t)u(0)$$

where $u(0), u_t(0) \in L^2(\mathbb{R}^3 \setminus K_n)$ are the initial position and velocity and objects like $\cos\sqrt{-\Delta_n}\, t$ are operators defined by the functional calculus for self-adjoint operators. (Note: $\Delta_n \le 0$ and $\frac{\sin t\sqrt{-x}}{\sqrt{-x}}$ is a smooth function on $(-\infty, 0]$.) Similarly for problem (2) we have

$$u(t) = e^{it\Delta_n} u(0).$$

With these formulas in mind we see that the solidification result is a consequence of the

THEOREM. *Suppose* $nr \to \infty$ *and the obstacles are evenly distributed in* Ω. *For every bounded continuous function* F *on* $(-\infty, 0]$ *we have*

$$F(\Delta_n)f \to F(\Delta_\infty)f \quad \text{in } L^2(\mathbb{R}^3) \tag{3}$$

for all $f \in L^2(\mathbb{R}^3 \setminus \Omega)$.

Note that $f \in L^2(\mathbb{R}^3 \setminus \Omega)$ implies $f \in L^2(\mathbb{R}^3 \setminus K_n)$ so that the assertion makes sense.

EXERCISE. This theorem shows that the solutions of (1), (2) converge to the solutions of $(1)_\infty$, $(2)_\infty$ for <u>fixed t</u>. Show that the convergence is automatically uniform on compact time intervals. (Hint: Show that the convergence in (3) must be uniform over compact sets of F

(in $BC(-\infty,0]$).)

PROOF OF THE THEOREM. The idea of the proof is to infer "resolvent convergence" from some elementary estimates, a compactness argument and the lower bound for $-\lambda_1$ found in Lecture #2. A dose of soft analysis then finishes the job.

STEP #1. If $f \in L^2(\mathbb{R}^3 \setminus \Omega)$ then $(1-\Delta_n)^{-1}f \to (1-\Delta_\infty)^{-1}f$ in $H_1(\mathbb{R}^3)$ (only convergence in $L^2(\mathbb{R}^3)$ will be needed.) To prove this we show first that $\{(1-\Delta_n)^{-1}f \mid n=1,2,\ldots\}$ is a bounded subset of $H_1(\mathbb{R}^3)$. As usual we extend $(1-\Delta_n)^{-1}f$ to $\mathbb{R}^3$ by setting it equal to zero on K_n.

Since $(1-\Delta_n)^{-1}$ is an operator of norm 1 we have $\|(1-\Delta_n)^{-1}f\|_{L^2(\mathbb{R}^3)} \le \|f\|_{L^2(\mathbb{R}^3)}$. Let $v_n = (1-\Delta_n)^{-1}f$, $v_\infty = (1-\Delta_\infty)^{-1}f$. Then

$$\|v_n\|^2_{H_1(\mathbb{R}^3)} = \|v_n\|^2_{H_1(\mathbb{R}^3\setminus\Omega_n)} = ((1-\Delta_n)v_n\ ,\ v_n)_{L^2(\mathbb{R}^3\setminus\Omega_n)}$$

$$= (f,v_n)_{L^2(\mathbb{R}^3)} \le \|f\|_{L^2(\mathbb{R}^3)}\|v_n\|_{L^2(\mathbb{R}^3)} ,$$

which proves the desired boundedness. We next show that v_n converges weakly to v_∞ in $H_1(\mathbb{R}^3)$, symbolically $v_n \rightharpoonup v$. By the weak compactness of the unit ball in a Hilbert space we see that $\{v_n\}$ is weakly compact. Suppose that w is a weak limit point and that $v_{n_j} \rightharpoonup w$. We will show that $w = v_\infty$ which establishes the convergence $v_n \rightharpoonup w$.

First we show that in the sense of distributions $(1-\Delta)w = f$ in $\mathbb{R}^3 \setminus \Omega$. For $\phi \in C_0^\infty(\mathbb{R}^3 \setminus \Omega)$ we have

$$(w,\ (1-\Delta)\phi) = \lim(v_{n_j}, (1-\Delta)\phi).$$

Furthermore $(1-\Delta)v_{n_j} = f$ so the right hand side is (f,ϕ), proving the assertion.

Next we show that $w = 0$ in Ω. We know that $v_{n_j} = 0$ on ∂K_n so

$$\frac{\int_{\Omega\setminus K_n} |\nabla v_{n_j}|^2}{\int_{\Omega\setminus K_n} v_{n_j}^2} \le -\lambda_1 .$$

We have a uniform bound on $\|v_n\|_{H_1(\mathbb{R}^3)}$, so the numerator is bounded as $j \to \infty$. Since $-\lambda_1 \to \infty$ we must have $\|v_{n_j}\|_{L^2(\Omega)} \to 0$ as $j \to \infty$ and therefore $w = 0$ in Ω. The three conditions

$$w \in H_1(\mathbb{R}^3),\ (1-\Delta)w = f \ \text{ in } \mathbb{R}^3 \setminus \Omega,\ w = 0 \ \text{ in } \Omega$$

identify w as $(1-\Delta_\infty)^{-1}f$.

Having established weak convergence the norm convergence is a consequence of $\|v_n\|_{H_1(\mathbb{R}^3)} \to \|v_\infty\|_{H_1(\mathbb{R}^3)}$. We have shown that $\|v_n\|^2_{H_1(\mathbb{R}^3)} = (f,v_n)_{L^2(\mathbb{R}^3)}$. Because of weak convergence this approaches $(f,v_\infty)_{L^2(\mathbb{R}^3)}$, which is equal to $\|v_\infty\|^2_{H_1(\mathbb{R}^3)}$, and Step #1 is complete.

STEP #2. (Proof of the theorem for bounded continuous functions on $(-\infty,0]$ which vanish at $-\infty$). Let A be the algebra of bounded continuous functions on $(-\infty,0]$ which vanish at $-\infty$. Let $\mathcal{A} \subset$ A be the set of functions for which the assertion of the theorem is true. A is a Banach space under the sup norm and $\mathcal{A}$ is easily seen to be a closed linear subspace of A. In addition $\mathcal{A}$ is a subalgebra. To see this suppose $F,G \in \mathcal{A}$; then for $f \in L^2(\mathbb{R}^3\backslash\Omega)$

$$(FG)(\Delta_n)f = F(\Delta_n)G(\Delta_n)f$$
$$= F(\Delta_n)G(\Delta_\infty)f + F(\Delta_n)[G(\Delta_n)f - G(\Delta_\infty)f].$$

Since $F \in \mathcal{A}$ the first term converges to $F(\Delta_\infty)G(\Delta_\infty)f$. The second term has norm dominated by

$$\sup_{(-\infty,0]} |F| \cdot \|G(\Delta_n)f - G(\Delta_\infty)f\|_{L^2(\mathbb{R}^3)}$$

which goes to zero since $G \in \mathcal{A}$. Thus $FG \in \mathcal{A}$. The assertion of Step #1 is that $F(x) = (1-x)^{-1} \in \mathcal{A}$ and clearly F separates points. By the Stone-Weierstrauss Theorem, $\mathcal{A} =$ A.

STEP #3. (Endgame) What we are proving is that $F(\Delta_n)$ converges strongly to $F(\Delta_\infty)$ as operators from $L^2(\mathbb{R}^3\backslash\Omega)$ to $L^2(\mathbb{R}^3)$. Since we have a uniform bound on $\|F(\Delta_n)\|$ it suffices to show that $F(\Delta_n)f \to F(\Delta_\infty)f$ for a dense set of f, for example for f of the form $e^{\eta\Delta_\infty} g$ for $g \in L^2(\mathbb{R}^3\backslash\Omega)$ and $\eta > 0$. (Note: $e^{\eta\Delta_\infty} g \to g$ as $\eta \to 0$.) Now

$$\begin{aligned}F(\Delta_n)e^{\eta\Delta_\infty}g - F(\Delta_\infty)e^{\eta\Delta_\infty}g \\ = [F(\Delta_n)e^{\eta\Delta_n}g - F(\Delta_\infty)e^{\eta\Delta_\infty}g] + \\ F(\Delta_n)[e^{\eta\Delta_\infty}g - e^{\eta\Delta_n}g] .\end{aligned}$$

By the result of Step #1 applied to the functions $F(x)e^{\eta x}$ and $e^{\eta x}$ the vectors in brackets tend to zero. The proof is complete. □

A final remark is in order on the interpretation of this result. It says that for fixed initial data the solutions of (1), (2) converge to those of $(1)_\infty$, $(2)_\infty$ provided $nr \to \infty$. How large nr must be before the convergence is evident will depend on the initial data. For example, consider the acoustic equation. No matter how large n is we may pose initial data which is an incoming wave of extremely high frequency, λ. If the wavelength is high enough the geometrical optics approximation becomes valid and one will not observe solidification. The solidification is intimately related to the failure of geometrical optics. It is caused by an overdose of diffraction.

A quantitative guess of how large λ must be can be made by dimensional considerations. It is evident both physically and mathematically that solidification does not depend on the size of Ω or the absolute number n of obstacles but on the density $n|\Omega|^{-1}$. For evenly spaced spheres if R is a measure of the distance between obstacles then $nR^3 \approx |\Omega|$. Thus in terms of r, R the number $n|\Omega|^{-1}r$ is essentially rR^{-3}, which has the dimensions $(\text{length})^{-2}$. A reasonable dimensionless quantity to replace nr is $r\lambda^2 R^{-3}$, where λ is a measure of the wavelength of the incident wave. In practical considerations I believe that this is the absolute number which must be large. Perhaps in the future this idea will find expression in concrete estimates.

REFERENCES

1. COURANT, R. and HILBERT, D., *Methods of Mathematical Physics* Vol. I, Interscience, New York, 1953.

2. FEYNMAN, R., *Lectures on Physics* Vol. II, Addison-Wesley, Palo Alto, 1964.

3. FRIEDRICHS, K. O., *Mathematical Methods of Electromagnetic Theory*, Courant Institute Lecture Notes, 1974.

4. KELLOGG, O. D., *Foundations of Potential Theory*, Dover Publications, New York, 1953.

5. LAX, P. D. and PHILLIPS, R. S., *Scattering Theory*, Academic Press, New York, 1967.

6. MAXWELL, J. C., *A Treatise on Electricity and Magnetism*, Dover Publications, 1954.

7. PROTTER, M. H. and WEINBERGER, H. F., *Maximum Principles in Differential Equations*, Prentice Hall, Englewood Cliffs, N. J., 1967.

8. RAUCH J. and TAYLOR, M., *Penetration into shadow regions and unique continuation properties in hyperbolic mixed problems*, Indiana U. Math. J. 22 (1973), 277-285.

9. RAUCH J. and TAYLOR, M., *Potential and scattering theory on wildly perturbed domains*, J. Functional Anal., to appear.

Vol. 277: Séminaire Banach. Edité par C. Houzel. VII, 229 pages. 1972. DM 22,–

Vol. 278: H. Jacquet, Automorphic Forms on GL(2) Part II. XIII, 142 pages. 1972. DM 18,–

Vol. 279: R. Bott, S. Gitler and I. M. James, Lectures on Algebraic and Differential Topology. V, 174 pages. 1972. DM 20,–

Vol. 280: Conference on the Theory of Ordinary and Partial Differential Equations. Edited by W. N. Everitt and B. D. Sleeman. XV, 367 pages. 1972. DM 29,–

Vol. 281: Coherence in Categories. Edited by S. Mac Lane. VII, 235 pages. 1972. DM 22,–

Vol. 282: W. Klingenberg und P. Flaschel, Riemannsche Hilbertmannigfaltigkeiten. Periodische Geodätische. VII, 211 Seiten. 1972. DM 22,–

Vol. 283: L. Illusie, Complexe Cotangent et Déformations II. VII, 304 pages. 1972. DM 27,–

Vol. 284: P. A. Meyer, Martingales and Stochastic Integrals I. VI, 89 pages. 1972. DM 18,–

Vol. 285: P. de la Harpe, Classical Banach-Lie Algebras and Banach-Lie Groups of Operators in Hilbert Space. III, 160 pages. 1972. DM 18,–

Vol. 286: S. Murakami, On Automorphisms of Siegel Domains. V, 95 pages. 1972. DM 18,–

Vol. 287: Hyperfunctions and Pseudo-Differential Equations. Edited by H. Komatsu. VII, 529 pages. 1973. DM 40,–

Vol. 288: Groupes de Monodromie en Géométrie Algébrique. (SGA 7 I). Dirigé par A. Grothendieck. IX, 523 pages. 1972. DM 55,–

Vol. 289: B. Fuglede, Finely Harmonic Functions. III, 188. 1972. DM 20,–

Vol. 290: D. B. Zagier, Equivariant Pontrjagin Classes and Applications to Orbit Spaces. IX, 130 pages. 1972. DM 18,–

Vol. 291: P. Orlik, Seifert Manifolds. VIII, 155 pages. 1972. DM 18,–

Vol. 292: W. D. Wallis, A. P. Street and J. S. Wallis, Combinatorics: Room Squares, Sum-Free Sets, Hadamard Matrices. V, 508 pages. 1972. DM 55,–

Vol. 293: R. A. DeVore, The Approximation of Continuous Functions by Positive Linear Operators. VIII, 289 pages. 1972. DM 27,–

Vol. 294: Stability of Stochastic Dynamical Systems. Edited by R. F. Curtain. IX, 332 pages. 1972. DM 29,–

Vol. 295: C. Dellacherie, Ensembles Analytiques Capacités. Mesures de Hausdorff. XII, 123 pages. 1972. DM 18,–

Vol. 296: Probability and Information Theory II. Edited by M. Behara, K. Krickeberg and J. Wolfowitz. V, 223 pages. 1973. DM 22,–

Vol. 297: J. Garnett, Analytic Capacity and Measure. IV, 138 pages. 1972. DM 18,–

Vol. 298: Proceedings of the Second Conference on Compact Transformation Groups. Part 1. XIII, 453 pages. 1972. DM 35,–

Vol. 299: Proceedings of the Second Conference on Compact Transformation Groups. Part 2. XIV, 327 pages. 1972. DM 29,–

Vol. 300: P. Eymard, Moyennes Invariantes et Représentations Unitaires. II, 113 pages. 1972. DM 18,–

Vol. 301: F. Pittnauer, Vorlesungen über asymptotische Reihen. VI, 186 Seiten. 1972. DM 18,–

Vol. 302: M. Demazure, Lectures on p-Divisible Groups. V, 98 pages. 1972. DM 18,–

Vol. 303: Graph Theory and Applications. Edited by Y. Alavi, D. R. Lick and A. T. White. IX, 329 pages. 1972. DM 26,–

Vol. 304: A. K. Bousfield and D. M. Kan, Homotopy Limits, Completions and Localizations. V, 348 pages. 1972. DM 29,–

Vol. 305: Théorie des Topos et Cohomologie Etale des Schémas. Tome 3. (SGA 4). Dirigé par M. Artin, A. Grothendieck et J. L. Verdier. VI, 640 pages. 1973. DM 55,–

Vol. 306: H. Luckhardt, Extensional Gödel Functional Interpretation. VI, 161 pages. 1973. DM 20,–

Vol. 307: J. L. Bretagnolle, S. D. Chatterji et P.-A. Meyer, Ecole d'été de Probabilités: Processus Stochastiques. VI, 198 pages. 1973. DM 22,–

Vol. 308: D. Knutson, λ-Rings and the Representation Theory of the Symmetric Group. IV, 203 pages. 1973. DM 22,–

Vol. 309: D. H. Sattinger, Topics in Stability and Bifurcation Theory. VI, 190 pages. 1973. DM 20,–

Vol. 310: B. Iversen, Generic Local Structure of the Morphisms in Commutative Algebra. IV, 108 pages. 1973. DM 18,–

Vol. 311: Conference on Commutative Algebra. Edited by J. W. Brewer and E. A. Rutter. VII, 251 pages. 1973. DM 24,–

Vol. 312: Symposium on Ordinary Differential Equations. Edited by W. A. Harris, Jr. and Y. Sibuya. VIII, 204 pages. 1973. DM 22,–

Vol. 313: K. Jörgens and J. Weidmann, Spectral Properties of Hamiltonian Operators. III, 140 pages. 1973. DM 18,–

Vol. 314: M. Deuring, Lectures on the Theory of Algebraic Functions of One Variable. VI, 151 pages. 1973. DM 18,–

Vol. 315: K. Bichteler, Integration Theory (with Special Attention to Vector Measures). VI, 357 pages. 1973. DM 29,–

Vol. 316: Symposium on Non-Well-Posed Problems and Logarithmic Convexity. Edited by R. J. Knops. V, 176 pages. 1973. DM 20,–

Vol. 317: Séminaire Bourbaki - vol. 1971/72. Exposés 400–417. IV, 361 pages. 1973. DM 29,–

Vol. 318: Recent Advances in Topological Dynamics. Edited by A. Beck. VIII, 285 pages. 1973. DM 27,–

Vol. 319: Conference on Group Theory. Edited by R. W. Gatterdam and K. W. Weston. V, 188 pages. 1973. DM 20,–

Vol. 320: Modular Functions of One Variable I. Edited by W. Kuyk. V, 195 pages. 1973. DM 20,–

Vol. 321: Séminaire de Probabilités VII. Edité par P. A. Meyer. VI, 322 pages. 1973. DM 29,–

Vol. 322: Nonlinear Problems in the Physical Sciences and Biology. Edited by I. Stakgold, D. D. Joseph and D. H. Sattinger. VIII, 357 pages. 1973. DM 29,–

Vol. 323: J. L. Lions, Perturbations Singulières dans les Problèmes aux Limites et en Contrôle Optimal. XII, 645 pages. 1973. DM 46,–

Vol. 324: K. Kreith, Oscillation Theory. VI, 109 pages. 1973. DM 18,–

Vol. 325: C.-C. Chou, La Transformation de Fourier Complexe et L'Equation de Convolution. IX, 137 pages. 1973. DM 18,–

Vol. 326: A. Robert, Elliptic Curves. VIII, 264 pages. 1973. DM 24,–

Vol. 327: E. Matlis, One-Dimensional Cohen-Macaulay Rings. XII, 157 pages. 1973. DM 20,–

Vol. 328: J. R. Büchi and D. Siefkes, The Monadic Second Order Theory of All Countable Ordinals. VI, 217 pages. 1973. DM 22,–

Vol. 329: W. Trebels, Multipliers for (C, α)-Bounded Fourier Expansions in Banach Spaces and Approximation Theory. VII, 103 pages. 1973. DM 18,–

Vol. 330: Proceedings of the Second Japan-USSR Symposium on Probability Theory. Edited by G. Maruyama and Yu. V. Prokhorov. VI, 550 pages. 1973. DM 40,–

Vol. 331: Summer School on Topological Vector Spaces. Edited by L. Waelbroeck. VI, 226 pages. 1973. DM 22,–

Vol. 332: Séminaire Pierre Lelong (Analyse) Année 1971-1972. V, 131 pages. 1973. DM 18,–

Vol. 333: Numerische, insbesondere approximationstheoretische Behandlung von Funktionalgleichungen. Herausgegeben von R. Ansorge und W. Törnig. VI, 296 Seiten. 1973. DM 27,–

Vol. 334: F. Schweiger, The Metrical Theory of Jacobi-Perron Algorithm. V, 111 pages. 1973. DM 18,–

Vol. 335: H. Huck, R. Roitzsch, U. Simon, W. Vortisch, R. Walden, B. Wegner und W. Wendland, Beweismethoden der Differentialgeometrie im Großen. IX, 159 Seiten. 1973. DM 20,–

Vol. 336: L'Analyse Harmonique dans le Domaine Complexe. Edité par E. J. Akutowicz. VIII, 169 pages. 1973. DM 20,–

Vol. 337: Cambridge Summer School in Mathematical Logic. Edited by A. R. D. Mathias and H. Rogers. IX, 660 pages. 1973. DM 46,–

Vol. 338: J. Lindenstrauss and L. Tzafriri, Classical Banach Spaces. IX, 243 pages. 1973. DM 24,–

Vol. 339: G. Kempf, F. Knudsen, D. Mumford and B. Saint-Donat, Toroidal Embeddings I. VIII, 209 pages. 1973. DM 22,–

Vol. 340: Groupes de Monodromie en Géométrie Algébrique. (SGA 7 II). Par P. Deligne et N. Katz. X, 438 pages. 1973. DM 44,–

Vol. 341: Algebraic K-Theory I, Higher K-Theories. Edited by H. Bass. XV, 335 pages. 1973. DM 29,–